CODES, CIPHERS & OTHER CRYPTIC
& CLANDESTINE COMMUNICATION

CODES, CIPHERS & OTHER CRYPTIC & CLANDESTINE COMMUNICATION

Making and Breaking Secret Messages from Hieroglyphs to the Internet

BY FRED B. WRIXON

Black Dog & Leventhal

Paperbacks

Published by Black Dog & Leventhal
151 West 19th Street
New York, NY 10011

Illustrated by Patty Hune

Designed by Martin Lubin Graphic Design

Typeset by Brad Walrod/High Text Graphics

Manufactured in the United States of America

ISBN: 0-7394-6013-7

ACKNOWLEDGEMENTS

My editor Elise Andaya and artist Patty Hune deserve special recognition for their contributions to this text. I would also like to thank Joshua Weitz for the section "Quantum Cryptography."

The following persons have provided both inspiration and information: Barry Carleen, Alex Chartier, Laura Clark, John Finnegan, Jim Foti, Rene Frank, David Gaddy, Jacque Guy, David Hatch, Jack Ingram, Rev. Kevin Massey-Gillespie, Richard Pekelney, Rebecca Raines, Ed Simmons, Gustavus J. Simmons, David Woods.

Prompt and efficient services were provided by the staffs and members of the following: The American Cryptogram Association; The American Philosophical Society; Carnegie Library of Pittsburgh, PA; Central Intelligence Agency; Certicom Corporation, Canada; Collier's Encyclopedia; Crown Publishing Group; Denison University Library; Dover Publications, Inc.; Encyclopedia Americana; Encyclopedia Britannica; Federal Bureau of Investigation; The Imperial War Museum, England; Kungl Biblioteket, Sweden; Library of Congress; Martins Ferry Public Library; National Archives; National Security Agency; Naval Historical Center; New York Public Library; Ohio County Public Library; Simon & Schuster, Inc.; U.S. Army Military History Institute; U.S. Coast Guard Academy Library; U.S. Naval Institute; West Virginia University Libraries.

CONTENTS

CIPHER DEVICES AND MACHINES   238

CRYPTOLOGY: FROM HIEROGLYPHS TO THE INTERNET

Codes, ciphers, signals and secret languages have concealed communications over the centuries, whether oral, written, gestured, audible or electronically conveyed. While the exact beginnings of these masking processes have been shrouded by time and by the very nature of their fundamental purpose—secrecy—cryptic knowledge has not always been kept hidden forever. Understanding has been reached through painstaking research, the sudden illumination of genius, the dark reality of treachery and the upheaval of war.

The fierce competition that developed between cryptographers and their solution-seeking rivals became known as cryptology, from the Greek *kryptē* < *kryptós,* meaning "secret, hidden" and either the word *logós*, meaning "word," or *ology*, "science." The term cryptology eventually encompassed the two competing skills: concealment by cryptography, (*kryptē* < *kryptós*, and *graphia* "writing") and revelation by cryptanalysis, from *kryptē* < *kryptós*, the very similar Latin *crypta*, the German *ana* "up, throughout" and *lysys*, "a loosing." The roots of cryptology are various, and scholars believe that as languages develop among different societies, so do the tendencies to seek means of communications concealments.

Historians believe that the first protocryptographic practices date to ancient Egypt about four thousand years ago. Scribes recording their rulers' lives gave varied forms to standard hieroglyphs on monuments and tombs to distinguish them from common characters and impart them with added respect. Temple scribes adopted this practice when transcribing religious texts in order make them appear more mysterious and powerful to the layman. The religious hierarchy took advantage of its powers of "translation" for the populace who were increasingly less informed and therefore more dependent on the powers of the priesthood. As the Egyptian civilization expanded, hieroglyphics became more involved. Symbols, pictographic representations and pronunciation became intertwined. With the increasing numbers of carvings on special edifices, the people lost interest. Egyptologists theorize that official scribes then modified some figures still further in an attempt to arouse curiosity and regain the public's interest. Though not a code or cipher by any means, these changes did involve two basic principles of cryptology, namely, transforming writing and masking it with secrecy. Despite these advances, there are no clear records indicating that they used these methods on a broad scale to conceal diplomatic, commercial or military plans.

The Mediterranean shores provide clearer cryptologic examples from the civilizations that prospered in Mesopotamia around the Tigris and Euphrates Rivers as well as in the eastern Mediterranean areas. Cuneiform writing was the prime means of recording the daily lives of Mesopotamian empire builders like the Babylonians, Assyrians and Chaldeans. In 1500 B.C.E. a cuneiform tablet held a carefully guarded formula for a pottery-glaze. The mixture of cuneiform figures defining the proper ingredients was purposely jumbled, making this the earliest true secret writing for which we have evidence.

From about 500 B.C.E. India also used secret writing ranging from spies' exchanges to types of script purportedly used by the Buddha (c. 563–c. 483 B.C.E.). These methods included phonetic substitutions, whereby the positions of consonants and vowels were exchanged; reversed letters aligned with one another; and styles of writing that were placed at odd angles. Various Indian texts, of which the Artha-´sāstra (c. 321–300 B.C.E.) is an example, indicate a broad awareness of obscuring methods. The knowledge of concealed writing is included as one of the more than 60 skills to be mastered by women in the *Kama-sutra*, a classic textbook on erotics and other forms of human pleasure by Vātsyāyana.

Indian historical texts also record the earliest known use of sign language for the hearing-and speech-impaired. Most versions of this method equated fingers with consonants and their joints with vowels. Some also assigned meanings to the spaces between the fingers. Variations on this simple sign language continued to be used in modern times in the signals used by beggars and moneylenders.

Though also partially obscured by time, biblical scriptures document many traditions of verbal and written records and elements of word transformation. Cryptohistorians suggest that the instances of Bible cryptography are better considered protocryptography, since they do not have the same aspects of secrecy that fully developed cryptography provides. Nevertheless, the examples are fascinating from a historical point of view.

In the Old Testament, two examples of word transformations use a traditional Hebrew alphabet letter substitution called *athash*. Athash is a reciprocal substitution process in which the first and last letters change places, the second and the next-to-last are substituted and so forth through the alphabet.

Jeremiah 25:26 and 51:41 contain transformations in which *Sheshach* replaces *Babel* (a name for Babylon). The second of these shows the apparent lack of a secrecy motive when Sheshach is followed by a phrase containing the word Babylon:

HOW IS SHESHACH TAKEN! AND THE PRAISE OF THE WHOLE EARTH SEIZED! HOW IS BABYLON BECOME AN ASTONISHMENT AMONG THE NATIONS!

The name Babylon also appears in a cryptic form in Revelation, in the form of a symbolic evil woman who bears that name. Babylon is interpreted as Rome and its empire in Revelation 17:5 as follows:

AND ON HER FOREHEAD WAS WRITTEN A NAME OF MYSTERY: BABYLON THE GREAT, MOTHER OF HARLOTS AND OF EARTH'S ABOMINATIONS.

The Bible also records an account of a man who could be called the world's first cryptanalyst. According the biblical story, Daniel was made a captive when the Babylonian Nebuchadnezzar conquered Palestine in 605 B.C.E. and grew up in Babylon under his reign. Credited with the ability to interpret dreams, Daniel became influential with a succession of rulers including Neriglissar, Labash-Marduk and Nabonidas. Daniel rose to prominence in Babylonian society, yet kept his faith and moral principles amid idolatry and a kingdom becoming badly encumbered with its own lavish excesses.

By 534 B.C.E. Belshazzar, son of Nabonidas, ruled a nation weakened by the corruption of unchecked power and decadence. Belshazzar had lost the loyalty of the people whose ancestors had fought bravely and well for Babylon. Daniel, now an old but still brilliant man, was in attendance at one of the king's feasts. In the midst of the revelry came the architypical prediction of doom; the message "Mene, Mene, Tekel, Upharsin" was written on the wall (Daniel 5:5–28).

The words *mene*, *tekel* and *upharsin* had equivalents in Aramaic names for money. A mina ("mene") was a coin worth one-sixtieth of a talent. A teke ("tekel") was quite similar to the Aramaic shekel and was one-sixtieth of a mina. A peres ("upharsin") was worth half a mina. These monetary types give credence to the belief that the words symbolized the financial separation of Babylon and its wealth and empire. Daniel interpreted the phrase this way (5:26–28):

MENE: GOD HATH NUMBERED THY KINGDOM, AND FINISHED IT. TEKEL: THOU ART WEIGHED IN THE BALANCES AND ART FOUND WANTING. PERES: THY KINGDOM IS DIVIDED AND GIVEN TO THE MEDES AND PERSIANS.

According to the Bible, Belshazzar was slain that very night. Darius, a Medean, assumed control. Not long thereafter, a powerful Persian commander, Cyrus, conquered the entire Babylonian empire. With his conquest, once mighty Babylon was no more; Daniel's prediction had been correct.

The Bible also mentions the use of a password, which is similar to a codeword. In Judges 12:5–6 the men of Gilead tried to stop the defeated Ephraimites from escaping across the Jordan River by requiring that all passersby say the word *Shibboleth*. They knew that the Ephraimites could not pronounce the initial *sh* in the word, and all those who could not pass the test were slain.

From a cryptological point of view these words were not particularly secretive. Although transformation in a literal sense was absent, these protocryptographic examples almost certainly influenced the later development of cryptology.

THE BEGINNINGS OF CRYPTOLOGY

The word *cipher* is derived from the Arabic *sifr*, meaning "nothing." When the Arabic civilization expanded through much of the Western world in the seventh century C.E., they brought with them a high culture, a flourishing scholarship in science and mathematics, as well as various secret writing practices. In addition, their mathematicians had developed decryption techniques through analyzing the repeating appearances of certain letters and words in their own texts. This became known as "letter frequency study" and eventually as cryptanalysis.

When Arabic practices brought a more organized approach to cryptology into the West, there was an already existing tradition of secret concealments. The early Greeks and Persians used simple cryptographic techniques to convey battle plans to soldiers in the field. In the fifth century B.C.E., the Greek historian Thucydides described what is believed to be the first complete system of transferring secret information. He credits the Spartans with developing a device called a *skytale* (rhymes with Italy), which transposed missives by wrapping parchment around rods of specific sizes and writing down the length of the rod. When unwrapped, the letters appeared in nonsequential order.

In later years, when Cyrus and other Persian leaders tried to spread the boundaries of their empire toward the Aegean Sea, they ran into staunch resistance from the Greek city-states, especially Sparta and Athens. As oral storytelling evolved into written record keeping, the Persian-Greek conflict provided more detailed facts about concealed messages and signals. As well as the time-honored use of couriers, both sides used steganographic techniques, which involved the physical concealment of a communiqué. According to the Greek historian Herodotus, among other techniques, the Spartans covered wooden tablets with wax to make them appear blank and also hid missives in the bellies of freshly killed rabbits.

Fire signals were also in use at this time. Greek mythology has tales of torch-sent messages; crude towers supposedly bore news, in the form of fire signals, of the fall of Troy from observers of that doomed city to Agamemnon's fortress at Mycenae; and Herodotus wrote of Greek vessels that sent flame warnings of the approach of a Persian fleet. Athens and the other Greek city-states also developed special signaling methods to help ships navigate the treacherous Greek shoals. Beginning with lamps and torches mounted in high vantage points and flames used to mark dangerous shores, more elaborate transmission systems developed. The Greek scholar Aeneas the Tactician compiled texts about warfare around 350–345 B.C.E. and described a signaling apparatus based on a combination of earth, air, water and fire.

The Greek historian and cryptographer Polybius is credited with a significant advance in signaling and ciphermaking. Based on an idea developed by the philosopher Democritus, the Democritus/Polybius system used various torch signals to represent the letters of the Greek alphabet. In the history of cryptography, Polybius is immortalized for his creation of a true alphabet-based system based on a five-by-five grid configuration, which came to be known as the Polybius checkerboard. This system was the first to provide a transferral of letters to numbers in a pattern that was relatively easy to recall, and became a foundation of later cryptographic techniques.

At the height of the Roman empire, Julius Caesar used a combination of signaling stations and a cipher to communicate with his generals during his military campaigns throughout Gaul. In his text *Gallic Wars,* Caesar described using Greek letters to mask his Latin communiqués. He also used rearrangements of the plaintext alphabet, and his basic shift substitution cipher still bears his name today.

Like other civilizations before it, Rome entered stages of decadence and decline, and when the empire finally crumbled, Western culture entered the chaos of the Dark Ages. The fragile lights of education as well as cryptology were nearly extinguished as widespread suspicion of scholarship swept over the European continent.

THE RENAISSANCE

After the suffering and violence of the preceding centuries, the fourteenth-century Italian revival of literature, art and scholarship seemed like a rebirth. In the intellectual atmosphere of the Renaissance, the scholars and cloistered monks who had protected fragile texts and scrolls from destruction and decay over the centuries slowly allowed access to their vast libraries. Among their parchments were documents that contained ancient ciphers and their solutions, as well as other insights into the ancient tradition of secret writing.

The growing interest in occult writings and practices also helped revive cryptology in Europe. At the same time, however, advances in cryptology were also hindered by its close association with mysticism. The Dark Ages had given rise to innumerable superstitions regarding the powers of magic as people sought relief through charms, incantations, amulets and omens. To avoid accusations of sorcery, which often meant death by fire and drowning, many soothsayers and alchemists used the signs of the zodiac to hide their experiments. During the expansion of Catholicism, secret writing was at first banned and then outlawed in many places. Eventually, centers of learning were restored and supported by the Catholic Church, and the Catholic hierarchy actually helped to keep cryptology alive as leaders found uses for concealing their correspondence amid doctrinal disputes and confrontations with powerful royalty and nobles. Papal secretaries in the early 1300s used some abbreviations for names in an early form of code: the letter *A* referred to a king and *D* to a pope. By the latter 1300s, correspondence among Italy's city-states had elements of letter substitution and listed names that equated with pairs of letters. These forms became known as nomenclators—from the Latin *nomen,* meaning "name," and *calator,* "caller"—concealment systems that combined code-words for words, syllables and names along with lists of cipher alphabets. Nomenclators continued to be the most widely used masking system until the 19th century.

The numerous divisions among religious groups during the 1300s led to the creation of some basic codes with name changes, abbreviations and some replacements of names with jargon. The Guelphs, a pro-Pope faction, were in conflict with the Ghibellines of Italy, who supported the Holy Roman Emperor. Their differences were rooted in policies and mixed loyalties between temporal and Vatican rule. As a crude jargon code, the Guelphs called themselves the Children of Israel and the Ghibellines the Egyptians. Simple codes used for name abbreviations also appeared during the Great Schism, when anti-Pope factions set up a church separate from Rome in Avignon, France. Soon two Popes were trying to lead their quarreling followers. The Avignon ruler, Clement VII, used a missive concealment wherein common names or words were replaced by two letters, for example, *queen* became "gh."

During the Church's fanatical crusade against heresy, groups who used codes and ciphers to conceal secret missives often came under suspicion and were persecuted for witchcraft.

One such society was the Rosicrucian brotherhood. A scholarly and religiously oriented group, their full name was the Ancient Mystical Order Rosae Crucis (AMORC). Their beginnings in the 15th century are somewhat shrouded in myth and legend, but many historic sources credit the founding to a Christian Rosenkreuz or Rosenkreutz, German for "rosy cross." The group purportedly chose their symbol and special seal, a cross bearing a rose, from the name of this sage.

The Rosicrucian doctrine teaches that everything is linked to a divine being. The cross is their central symbol and their goals include earthly peace and the pursuit of ethical scientific knowledge. They believe that once humans are aware of the divine nature within themselves, they can become a microcosm of the universe. With such wisdom, they can begin to affect the energies and forces around them, using these powers to work positively against evil.

Even within the brotherhood itself there are many debates surrounding the origin of the fraternity. While some historians claim that Rosenkreuz was a mythological figure, others in the order believe that he may have existed during the latter 15th century when the German-speaking duchies showed a growing interest in the movement. They believe that Rosenkreuz founded the first lodge in a principality that one day would become part of a unified Germany.

Other Rosicrucians place their origins much earlier in history. Because several of their symbols, beliefs and rituals contain references to the ancient Egyptians, some followers draw connections between the Rosicrucians and the mysteries of the Sphinx. Though no one forefather or single family tree is mentioned, references to Pharaohs and wise men alike are given in Rosicrucian texts.

According to this sector, the original Egyptian organization was known as the Great White Lodge; however, AMORC studies have placed it at more than one location among the ancient sites. They believe that special knowledge, known as the Osirian mysteries, was discussed and perhaps actually revealed for the first time at this sacred site. These teachings relating to the god Osiris were believed to contain the answers to many of life's questions, but were revealed only to a chosen few illuminati. Chief priests, referred to as Kheri Hebs, recorded these mystical facts on carefully guarded papyri and handed down this secret knowledge to apprentices known as adepts or initiates. As the order expanded, illuminati spread across what is now the Middle East, founding worship centers and schools as they moved.

Historians and archivists not associated with AMORC give a different account. They believe that the *Fama Fraternitatis,* a manifesto supposedly written by Rosenkreuz, was actually an anonymous pamphlet. The *Fama* text was printed in 1614 in Kassel, Germany, and is said to contain the first reference to the Rosicrucians. According to this pamphlet, Rosenkreuz was taught in a monastery and traveled through Islamic countries to learn the secrets of the universe. He then started the Rosicrucian monastic order and swore its members to secrecy by sacred oath. His followers promised to devote themselves to world renewal and peace through a joining of religious faith and scientific fact.

Two years later another account, the *Chymische Hochzeit*, appeared. This was reputedly the story of Rosenkreuz as an elderly prophet and was written by Rosenkreuz himself. Some years after, the Lutheran Johann Andreä confessed that he had created this material as a young man's prank, and some researchers believe he may also have written the *Fama* texts.

Doctrines similar to that of the Rosicrucians were both known and practiced in varied ways by Greek scholars, Hermetic sages, Gnostic philosophers in Egypt and Hindu priests. In spite of charges that occult and pagan rituals were being fostered, the Rosicrucians grew in membership. Chapters spread as formal education made its way from Greece, north and east and across the Balkans to Italy. Rosicrucian records and other histories document a series of persecutions, raids upon chapter dwellings and social ostracism. Like many other such groups, they were frequently the target of prejudice and misunderstanding, yet loyal members kept their philosophies alive through the Dark Ages. Using a simple cipher to correspond with each other and to conceal their sacred texts from prying eyes, the Rosicrucians emerged from this time of upheaval stronger and more unified than ever. The Rosicrucian cipher remained a secret until some members left the order and broke their vows of secrecy.

With the passing of the Middle Ages, the Rosicrucian brotherhood had become an influential society, listing philosophers such as Descartes and royalty such as Louis XIII as believers. By the 1700s the order had moved to the Americas, where Thomas Jefferson was among its supporters. Rosicrucian societies continue to flourish today in America and Europe, although they no longer use their secret cipher.

A later fraternal organization that derives many of its beliefs and practices from the Rosicrucians is the brotherhood of the Free and Accepted Masons. The origins of the Freemasons, or Masons, are more clearly defined than those of the Rosicrucians. The Masons were originally stoneworkers who came to prominence by building cathedrals and other such edifices between 850 and 1650 C.E. They joined with other builders in guilds to ensure mutual protection and set a quality standard among their members. This connection to medieval times and the attempts by some historians to link the Freemasons to the Egyptian pyramid builders caused confusion in determining the starting point for the organization. Unlike the Rosicrucians, however, the Masons have a specific date by which they chart their formal association.

On June 24, 1717, four builders' associations met in London and as some of them referred to their memberships as lodges, the four combined and called themselves the Masonic Grand Lodge. This fraternal body chose builders' tools as symbols. They believed in the brotherhood of mankind within God's guidance and used the precepts derived from many different faiths. Though not a specific denomination, the Masons fostered charitable efforts and gave instructions for self-improvement.

Freemasonry gained a strong foothold in the British colonies among the craftsmen who immigrated to the New World. By 1730 lodges had been set up in both Philadelphia and Boston. Benjamin Franklin was a prominent member of the Pennsylvania colony and in 1734 he printed the first Masonic text for the New World. Among the leaders of the American Revolution, George Washington, John Hancock and Paul Revere were all Masons.

While the Free and Accepted Masons maintained open relationships in their public dealings, they were more secretive when it came to membership initiations and certain rituals. Texts describing history and rites, as well as correspondence between lodge leaders, were often concealed using a cipher similar to the Rosicrucians'. The mystery surrounding Mason membership, while causing many people to link them to black magic practitioners, also lent a certain allure to the society and it flourished through the centuries. Eventually, time, carelessness on the part of some Masons and the loose tongues of former members combined to reveal the Masons' cipher.

Like the Renaissance as a whole, cryptological advances of 14th and 15th centuries owed a great debt to Italian literature, art and science. It was the internal security needs of continually feuding Italian city-states that initiated the growth of cryptology as leaders tried to ensure secrecy of commercial, diplomatic and military exchanges. The emphasis on science and learning fostered by the Renaissance also provided a nurturing atmosphere for a number of mathematicians and inventors who contributed significantly to cryptographic advances.

As competition increased, the leaders of Venice began to employ people to surreptitiously read the mail intended for other cities' ambassadors. One of the earliest examples of this postal surveillance occurred in the early 1500s at the Doges' Palace where, in a special chamber, three paid cryptanalysts studied everything from the mail of important merchant families to the dispatches of ambassadors. The secretaries' discoveries enabled the leaders of Venice to plan successful military and economic strategies for decades. Such chambers became a standard aspect of court influence and intrigue. The residents of Florence, Genoa and Naples were soon doing the same with their own trained secretaries. This led to official sanction of what the French called the Cabinet Noir, or Black Chamber.

Throughout the Renaissance substantial advances were made in the field of cryptography through the work of pioneer cryptographers such as Leon Alberti, Johannes Trithemius, Giovanni Porta, Girolamo Cardano and Blaise de Vigenère. Cryptography moved from simple substitutions and symbol crytography into polygraphic substitution, grilles and key progressions. Cryptanalysis, or the decryption of ciphered or coded missives, also made strides, incorporating ideas of repetition and linguistic probability to create much more reliable code and cipher solving.

MONARCHIES AND REVOLUTION

The 16th century in England was marked by a violent schism between the Catholics and Protestants, with the throne of a growing empire as the prize. The political turmoil taking place at all levels of English society was reflected in the series of plots and intrigues in the struggle between Mary, Queen of Scots and Elizabeth I of England, a drama in which ciphers played a crucial role.

The Protestant Elizabeth ascended to the throne in 1558 amid numerous rumors of schemes to depose her and place a Catholic on the throne. Fully aware of her tenuous position, Elizabeth gathered a very capable group of devoted followers. Among them was a contingent of fifty or so agents dedicated almost exclusively to security intelligence. The notorious master spy Sir Francis Walsingham carefully assigned these agents according to their personal skills.

Walsingham was born in Kent in 1530 and studied at King's College, Cambridge. He made detailed studies of foreign languages, customs and diplomatic traditions, and his demonstrated abilities in Europe's courts gained him a knighthood in 1577 and the position of secretary of state soon thereafter. From this vantage point he was able to compile a list of all those who could endanger his queen. First among the names of threatening persons was that of Mary Stuart.

Born in 1542, Mary was the daughter of Scotland's King James V and Mary of Guise. As the cousin of the childless Elizabeth I, she was the direct heir to the throne of England. After the death of one husband, the murder of a second and the flight of a third, Mary was forced by influential nobles to leave Scotland. Seeking Elizabeth's sympathy and protection, she was instead placed in ever closer custody for 19 years.

For reasons both personal and political, Elizabeth permitted her rival to remain alive. While in virtual imprisonment, Mary became a symbol of Catholic oppression for her many followers, who believed Elizabeth to be a usurper. Trusted followers of Walsingham indicated that plans were being made to seize the crown for Mary. Yet Elizabeth insisted on proof before she signed a warrant for her kinswoman's death, and Walsingham decided that he had to move directly to save Elizabeth and Protestantism as England's religion. In 1586, he provoked a circle of conspirators to implicate themselves in what came to be known as the Babington Plot.

Among Walsingham's close associates in this effort was Thomas Phelippes, England's first noteworthy cryptanalyst. As the preparations for the entrapment began, Phelippes was ready to apply his uncanny skills with foreign tongues and messages.

The plot's namesake, Anthony Babington, was a boastful, wealthy adventurer who was born of a family with lineage to the Normans. He was secretly also a devout Catholic, who had once been a page to Mary. Financially able to support a number of schemes, he vainly had begun to envision himself as Mary's champion and had already formulated the groundwork of a plan to overthrow the government. He and his friends, including a zealous priest named John Ballard, sought overseas support from France and Spain for a scheme that would free Mary and perhaps assassinate Elizabeth.

One of Babington's confidantes was Gilbert Gifford. The son of a prominent Catholic family, Gifford had once trained for the priesthood but had gone astray and was facing a prison sentence when Walsingham intervened on his behalf. Quite willing to comply with his benefactor's wishes, Gifford agreed to use his religious background to gain the trust of the priests and Mary's followers. Pretending to be an ardent Catholic, Gifford gained the acquaintance of Catholics on the Continent and in England and fueled Babington's dream of gathering those devoted to securing freedom and the crown of England for Mary.

Mary was being held at Chartley Castle in Staffordshire. Here Walsingham had Gifford pose as a cleric and patiently ingratiate himself with the lonely Chartley resident. He offered to help Mary exchange secret letters with her friends and even devised a means to disguise this transfer of missives.

With a sympathetic brewer acting as courier, the messages were wrapped in leather and then placed in a corked tube in the stoppers of beer barrels. Once the container was inside the castle, Gifford had one of his helpers bring the disguised correspondence to Mary. She in turn would place her letters in the container, which would then be taken beyond the walls in another barrel. Still wary, Mary enciphered her messages using a nomenclator. As Walsingham's agent provocateur, Gifford intercepted Mary's correspondence and made sure that cryptanalyst Phelippes had a chance to decipher them. The encryptions of the Scots' queen, a very simple partially disguised mixture of Greek letters, numerals and symbol cryptography, were no challenge for Phelippes' skills.

Meanwhile, the plans of Anthony Babington, John Ballard and 11 other prime collaborators began to coalesce into a full-scale conspiracy. The complicated scheme included the assassinations of Elizabeth, her minister William Cecil and Walsingham; a ruse to free Mary from Chartley and take her to a safe haven; the seizure of the Royal Navy's ships guarding the Thames; and a call to arms for all true Catholics. During the general confusion and upheaval, help was to be sought from Philip II of Spain. Once the Protestant monarchy was completely toppled, Mary would be crowned Queen of England.

Caught up in the drama, Babington wanted Mary to know that he was the chief conspirator. Naively depending on the barrel ploy and the disguised notes, he insisted that she recognize him by writing directly to him. In a damning message on July 12, he explained the details of Elizabeth's murder and Mary's deliverance. Again he asked for a personal reply from the Scots' queen.

On July 17, Mary replied in cautious agreement and in so doing clearly implicated herself. After Gifford had intercepted the barrel transfers long enough to copy them, Walsingham and Phelippes were able to decipher them in a matter of hours. What had appeared to be a trustworthy mask to the schemers was a thin veil to the cryptanalysts. Walsingham and Phelippes had had many opportunities to see messages with very similar word phrases, especially in the opening greetings and closing remarks. Both men knew about studies of letter frequency and looked for examples of repetition.

Once they had found the tell-tale sequences, Walsingham and Phelippes had untied the first knots of the cipher and the rest unraveled quickly. When Mary replied, she fully incriminated herself in the following words: "When your preparations both in England and abroad are complete, let the six gentlemen who have undertaken to assassinate Elizabeth proceed to their work, and when she is dead, then come and set me free."

As an added strike against the plotters, Walsingham had Phelippes pen a cipher postscript on the letter making it appear that Mary sought the names of the six would-be assassins of Queen Elizabeth. With all the parts in place, Walsingham moved quickly to snuff the first flames of the uprising. Babington and the other conspirators were arrested, forced to confess, and sentenced to death. In a separate trial, the evidence of the cipher missives weighed heavily against Mary, and Elizabeth could extend no special mercy to her cousin. Mary was found guilty of high treason and her complicity with Babington became public knowledge. Beyond Elizabeth's own mixed emotions, political pressures and the laws governing traitorous acts forced her to do what was expected of a threatened ruler. Mary was beheaded on February 8, 1587, at Fotheringhay Castle, Northamptonshire.

As fate would have it, another Stuart was to be undone, and ciphers were to play a key role once more. This Stuart was Mary's grandson, King Charles I, who, like his grandmother, ruled at a turbulent time in English history. Charles was born on November 19, 1600, and wore the crown when England's civil war (1642–52) began.

This conflict was the result of religious bickering, class differences and political rivalries. The primary incidents revolved around Charles's disputes with Parliament and the rise of powerful royal competitors. These rivals were called the Puritans because of their desire to "purify" the Church of England. They were also derisively labeled the Roundheads because they chose to cut their hair shorter than the long-haired supporters of Charles, the Cavaliers.

When armed conflict began on August 22, 1642, the Puritans appeared to be outnumbered and in a weak position. The Cavaliers controlled most of England's large cities except London and much of the countryside. The Puritans, however, had two important secret weapons. Their first was soon revealed: the brilliant leader and tactician, Oliver Cromwell. Using strict discipline, detailed training and a keen mind for strategy, Cromwell molded the divided Puritans into an efficient fighting force. Soon the mocking nickname, Roundheads, began to strike fear into the Royalists.

The Puritans' second weapon was Sir John Wallis, a mathematician who would come to be known as the father of English cryptology. An ordained clergyman, Wallis amused himself with problems of numerical complexity. As with others skilled in arithmetic or music, his talents seemed to lead to a special understanding of cryptic messages.

Wallis had been earning a living as a chaplain to a wealthy widow during the civil war, when a letter belonging to the supporters of Charles I was found after the battle of Chichester. Knowing of Wallis's Puritan leanings, one of Cromwell's men brought him the captured Cavalier message. Asked to solve it, Wallis lifted the veil of this rather easy cipher. Cromwell recognized Wallis's talents and quickly befriended him. Soon Wallis was reading intercepted letters and military orders written by the Cavaliers, beginning series of revelations that helped Cromwell reverse the trend of the war.

Throughout the struggle the Puritans had wanted to connect Charles to direct plots against members of Parliament, but his Cavalier supporters had always managed to wage war without having their king appear to be directly involved. After the discovery of some private correspondence, Charles's fortunes changed dramatically.

On June 14, 1645, the parish of Naseby in Leicestershire was the site of a Puritan victory. Amid their rejoicing, Cromwell's men found a most interesting collection of letters among military dispatches captured from the Royalists and entrusted them to Wallis. Hoping for some clues to the Cavaliers' next moves, the Roundheads were most pleasantly surprised by the news Wallis had for them. In this case the missives were much more valuable than the rival army's plans. Wallis had deciphered letters written by King Charles to his wife. In them he made statements that clearly defined his involvement in anti-Parliament schemes. As with his ancestor, Charles's false trust in ciphers initiated his demise. Four years later, largely on the basis of this evidence, the grandson of Mary Stuart was also beheaded.

Even after the civil war ended in 1652, the victorious Roundheads continued to bring missives of all types to Wallis. His mathematical achievements during the Restoration years included his classic *Arithmetica Infinitorum*, which provided Sir Isaac Newton with the foundations for his own discoveries.

Across the English Channel, the vicious battles between Catholics and Protestants continued into the next century, and set the stage for the entrance of France's great cryptologist of the 17th century, Antoine Rossignol. In 1628 a Catholic army led by Henry II, the prince de Condé, had surrounded Réalmont, a Huguenot (French Protestant) bastion in southern France. The Protestant defenders appeared to be well entrenched and prepared for a long and costly siege.

Catholic soldiers captured a citizen of Réalmont carrying an encrypted missive to his confederates. After military advisers failed to solve the cipher, Henry sent the cipher to young Rossignol, who lived in a nearby town. Rossignol quickly decrypted the contents, revealing that the Protestants had a very low supply of weapons and ammunition. The Catholics showed the decryption to the defenders of Réalmont; their weakness exposed, they abruptly surrendered.

Rossignol's success earned him a position with France's powerful Cardinal Richelieu. The cardinal had sent another Catholic army to besiege the Huguenot fortress of La Rochelle. Catholic forces intercepted some enemy letters and Rossignol solved the encryption, revealing that many in La Rochelle were facing severe food shortages. The leaders of La Rochelle were holding out for support from British naval forces that were expected to bring relief and extra firepower.

With this new information provided by Rossignol, Cardinal Richelieu alerted his commanders and ordered extra vessels into the area and strengthened the armaments of nearby forts to thwart the Royal Navy. When the British ships arrived, they were faced with cannons primed to fire. Unable to anchor in the harbor, the vessels were forced to retreat. Threatened with starvation, the inhabitants of La Rochelle had to lay down their arms.

Rossignol became the original cryptologist of the Bourbon dynasty and was showered with wealth and honors. He was a favorite of Louis XIII and was often treated better than many of the nobles at court. His position was maintained when Louis XIV became king, and he worked for this influential monarch during a golden era of French art, science and philosophy.

Rossignol made an important contribution in his own field by directly altering the predominant nomenclator method, creating the first two-part nomenclator. The concealment he made for Louis XIV came to be named the Great Cipher. It was considered unbreakable and indeed confounded other nations' experts for years.

With the death of Rossignol in 1682, the Great Cipher was not always applied to royal correspondence and this carelessness would prove costly. For poised to decrypt France's most guarded secrets was England's master cryptanalyst, Sir John Wallis.

Wallis's services were valued so highly that he was retained by Charles II after the end of Puritan rule in 1660. During the reign of William and Mary, Wallis continued to serve as a secret-writing analyst, reporting to the earl of Nottingham, King William's secretary of war. In 1689, Wallis solved the cipher covering the correspondence between Louis XIV and the French ambassador in Poland. Since it was not protected by the Great Cipher, this accomplishment arguably could be considered average for Wallis. The results, however, were of high-level importance.

When Wallis uncovered the French communiqués, he realized that he was holding powerful information. One of the letters revealed that Louis was scheming with Poland's king to form an alliance and declare war on Prussia. Because this union threatened the balance of power in Europe, it was also dangerous to England. Wallis knew that William wanted the European nations divided and France, England's ancient enemy, isolated. Nottingham went directly to William.

A shrewd practitioner of court intrigue, William found ways to make the secret diplomacy public. The results were extremely damaging to Louis; not only did his scheme to attack Prussia fail, but the French ambassador corps was expelled from Poland and the Polish were alienated for several years thereafter. Prussia, rarely unprepared for war, became even more militant. To William's great pleasure, the continental powers were at odds and Louis XIV was internationally disgraced.

The 17th and early 18th centuries saw a variety of approaches to codes, ciphers and secret languages. One of the major advances in cryptology was the spread of institutionalized cryptanalysis in the form of the Black Chambers. While some mail interception was taking place during the16th century, by the 1700s the Black Chambers' occupants were well-organized and well-paid employees of their respective nations. The development of these chambers was extremely important to the advance of cryptography as each side attempted to break each others' ciphers. Aware of these practices, senders worked to improve their cryptographic methods, and a seesaw battle was vigorously contended.

The most famous of the Black Chambers was Austria's Geheime Kabinsets-Kanzlei, established under the rule of Empress Maria Theresa. In the 1700s Vienna was a hub of European commerce and a center for diplomacy and mail transfer. Austria's rulers and generals alike were well-informed about goods, travelers and mail passing through the city to and from western Europe's capitals and more distant cities such as Constantinople and St. Petersburg. Under the empress the Black Chamber thrived. Financial rewards, career advancement and honors were freely bestowed among talented cryptanalysts. Maria Theresa is said to have known enough about cryptographic practices to have advised Austria's envoys to change their nomenclators and encryption keys lest they be overused and thus vulnerable to other code-breakers.

At the height of its productivity, from the 1730s to the 1760s, the Kabinets-Kanzlei was considered by many to be the finest such organization in Europe. It is interesting that historians praise its finest director, Baron de Koch, not for the particular skills of a Wallis or a Rossignol but for his administrative abilities and respect for his personnel. In the 1700s other nations had also adopted this practice: England had its Decyphering Branch, France its Cabinet Noir and Russia its secret police.

Yet even the Kabinets-Kanzlei's smooth system was not free from espionage. In 1774 Abbot Georgel, the secretary of France's ambassador in Austria, purchased a bundle of decipherments from a spy and sent them by an experienced courier to Louis XV. When the king opened the package he discovered the neatly ordered and efficiently compiled results of very recent Kabinets-Kanzlei deciphering successes, including some copies of some of his own missives. The very compact, ordered and accurate records of the Kabinets-Kanzlei had made them quite tempting and open to theft by an opportunistic employee. The man who contacted Abbot Georgel was apparently never identified. The abbot was rewarded for his efforts and Louis XV decided to apply the Great Cipher more frequently to protect himself from England to the west and the efficient Austrians to the east.

In the latter 18th century England was rocked by news of the rebellion in the young American colonies and was soon sending troops to suppress the uprising. Although the American patriots did not have the type of organized cryptography or cryptanalysis that was prevalent in Europe, many of their betrayals and achievements are worthy of inclusion in the history of cryptology.

In the early months of the conflict with England, Benjamin Franklin recognized the talents of Charles William Dumas of the Netherlands, a German-born sympathizer of the Colonial cause. It was Dumas who suggested to Franklin that Britain's rivals Spain and France should be approached for assistance.

In 1776, as a secret agent for America, Dumas took advantage of Holland's neutral status and his friendship with ship captains to transmit letters to the patriots containing important facts about the Colonial struggle. He concealed his messages using a mixture of codewords and an alphanumeric cipher.

Franklin also used this system to write from the Colonial ministry at Passy near Paris when he was conducting high-level diplomacy for the rebel cause. His sagacity and charm are justly credited with winning many friends for the colonies. Franklin's timely diplomacy was largely responsible for securing the pivotal alliance with France that turned the war in favor of the rebellion.

Franklin's judgment of others may not have been foolproof, however. His trusted friend and protégé Edward Bancroft was hired as the secretary of the Colonial legation in France, in which position he had access to official records and encrypted dispatches. He had served Franklin as a spy and knew the espionage craft well. In the 1880s it was finally revealed that Bancroft was actually working as a double agent for the British and that he even managed to turn Franklin's colleague Silas Deane toward the British cause. Several historians believe, however, that the wily Franklin knew of Bancroft's and Deane's duplicity and may have been using their treason for his own purposes by allowing them to pass along false documents, rumors and misdirections.

Another lesser known double agent of the Revolution, Benjamin Church, used a type of symbol cipher in his communications. A trained physician and a member of Paul Revere's Boston spy ring, Church used his position to gain information that he then passed to Thomas Hutchinson, the royal governor of Massachusetts. He later reported to Hutchinson's successor, British General Thomas Gage, and some historians believe that it was Church who told General Gage of the Colonials' cache of military supplies in Concord during the fateful spring of 1775.

It was Church's mistress who began his downfall by entrusting one of his ciphered letters to a former lover named Wenwood in Newport, Rhode Island. Instead of giving the missive to British officers as requested, Wenwood became suspicious and gave the letter to General George Washington. The mistress was summoned, questioned and, after protracted disclaimers, finally divulged the physician's name. Washington sought help from amateur cryptanalysts Reverend Samuel West, Elbridge Gerry and Massachusetts Militia Colonel Elisha Porter, who solved the doctor's cipher to reveal facts about Colonial privateers, rations, ammunition and currency, among other damning revelations. Church was imprisoned, then expelled from Massachusetts. Aboard a schooner bound for a West Indies exile, he was lost at sea.

The British and their Loyalist allies used several cryptography and steganography systems, with varying levels of success. Early in the war the physicist Benjamin Thompson sent a missive concealed with invisible ink to British headquarters in Boston, describing some of the plans of the rebellious New England troops. Later, Sir Henry Clinton, British commander of New York, rendered his correspondence more secure with an alphabet table, a nomenclator, a number substitution and a type of grille. Clinton used the grille in a letter sent to General John Burgoyne in 1777. Their correspondence was linked to Burgoyne's invasion from Canada through the Lake Champlain region and into the Hudson Valley.

At the time of this letter, neither general could foresee the fateful days of October 1777. During a series of confrontations in the battle of Saratoga, Burgoyne's army was soundly defeated. This rebel victory sent a shock wave through the royal courts of Europe. General Horatio Gates was given credit as overall commander, yet many observers believed that a heroic charge led by Benedict Arnold saved the day. Theorists have since speculated that Arnold's not receiving proper credit helped lead to his eventual shift in loyalty.

Perhaps the best-known case of treason in American history, the infamous actions of Arnold involved the use of an awkward type of code known as book code. With greed exacerbating his wounded pride, Arnold turned on the patriot cause and conspired to betray the strategic West Point fortress and surrounding sites, including Stony Point and North Castle. His counterpart in the British army was Major John André. A young, cultured gentleman, André's sole motive in the plot seems to have been loyalty to the Crown. Apparently Arnold performed his own encoding duties, while redcoats relied on cryptologists Jonathan Odell, a New York pastor, and a Philadelphia merchant named Joseph Stansbury.

Ironically, the strength of the plotters' code was never tested. Because of the trust and respect that Arnold had previously achieved, he was not being watched nor was his correspondence being studied. Not a single one of his missives was ever intercepted for analysis and the actual strength of his code was never tested. André was captured by three rebel militiamen whom he mistook for British soldiers, and it was this wary rebel patrol that foiled the nearly successful conspiracy. After appeals to spare his life were rejected and a deal to exchange him for Arnold failed, André was hanged as a spy at Tappan, New York. Arnold, escaping to the British lines, lived to see his name carved in the corridors of infamy.

The near disastrous loss of the vital Hudson fortress caused General Washington and other Colonial leaders to urge an increase in codemaking and spying efforts on every front. One of the sites where the rebels could count on efficiency was in a rather unusual place, Loyalist-dominated New York City. Three of the best rebel agent/codemakers were residing in that area: Major Benjamin Tallmadge, originally of the Second Connecticut Dragoons; New York's own Robert Townsend; and Samuel Woodhull of Setauket, Long Island. Townsend was given the codename Culper Junior, and Woodhull was designated Culper Senior. These three men and their friends had the unenviable task of gathering information in a city that the redcoats had turned into a virtual armed camp. Unable to send messages by visual signals such as lanterns or objects mounted in doors and windows, Tallmadge and the Culpers corresponded with Washington using a nomenclator and a dictionary-based code provided by the spymaster Tallmadge.

Their messages were given an extra level of concealment with stains, or invisible inks. Some were made from lemon and onion juice applied with quills to parchment. To the average eye, the words written with this base were invisible. They were "developed" at Washington's headquarters by holding the letters above a source of heat. The added temperature revealed the hidden and encoded words. Different and very effective stains created by John Jay's physician brother, Sir James Jay, were so secret that their chemical makeup was never revealed. The combination of the dictionary code and the invisible inks enabled the rebel agents to compile valuable facts about British troop strength, morale, supplies and even Royal Navy ships in the harbor.

By 1780 Clinton commanded the British war effort from his base in the New York garrison and, using information provided by Robert Townsend, General Washington learned that Clinton was planning to attack French troops and ships in Newport, Rhode Island. By carefully moving his own troops into key positions around the city, the Virginia squire kept Clinton off guard. Sending missives disguised in the dictionary code, Townsend passed important details on troop strength and vessel movements along a circuitous 150-mile chain to General Washington, who then warned the French.

When Clinton's forces were poised on Long Island to begin the attack on July 30, he received a double shock. First, reports from the Newport area showed it to be heavily fortified. Second, news came that General Washington had moved a large body of men to the eastern bank of the Hudson River, close to New York City. Though he had a much more powerful force, Clinton dared not lose the key port with its harbor for the navy. The strong defenses ahead and the potential threat to the city behind him forced Clinton to cancel his plans. Clinton ended up on the defensive, and his men often came under sniper fire when they wandered too far from the city.

A great deal of conjecture exists among military analysts regarding Clinton's inaction during this period. Many have offered examples demonstrating how he could have successfully moved against the thin arc of Colonial troops around the city on a number of occasions. However, hindsight in warfare is always much clearer than the vision afforded to anyone using a long glass to peer at woods or shores on the eve of battle. What is certain is that Washington was clever enough to turn a very tenuous defensive position into a virtual stalemate. He developed a system of spies and codes and was not too vain to heed the knowledge it generated.

The standoff in this region was to prove even more important as events unfolded in the southern colonies. Because Washington kept the pressure on Clinton, the British commander could not move his formidable garrison at will or send large numbers of reinforcements to the hard-pressed redcoats in the South.

During the final campaign of the war, cryptanalysis again benefited the Colonial forces at a crucial time. By autumn of 1781, Cornwallis had led his army to Yorktown on the Virginia coast to await reinforcements and resupply from New York. Hoping to receive Royal Navy support from the sealanes to New York, he was instead hemmed in by a Colonial army led by Nathaniel Greene and the Marquis de Lafayette. An important revelation occurred when a British water-borne courier and his dispatches were captured en route from General Clinton in New York City to Lord Cornwallis. The intercepted dispatches were sent to Rebel cryptographer James Lovell, who deciphered them and revealed a British plan to relieve Cornwallis by sea.

Some of the cryptograms were cloaked in monoalphabetic substitution. Others combined aspects of single and multiple alphabets, with digits serving as encipherment. Lovell noticed that these styles appeared to be in general use among the enemy commanders. He also discovered that when changes in the cryptomethods occurred, the resulting differences were only ones of position in the same alphabet. After breaking the cipher, Lovell informed General Washington of his findings, allowing Washington to formulate his next plans.

In a strategically brilliant decision, Washington cleverly pretended to continue encircling New York. In fact, he took the majority of his forces to join Lafayette and Greene. Fooled by Washington, Clinton held his reserves too long in New York. General Washington and Admiral de Grasse, his French ally, closed a land and water ring around Yorktown and the Chesapeake Bay. Cornwallis surrendered on October 19, 1781, and when Admiral de Grasse chased the Royal Navy from the area on October 30, the Colonial victory was complete.

By the 1780s America's wartime ally, France, was herself in a state of ever-deepening social and economic strife. Years of royal excesses had led to an extremely volatile situation and the formerly well advised Bourbon dynasty was in the questionable hands of Louis XVI. Although not an evil man, Louis was plagued by almost every kingly malady, from indifference to ignorance of the military chain of command.

His bride, Marie-Antoinette, is often described as a foolish, giddy squanderer and was extremely unpopular with her subjects, yet history suggests she was more resourceful and practical under pressure than this portrayal shows. Her mother, Maria Theresa of Austria, had, as we have seen, a keen interest in secret correspondence and this seems to have influenced Marie, who used one cipher similar to Giovanni Porta's and one based on a popular novel, *Paul et Virginie,* to correspond with her various lovers.

Retreating from public life, Marie surrounded herself with sycophants adept at befriending the lonely and powerful. Amid the intrigues and backstabbing jealousies of the increasingly decadent Versailles court, Marie's spending binges and disputes with Louis were cleverly encouraged by her entourage, who benefited both from her shared opulence and by being included in the influential royal favor.

During times of increased alienation, both Louis and Marie had affairs. One of Marie's relationships is of interest here because it involved both enciphered love letters and political intrigues. Her lover and cryptographic adviser was Axel Fersen, a handsome Swedish count who had been a hero of the American Revolution.

Though Louis XVI and Marie-Antoinette were divided in their personal lives, they were drawn together as the year 1789 brought a national crisis to the gates of Versailles. France's long-festering ills quickly escalated into revolution against the Bourbon regime. The storming of the Bastille was more symbolic than dangerous to the monarchy, but the sacking of estates, the disloyalty of army units and the murders of numerous aristocrats were signs of serious political trouble.

Louis and Marie were forced to leave Versailles and were escorted to the Tuileries Palace in Paris, supposedly for their safety. As Louis became more confused and ineffective, Marie demonstrated an inner resolve that surprised her critics. With Count Fersen's help, Marie used ciphers to exchange plans with loyal monarchists. They hoped to unite the numerous but disorganized provincials who still held the royal family in awe. It was Fersen, with his war experience, who took care of the clandestine meetings with their followers. Marie also dispatched requests for help from monarchist relatives and friends throughout Europe which were concealed by a cipher system of Fersen's design—a polyalphabetic substitution with keywords.

Though this multiple alphabet did not compare to other types in difficulty, it was enough to fool the increasingly violent Paris mobs. The leaders of various factions, including Georges-Jacques Danton and Maximilien-François de Robespierre, were too busy trying to consolidate their own shaky positions among the many shifting allegiances to be able to watch the Royalists' every move. Nor were there men of Rossignol's caliber who could be hired as cryptanalysts.

Louis and Marie did hire the compte de Mirabeau, a popular orator, in an attempt to steer the nation toward a constitutional monarchy. However, when Mirabeau died in 1791 and the threats to their personal safety increased, the queen persuaded the king to flee to France's eastern provinces.

An elaborate scheme was arranged in which the royal family appeared to be retiring for the evening at their palace, the Tuileries. Instead, they disguised themselves as servants and began their departure in an enclosed carriage, with the loyal Count Fersen posing as their driver. They hoped to meet supporters near the border, and soon seemed to have their goal within reach. But at the entrance to a small bridge in the village of Varennes a local man's wagon blocked their access. The family was recognized, arrested and returned to Paris.

Count Fersen managed to avoid punishment and remained loyal to the royal family, although separated from the queen.

More purposeful than ever before, Marie sought help from overseas. She encouraged her brother, Holy Roman Emperor Leopold II, to challenge the Jacobin revolutionaries. In more secret correspondence, she sought a union of other royal houses against the spread of radicalism.

As Louis agreed to abide by a new constitution formulated by moderates, the queen remained in touch with other governments. Marie's attempts were nearly successful when a confrontation between France and Austria seemed imminent in 1792. Hoping to guarantee the defeat of the rebels' forces, she took the fateful step of divulging the French army's plans to her friends in Vienna.

An outbreak of renewed fighting within France toppled the constitutional monarchy in August 1792 and the moderates lost their authority. Leopold's threats faltered and the Jacobins gained full control. Seeking revenge for years of real grievances, the Jacobins were soon corrupted by their own power. Eventually losing their grip on the populace, France was turned over to the mob rule of the Reign of Terror.

By this time Marie's cipher had been broken and the royal family's complicity with other monarchs had become public knowledge. These attempts led to accusations of counterrevolutionary plots and the reigning Jacobin government charged Marie and Louis with treason.

Almost eight months to the day after the king's execution, on January 21, 1793, Marie was found guilty of high treason and executed at the guillotine. Although he escaped the violence of Paris, Axel Fersen was falsely accused of poisoning Sweden's crown prince and murdered by a Stockholm mob in 1810.

The anarchy of the French Revolution had disgusted even the most embittered antiroyalists in the United States. The former allies who had defeated Britain in the American Revolution grew dangerously at odds, and by 1797, the military and diplomatic ties between the United States and France had declined to mutual recriminations and threats. Added to this background of social upheaval was U.S. distrust of the French diplomatic ministers, Edmond Genet and Pierre Adet. When Genet assumed powers far beyond those of a foreign envoy, initiating unauthorized meetings with U.S. government officials and offering bribes for favors rendered, George Washington demanded that France recall him. Adet, the succeeding French envoy to the United States, later added fuel to the fire when he announced that France would consider U.S. seamen serving on British vessels to be buccaneers, sea raiders without special rights. As the new president, John Adams, assumed office, Americans demanded that he take action.

Adams reacted by sending a special commission to France that included Charles C. Pinckney, John Marshall and Elbridge Gerry. The first two of these men represented the conservative Federalist point of view on foreign relations, while Gerry was partial toward the liberal Jeffersonian view.

The United States trio arrived in late September 1797 and became involved in a dangerous political and financial intrigue which became known as the XYZ Affair. A Swiss banker, Jean Hottinguer (X), requested a 250,000-dollar "gift" for French officials. Soon a Hamburg merchant, Mr. Bellamy (Y), offered a secret treaty in exchange for a sizable loan. Not long thereafter, Lucien Hauteval (Z), a messenger for Charles Maurice de Talleyrand-Périgord, French minister of foreign affairs, also sought monetary commitments from the commissioners.

Marshall, Pinckney and Gerry used a nomenclator to send the account of these affairs to Secretary of State Timothy Pickering. When told by Hottinguer that monetary payments were an expected part of negotiations, the commissioners sent an answer to Pickering, encoded in their nomenclator, that would become a piece of U.S. diplomatic history: "No, no, not a sixpence." The representatives of a young but proud nation had answered. The outrage of many U.S. citizens forced a more moderate French position over time. Though problems with France continued, the United States had made its claim to a place of respect in the world community.

As the 18th century drew to a close, America came into its own with Thomas Jefferson as its president. As America's third president, primary author of the Declaration of Independence, architect and inventor, Thomas Jefferson had a life filled with many varied and significant accomplishments. Rarely mentioned, however, is his interest in cryptology.

In the period between 1785 and 1793 Jefferson used different nomenclators to correspond with James Madison and U.S. Secretary of Foreign Affairs Robert Livingston.

Jefferson's best-known contribution, however, a cipher device he called a "wheel cypher" was not recognized during his lifetime. This simple and practical device was well ahead of its time and in fact superseded a number of similar inventions in Europe. Yet neither the American military nor the fledgling diplomatic corps was to benefit from the wheel cypher because Thomas Jefferson did not apply it for practical use. Curiously, Jefferson apparently put the idea aside and eventually forgot about it. Not until 120 years later was a similar version of the wheel cypher made available to the U.S. armed forces, but its worth was verified by the fact that the U.S. Navy continued to use similar mechanisms for decades after its introduction.

Thomas Jefferson was also indirectly involved in a scandal involving codes and ciphers and a candidate for the presidency, Aaron Burr. Born in 1756, Burr matured during the tumultuous early years of America. He secured command of a Colonial regiment in 1777 and fought bravely in the Battle of Monmouth a year later. He even served for a time on Benedict Arnold's staff. Burr was no longer associated with Arnold during his West Point treason plan, but this service with Arnold did bring Burr into contact with another young officer named James Wilkinson, who was to be linked with Burr at a later date.

In busy postwar New York, Burr became a lawyer with political organizing skills and a polished style. He turned the city's Tammany Social Club into one of the first American political "machines." This led to a strong position in the New York

State legislature and eventual control of the presidential electoral votes for New York.

Realizing his influence, the Jeffersonian Republicans picked him to be Thomas Jefferson's vicepresident in 1796 and 1800. The first time on the ballot, Burr finished third. But in 1800 the variables of the Electoral College led to an unusual situation. The electors cast an equal number of votes, 73 for each man, compared to 63 for Federalist rival John Adams. The election had then to be settled by the House of Representatives.

Because Jefferson's backers were also the party's leaders, they did not look favorably upon Burr's sudden prominence. Though Jefferson did not personally compete with Burr, accounts indicate that he followed his advisers' words to place some distance between himself and the fast-rising New Jersey native. As the House started deliberating, Burr began to feel this alienation.

The crushing blow to Burr's presidential hopes came not from the Jeffersonians but from his New York political rival, Alexander Hamilton. After a week of House indecision and 35 ballots, Hamilton persuaded some of his New York associates to cast blank ballots. Jefferson's men held firm, repeated their votes on the 36th ballot, and their candidate was declared president.

Burr continued this term as vice president, though he was understandably disillusioned, and by 1804 he was actively seeking a stronger political base. To do so, he entered New York's gubernatorial race with private Federalist backing. At the same time some New England Federalists, such as the "Essex Junto" and the "River Gods," disagreed so strongly with Jefferson's policies that they seriously considered forming a New England–New York confederacy.

Hamilton learned of this plan, publicly linked Burr with it and denounced the idea as a plot against the nation. These accusations effectively destroyed Burr's support in upstate New York and cost him the governorship. Enraged by Hamilton's action, Burr challenged him to a duel. On July 11, 1804, the pistol exchange at Weehawken, New Jersey, ended with

Hamilton being mortally wounded. In gaining what he considered to be a gentleman's revenge, Burr sounded the death knell for his political career.

Hamilton's death caused a public outcry that Burr could not have envisioned. As even his closest Tammany friends turned against him, he reacted by creating one of the oddest schemes in American history. Reports about his complicated plot included foreign money, a private army, territorial acquisitions and ciphers to link the grandiose venture.

Burr had resumed his association with James Wilkinson, now a general in charge of the newly acquired Louisiana Territory. With funds apparently supplied by his son-in-law, Joseph Alston, and by a wealthy Irishman named Harman Blennerhassett, Burr purchased the title to more than a million acres of Orleans Territory. Various sources tried to connect his monetary support, rumored to be as much as a half-million dollars, to British and Spanish interests who sought a breakup of the recently united colonies. Burr also had a strategic base of operations on an Ohio River island owned by Blennerhassett and retained a small army around 1806, leading to speculation that he was planning to seize more territory and establish an independent country in the Southwest. From this site in 1806 Burr sent messages to Wilkinson in a dictionary code from John Entick's *New Spelling Dictionary,* combining it with a symbol cipher. In the late summer of that same year he left Blennerhassett's island with 60 men on 13 flatboats.

All of Burr's planning was nullified when Wilkinson turned himself in and submitted deciphered versions of Burr's letters to President Jefferson. It was later revealed that Wilkinson was a paid agent of Spain, a nation very interested in the Louisiana lands bordering on Mexico and its other southwest possessions. Upon reading it, the man who had once shared the Republican ballot with Burr now ordered the suspected expedition to be disbanded and Burr to be arrested. After some months, Burr was caught. His mysterious plans halted, he was brought to Richmond, Virginia, and was indicted for treason.

The outcome seemed certain, but in yet another twist, the judge presiding at the trial surprised everyone. Chief Justice John Marshall, wishing to uphold strict legal principles, was not as eager as were many to find Burr guilty. Despite the evidence of the deciphered missives, Marshall's careful instructions called for the testimony of two witnesses regarding treasonous acts. The prosecution could not comply, and Burr was acquitted on September 1, 1807. In spite of this reprieve, Aaron Burr's fortunes never improved. Ruined in public life and shunned by creditors, the man who had been a vote from the presidency slipped into ignominious obscurity.

Burr's adventures, schemes and treason trial claimed national attention and the interest of the public for a while. Soon afterward, however, the headlines were being filled by events in Europe and by news of another individual whose actions were affecting entire nations. French emperor from 1804 to 1815, Napoleon, or as he was often nicknamed, "the Little Corporal," became a military giant bestriding Europe for a generation.

Born on the island of Corsica in 1789, the spirit of revolution flooding through the United States and France directly affected Napoleon's early life. He had joined the French army and was beginning to rise through its ranks even as the Terror was upsetting France's social order. While Napoleon was achieving his first successes abroad, the French people were crying out for an end to the bloodshed at home. Napoleon was a natural leader, and the people turned to him as a central figure to restore order and stability. Within a generation, the French had toppled a monarch only to welcome a man who crowned himself emperor.

While Napoleon was a wise strategist and arguably the best tactician under pre-1860s battlefield conditions, his use of French inventor Claude Chappe's aerial telegraph, a type of semaphore system, remained his hidden ace for years. Opposing generals both feared and hated him for his apparent prescience. In actuality his battle plans depended on the aerial telegraph to convey orders for troop movements and to secure reinforcements and supplies. As his conquests extended throughout Europe, the emperor replaced this visible semaphore-type system with a more secretive means of communication.

Although more secure than the aerial telegraph, Napoleon's *petit chiffre* (little cipher) was a very simple nomenclator that consisted of only two hundred or so letter groups. His own pride and his officers' prejudices all caused him to frown upon complex secret writing. His tactical skills seemed to require only a few encrypted messages through his many years of victories. But the ill-fated invasion of Russia in 1812 was to change history's course.

Along with the notorious weather conditions and Russia's immense size, Czar Alexander I had another ally—cryptography. Many of the cryptologists of the Grande Armée were lost in Russia's vast frozen expanses. Furthermore, after the reign of Peter the Great, the Russians' cryptographic skills had increased immensely. The Corsican's "little cipher" would have been no match for Alexander's cryptanalysts. While his cipher was almost certainly broken during this invasion, Napoleon's application of his cipher suffered another setback. The fire that swept Moscow around Napoleon's occupying forces destroyed copies of the *petit chiffre*, forcing him to write a number of command messages in cleartext. It is thought that some of his unconcealed orders were intercepted and used against his already decimated army in its disastrous retreat across Russia.

A cipher failure was a direct factor in Napoleon's second major defeat a year later in the "Battle of Nations." From October 16 to 19, 1813, near Leipzig, Napoleon engaged an allied force of English, Belgian, Prussian and Austrian troops. He hoped to make an orderly withdrawal from Leipzig and sent ciphered orders to his reserve subordinate, Marshal Augereau. The marshal was to advance to the town, hold the main Elster River bridge, and build temporary bridges. Augereau's garbled cipher reply was understood by some of Napoleon's officers to be an affirmative response. Instead, the marshal's forces not only arrived in disarray but the Elster Bridge was inexplicably blown up. Trapped, Napoleon and the remnants of his shattered army had to flee for their lives.

The Little Corporal's time of contemplation while in exile on Elba apparently did not increase the complexity of his cipher systems. After escaping from the island and gathering a new army, he cleverly attacked the allies where they least expected an assault—in Belgium. But two major setbacks occurred in the form of confused messages and orders.

On June 16, 1815, at the crossroads town of Quatre-Bras, French Marshals Michel Ney and Jean d'Erlon exchanged orders that were misunderstood. Instead of supporting Ney, d'Erlon's corps lost time marching between Ney's forces and Napoleon's army at Ligny. D'Erlon's corps helped neither commander and Ney lost his advantage at Quatre-Bras against the great General Arthur Wellesley, the duke of Wellington. Two days later, Wellington was defending stronger positions near a town called Waterloo.

The historic battle of Waterloo hinged on many factors beyond the scope of this text. Nevertheless, at a very crucial point, Napoleon's army had an acute communications lapse. Napoleon had sent French Marshal Emmanuel de Grouchy in pursuit of Prussian General Gebhard von Blücher. But by 4:00 P.M. on that fateful June day, Grouchy had not yet made contact. Neither by courier nor cipher did Napoleon learn about Blücher's location until the Prussian's advance units appeared on Napoleon's right flank. As the French elite Imperial Guard was checked by Wellington and his thin red line, Marshal Grouchy was engaging Blücher's rear guard in the town of Wavre, 8 miles to the east. Grouchy never entered the struggle at Waterloo, where the Corsican's last vestiges of empire were lost in the bloody Belgian fields.

THE MECHANICAL AGE

In February 1838, Samuel Morse sent the world's first recorded telegraph message: "Attention, the Universe, by Kingdom's Right Wheel," and heralded the entrance of a new mechanical age. The development of Morse code was extremely significant in the history of cryptography as it provided a system by which numbers and letters could be converted into a simple dot-dash pattern. Morse code, as it came to be known, reached its heights with the wide-spread acceptance of the electric telegraph, which became a crucial means of communication among the military and civilians alike.

In the same period, the development of the Playfair cipher signaled an advance in cipher making. In the early 1850s, Charles Wheatstone created what he called a rectangular cipher for use in telegraphy. It was Wheatstone's close friend Lyon Playfair, the first baron of St. Andrews, who introduced the cipher to Home Secretary Lord Palmerston and Prince Albert, at a formal dinner in 1854. As Playfair was widely known in military and diplomatic circles, Wheatstone's creation became immortalized as the Playfair cipher.

The cipher was extremely practical for military purposes, and given more time, might have become an integral part of the British diplomatic and military planning. But by the mid-1850s the attention of Great Britain's political and foreign service hierarchy was being centered on the Crimea. The first indications of war were causing Britain's leaders to rely on familiar, trusted tactics and methods.

Crimea is a peninsula of Russia on the northern shore of the Black Sea and, in the mid-1850's, became the focal point of a series of disputes regarding territorial claims and religious interests between Russia, Turkey and provincial regions around the Black Sea. One central factor in these disputes was Turkey's general distrust of Russia and the fermenting ill-feeling between the two nations. Indeed, Czar Nicholas I had called Turkey "the Sick Man of Europe." As the once united Ottoman holdings were beginning to separate into independent nations for religious and economic reasons, Russia began to annex some of this territory near her own borders. While Czar Nicholas claimed to represent and protect the nearly twelve million Orthodox subjects within the Moslem-dominated regions, the Turkish leadership believed that he intended to invade Turkey in order to acquire access to more warm-water ports.

France entered the fray primarily because of the influence of Napoleon III, whose throne was built upon the unsteady rubble of the Second Republic. Embroiled in political and economic problems at home and wishing to direct his subjects' interests elsewhere, he fixed upon the possibility of a foreign war as a means of increasing his popularity even though this involved a patchwork alliance with France's centuries-old rival, England.

Although Great Britain had no direct concerns in the Black Sea area, because of its involvement with India, Queen Victoria and her advisers were wary of shifts in the balance of power in western Asia. Turkish landholdings formed a barrier to potential Russian advances in the region, so it was indirectly beneficial to England to maintain the status quo.

There had been a long history of Russo-Turkish hostility, but the inclusion of the European powers changed the entire situation. By early 1853 the alliance against Russia included Turkey, England, France and Sardinia, an island kingdom near Italy. While some attempts were made at a peaceful resolution, entrenched distrust between each side hobbled negotiations.

By July, Russia had occupied the Turkish principalities of Walachia and Moldavia, and in response, Britain ordered their fleets to the strategic Dardanelles entrance to the Black Sea. The Czar did not answer an October ultimatum to leave the principalities, leading to intermittent land fighting. In November the Russians defeated the Turkish fleet at Sinope, Turkey, and rumors of the slaughter at Sinope resulted in a feverish clamor for action in England and France. In March 28, 1854, both nations declared war on Russia.

Relying heavily on England's naval power, the new England-France alliance launched a major offensive in the pivotal Black Sea area. By attacking the Crimean port fortress of Sevastopol, the allies hoped not only to disrupt Russia's sea lanes but also to bring the fighting to the Czar's own soil.

Within months, thoughts of a glorious quick victory proved unattainable due to poor war preparations and worsening weather conditions. With supply wagons mired in the mud, soldiers suffered from lack of food, ammunitions and medical supplies. Though the Royal Navy controlled the waterways, the land war became a costly stalemate. Almost inconceivably, the alliances' military leaders had failed to heed the lessons of Napoleon's disastrous invasion of Russia just 42 years earlier. Once more the terrain and the increasingly cold weather affected and prolonged the struggle.

Telegraphy began to play a key role in the Crimean War, although not in tactics. The telegraph enabled the press to cover this war more extensively than had ever previously been possible. The news of losses due to battle, disease and bitter cold reached Paris, London and Moscow with shocking regularity. As the war dragged on, the increasing casualty totals had a chilling effect on the once boisterous war fever.

Despite the speed and the general reliability that the telegraph provided, the construction of telegraph stations was a luxury often not possible during its early years. The results of many battles such as those in Inkerman, Malakhov and Balaklava could have been very different if missives had been rapidly conveyed and capably enciphered.

The Balaklava confrontation, the scene of the tragic charge of the British Light Brigade, caught the imagination of a generation and was immortalized in Alfred Lord Tennyson's classic poem. After years of debate and conjecture, the actual events of the Charge of the Light Brigade remain a mystery of confused orders and ill-conceived tactics.

On October 25, 1854, Russian general Prince Menshikov attacked the English base at Balaklava. Several important artillery batteries were taken, and their loss endangered the entire British position in that part of the front. A series of indecisive choices led to a fateful decision: a swift attack was needed to regain the lost cannons. Renowned for the speed provided by their fast steeds and light armament, the Light Brigade was chosen to accomplish the task.

The seventh earl of Cardigan, James Brudenell, received orders to advance through a nearby valley, but no other units nor adequate cannon fire were provided in support. Many questions have been raised regarding the intent and timing of this command and no telegraph linkup existed to rescind the order after it had been put into motion. Several historians and writers have even speculated that Cardigan's old enemy, his brother-in-law Lord Lucan, had conspired against him by misinterpreting or misdirecting the orders. Lord Lucan led the division of which the Light Brigade was a part and was later severely castigated for not coming to the aid of Lord Cardigan and the doomed cavalry. No doubt realizing the danger of the situation, the Light Brigade nevertheless rode into a heavy concentration of Russian firepower and were decimated.

After other costly sacrifices at the Redan and the Chernaya River, the allies finally captured Sevastopol in September 1855. Under the Treaty of Paris, Russia was forced to accept a neutral Black Sea and agreed not to interfere with Turkey's internal affairs, finally bringing an uneasy peace to the Black Sea.

The American Civil War (1861–65) was the first major confrontation in which the telegraph's potential was fully realized, especially by the industrially advanced Union forces. Along with telegraphy, other concealment systems played a significant role in concealing both battle plans and missives sent by spies across the country.

Union General George B. McClellan, for example, used a type of word transposition in 1861 during his victorious campaign in the Ohio Valley and western Virginia. The cipher was created by Anson Stager, first superintendent of Western Union. Stager's cipher grew in importance as Union commander John C. Frémont and detective-turned-spymaster Allan Pinkerton applied it in the west and east, respectively. In 1862 the Stager transposition came into use throughout the Union army, thanks to the first large-scale wartime use of the telegraph. In fact, President Lincoln frequently visited the telegraph room in the War Department to keep informed of unfolding events.

The Stager cipher was readily adaptable to telegraphy because of its easy arrangement. A type of word transposition, the message was written in lines and transcribed using the columns that the lines formed. Secrecy was provided by the order in which the columns were read, namely, down one column and up another. As time passed, further security was added by nulls, codenames for special terms and multiple directions that formed a larger maze of routes necessary to trace the message.

The various branches of the Union army did not all use the Stager cipher. Some forces continued to use other elementary word transpositions and members of the new Signal Corps, notably Sergeant Edwin Hawley and Major Albert Myer, devised other methods. Hawley's creation was a set of 26 wooden pieces bearing different ciphertext alphabets used with a keyword, and became the first American cipher method to be patented. Myer, the Signal Corps' commander, developed a cipher disk as well as a signaling system for a flag, torch and other hand-held objects which soon became affectionately known as "wigwag."

Wigwag, or flag telegraphy, is a system of positioning a flag at various angles in order to relay a message. This method was created in the mid-1800s by three men working in separate locations: in England, Navy Captain Philip Colomb and Army Captain Francis Bolton, and inventor-surgeon Albert J. Myer in America. Myer termed his banner method, which he developed around 1856, flag telegraphy. Some accounts, including that of *Traditions of the Signal Corps* (1959), credit an unnamed Civil War general with the name wigwag. First used as a nickname for the signals' movements, the term eventually supplanted Myer's own choice.

Myer prepared a report on his experiments, particularly the wigwag system, for his superiors in 1856. Though mildly complimentary, the ranking officers did not incorporate his creations. Nevertheless, Myer continued testing and improving the system and its variations through 1859 with one of his chief assistants, Edward P. Alexander. Although both men were employed by the U.S. Army, the Civil War was soon to separate them. When it began, Alexander joined the Confederate army and it was he, not Myer, who first used wigwag in actual combat.

On Sunday July 21, 1861, Alexander, then a captain of the Confederate army, was serving on one of four observation towers that he and General P. T. Beauregard had constructed and equipped with signal flags. These towers overlooked the countryside around a Virginia rail depot called Manassas Junction. Fresh water was provided by a nearby creek, Bull Run.

During the previous night, divisions of Brigadier General Irvin McDowell's Union army had moved to flank the Confederate left. After delays caused by the rough terrain, advance units of McDowell's force had reached Sudley Ford by 8:45 A.M. on July 21. On the verge of surprising the southern troops at an important crossing—Stone Bridge—the Union army did not notice the sun glinting on one of their fieldpieces.

This reflection and the gleam from bayonets drew Alexander's looking glass toward the ford. Danger was clearly imminent. With the swiftness and skill born of his training under Myer, Alexander sent the following flag warning to Confederate Colonel Nathan Evans at the Stone Bridge defenses: "Look out for your left, you are turned."

Colonel Evans ordered cannon and musket fire toward the Federal troops and the attack was halted before it had even begun. The application of this new signal system had played a part in the shocking Union defeat that eventful July day. At that moment none of the participants in this conflict could know what the successful warning would mean. Analysts now say that Alexander's vigilance led to changes in the tactics of the entire struggle around Manassas Junction.

Amid the confusion and accusations of blame following the first battle of Bull Run, Myer continued his struggle for better military communications. His fortunes took a turn for the better on July 26, 1861, when McDowell was replaced by McClellan. Myer presented his proposals to the general, who was himself seeking solutions to the Union's inefficient communication systems. Aware of the uses of telegraphy in the Crimea, McClellan found in Myer a resourceful and inventive aide, but both the Congress and the powerful secretary of war, Edwin Stanton, proved to be formidable barriers. It took more battlefield evidence to convince them that changes were needed.

McClellan was beset by military problems through the summer of 1862. His efforts to capture Richmond, known as the Peninsula Campaign, became a series of stalemates, then costly reversals. As McClellan's tactics began to fall into disfavor, Myer was more successful in persuading some commanders to use more efficient battlefront communications.

In September 1862 Robert E. Lee initiated his first invasion of the North, ordering his troops to cross the river at a remote site about 30 miles north of Washington, D.C. Lee's army was near the town of Frederick, Maryland, and advancing unchecked when one of its wagon supply units was spotted.

This support column was not discovered by regular Union scouts; it was sighted by one Lieutenant Miner, a flag signalman stationed at a post on Sugar Loaf Mountain. On September 6 the lieutenant flagged this news to the Signal Corps' base on Point of Rocks. From there, telegraph lines that ran along the Baltimore and Ohio Railroad track conveyed the alert to Washington. The Army of the Potomac was hastily ordered to march and counterattack. Although the Union was to suffer more setbacks in this campaign, General Lee's surprise attack was thwarted by the alert signalman and effective signaling systems.

This evidence jarred Congress and the War Department into allocating much-needed funds, which were used to purchase everything from copper wire for telegraphy to real flags for men previously training with broomsticks and tattered shirts. The benefits were finally realized almost a full year after Lieutenant Miner's warning when, at a town in southern Pennsylvania, sightings of epic proportions occurred.

In June of 1863, General Lee had again invaded northern soil and on June 24 a vastly improved flag system relayed word of Lee's most recent Potomac crossing. In spite of cut telegraph lines, the message moved rapidly from the observation post at Maryland Heights to the War Department in Washington.

The Union's high command was in disarray. On June 8 General Joseph Hooker had resigned and George Meade was made commander of the Army of the Potomac. While General Meade formed his own chain of command, he set up temporary headquarters at Taneytown, Maryland. In the midst of calling in scattered army units, startling news came to Taneytown through the watchful signalmen. A patrol of Union General John Buford's command had skirmished with Confederates at a site near the Maryland border and soon both sides were committing larger numbers of troops to the fray. The wigwag messages made the situation clear: a major force was facing Buford and not in Maryland, but in Pennsylvania. In fact, it was Lee himself who was commanding the Confederate troops at the town of Gettysburg.

Signalmen of both armies sent out calls for reinforcements. Support units were force-marched from locations dozens of miles away and by July 1, 73,000 gray and 88,000 blue met in one of America's most decisive battles. Although history books rarely credit signaling systems during this battle, crucial sightings made by Union observers directly tipped the scales against Lee's best tactics.

On July 2 Union signalmen, positioned on the two high points called Big Round Top and Little Round Top, discovered rebel troops in the woods near the strategic Emmitsburg Road. Reinforcements were rushed to the North's defenses and they thwarted a flanking attack that could have won the day for the South.

On July 3 the bulk of Lee's forces attacked the Union's center positions in a desperate bid for victory. Artillery and rifle fire were extremely intense and the high-ground observers could not use their flags lest they expose themselves to sharpshooters. Nevertheless, these brave watchers continued to relay missives by courier. One such rider brought vital news to the Union's battered signal command post.

On cemetery Ridge, Captain David Castle received word from a courier that rebel General George Pickett was beginning a massive charge against the middle of the Union line. Confederate artillery barrages along the ridge sent other flagmen under cover, yet Captain Castle used a wooden pole and a bedsheet for a makeshift flag to alert Union headquarters. The message arrived in time for General Meade to order increased cannon fire and shifts of troops. Thanks to crucial warnings by previously disregarded flagmen, Pickett's charge was stopped short of breaching the Union line; General Lee's desperate gamble had failed.

Women also played a vital role outside the battle lines, in the shadowy world of spies and their secret codes. Though neither side maintained training centers for special agents, both benefited from espionage. As the South had first achieved success with flag signals, so did the rebels' cause prosper early because of ambitious spies.

One Southern agent, Rose Greenhow, gave the Union trouble from the outset. In her mid-forties, she was the charming widow of a prominent Washington physician. Greenhow used her position in the elite circles of Washington society to garner valuable information. While men and women gathered for a respite from the war, she kept her senses keen. Her eyes noted officers' faces, and her ears picked up the slightest slip of a loose tongue. She was adept at ingratiating herself with the relatives of wealthy industrialists and powerful congressmen, and her sharp mind held and linked rumors, gossip and carelessly revealed facts in a mosaic of information. After much careful checking, she sent reports to rebel commanders disguised in a number and symbol substitution cipher.

During the battle of Bull Run in 1861 "Rebel Rose" Greenhow sent a missive to the Confederate defenders. Historians credit her with providing information about Union troop movements and numbers that helped General Beauregard and his allies that fateful day. This early success was followed by a series of communications that Greenhow conveyed using everything from pockets hidden in clothing to coded designs embroidered in dresses. The information she provided for the Rebels was so valuable that Federal authorities began their own counterespionage mission. Increasing their number of agents, they began detailed tracking of leaks in military information to their sources in party and parlor gossip.

Washington social circles were close-knit and Greenhow had an increasingly difficult time hiding her double life. Her chief nemesis was none other than that pioneer of detectives, Allan Pinkerton. By a combination of luck and determined effort,

Pinkerton was eventually able to trap Greenhow, arrest her and place her in Old Capitol Prison.

Undaunted, she created new links to her old network from her prison cell. Eventually, Union officials granted her amnesty and sent her to Richmond. Her bravery and resourcefulness were rewarded when Confederate President Jefferson Davis gave her the assignment of gathering support for the Secessionist cause in London and Paris.

Belle Boyd, another Confederate spy, was only 18 years old when the Civil War broke out. A West Virginia native, her adventurous courier missions through the Shenandoah Valley have been immortalized in stories and campfire songs. Her messages are credited with one direct military result: her report about Union troop activities ensured the success of General Stonewall Jackson's attack near Front Royal, Virginia.

The Federal government also had its secret agents as well as its sympathizers south of the Mason-Dixon Line. A spinster in her early fifties, mild-mannered Elizabeth van Lew moved quietly about Richmond, Virginia, with her heart directed northward. In addition to providing facts about Richmond's defenses, she helped Union prisoners escape from the infamous prisons of Belle Isle and Libby. While these men recuperated, van Lew hid them in the basement of her Church Hill mansion. Upstairs she compiled detailed facts that contributed to General Grant's strategic plans in his siege of the Rebel capital, hiding her missives in a substitution cipher.

A second Union loyalist was Pauline Cushman. An actress of New Orleans Creole ancestry, Cushman was in her late twenties when she was asked to help the North with her theatrical abilities. While performing in Nashville in May 1863, she traveled to Shelbyville, Tennessee, where she pretended to be a Rebel sympathizer evicted from her former surroundings and gained favor with residents and military staffers alike. Using her memory honed from years of acting, she conveyed news about General Braxton Bragg and the Confederate forces based in the area.

The intervening years and the secrecy of their missions have obscured some of the methods used by both the military and their spies during the Civil War. Most of the ciphers were simple nomenclators consisting of homemade symbol, number and letter combinations. Both sides also used a type of route cipher known as the rail fence, as it was relatively easy to learn and apply. Rather basic in their structure, both route and substitution ciphers became popular during such trying times when the ability to encipher and decipher missives quickly was more expedient than total secrecy.

For the most part, the South's cryptographic efforts were not as advanced as the North's. General Albert Sidney Johnston used a Caesar substitution to communicate with General Pierre Beauregard during the devastating Battle of Shiloh, and Johnston and Jefferson Davis used a dictionary-based book code. The South's cryptographers also used ciphers such as the Vigenère table, yet because of transmission errors by their own cipher clerks who were translating the ciphertext into Morse code, Confederate officers often had a more difficult time deciphering their own communiqués than did the Union cryptanalysts.

Despite the variety of codes and ciphers applied during the Civil War, none affected the outcome in the same way that telegraphy did, which the more industrialized North used to a better advantage. By 1864 Union commander Ulysses S. Grant was initiating a grueling war of attrition with General Lee, forcing Lee to protect his beloved Virginia soil and its capital, Richmond. No longer was he able to move with the quick strikes and clever flanking movements of his earlier offensives. Struggles in places like Spotsylvania Court House, the Wilderness and Cold Harbor led to a mounting casualty list. Lee was on the defensive, hemmed in by Union numbers, firepower and technology.

General Grant combined massive factory production, miles of railroads and a virtual forest of telegraph poles to encircle his opponent. Messages sent by Grant's telegraphy bases outraced the fastest rebel couriers as Federal troops arrived at strategic points before Lee's best strategies could be enacted. The dual advances of telegraphy and Morse Code enabled Grant to utilize a broad-scale strategy that was unknown in previous warfare. He could literally be commanding in one battle while he sent and received ciphered messages from other fronts. Only a similarly equipped army would have had an equal chance against these factors, but the decimated South could not match the technology and the cipher power of the Union.

On April 9, 1865, surrounded by Grant's army, Lee surrendered at Appomattox Courthouse and his 30,000 troops were allowed to return to their homes. Jefferson Davis was captured in Georgia on May 10, and on May 26, the last Confederate army surrendered. Almost 360,000 Union and 258,000 Confederate lives had been claimed in the four years of the Civil War.

The painful legacy of the Civil War continued to haunt the nation long after Appomattox. In the 1876 race for the presidency, some 11 years after the Civil War's conclusion, both festering ill will and secret communications about vote buying subverted the election.

Samuel J. Tilden, governor of New York, was the Democratic party's nominee. He had built a fine reputation and gained nationwide fame as the man who had led the fight to stop William M. "Boss" Tweed and the corrupt Tammany Hall political machine in New York City. The Democrats had been successful with their accusations of Republican misconduct in high places and in the condemnation of their opponents' controversial Reconstruction policies.

Tilden opposed Ohio's Governor Rutherford B. Hayes, who also had a reputation for personal honesty. However, the political spoils system and its corruption had spread like a plague through the state capitals and Washington, D.C. A number of southern states were still being controlled by self-serving profiteers and "carpetbaggers," northern politicians who went to the South to take advantage of the unsettled conditions. Voting procedures and regulations nationwide had not been standardized. All of these factors led to the most bitterly disputed presidential election in American history.

When the ballots were tallied, it appeared that the Democrats had won the presidency for the first time in 20 years. Conflicting returns from Oregon, South Carolina, Florida and

Louisiana placed the final number in doubt. Tilden's popular margin of "victory" was some 250,000 votes, giving him 184 electoral college votes—one short of complete election success—and leaving the Electoral College in dispute. Local election boards, favoring the Republicans, negated enough Democratic votes to declare Hayes the winner. Hayes' close victory was credited to fraudulent vote counts and bribery, tainting Hayes, a previously unsullied three-time governor. Democrats claimed the election had been stolen, and the boiling controversy overflowed into Congress, where a Republican-controlled Senate and a Democrat-dominated House were engaged in a standoff.

An election commission of five members each from the Supreme Court, the Senate and the House was chosen to settle the issue. This group was anything but impartial during its investigation. For about four months claims of improper conduct and influence-peddling were regularly heard or printed. But when one of the Supreme Court justices was replaced by a loyal Republican, this led to a straight party vote of eight to seven for Hayes.

All of the disputed 22 votes were given to Hayes and made his the winning total of 185. Though the Democrats raised an outcry of fraud, the commission's decision stood, and Rutherford B. Hayes became America's nineteenth president. Though never directly implicated in anything illegal, Hayes's otherwise productive administration was irrevocably tainted by the election dispute.

Many believed that Samuel Tilden had been cheated. Then in 1878, the *New York Tribune* and various cryptanalysts broke a story with startling revelations. The *Tribune* was a Republican paper, but overcame initial skepticism about its bias by supporting its story with facts uncovered in encrypted telegrams. These telegrams had been obtained from Western Union by a Congressional investigative committee, whose Republican members provided the wires to the *Tribune*.

Twenty-seven still-enciphered messages were printed in the *Tribune* during the summer of 1878 and caused a nationwide scandal. Readers and amateur cryptanalysts began sending in attempted solutions. Eventually, *Tribune* editor John Hassard, its economic writer, William Grosvenor, and Edward Holden, a mathematician at the Washington, D.C. Naval Observatory, each independently broke the ciphers.

The cipher types included word transpositions, dictionary-based code words and monoalphabetic substitutions using pairs of numbers. Their solutions provided irrefutable evidence that William Pelton, Tilden's nephew, had bargained and offered bribes for votes. Most damning was the proof that some of Pelton's correspondence had been mailed to 15 Gramercy Square, Tilden's New York home.

The result was a backlash against the Democrats and big Congressional gains for the Republicans, especially in the Northeast. Tilden denied any direct involvement in the scheme and while no charges were ever brought against him, his reputation was ruined. The man who had been one electoral vote from the presidency was denied a second chance by the cipher revelations.

While these political disputes held the nation's interest for a time, attention inevitably turned toward America's expanding borders. The advance into the West was facilitated by the civilian use of "talking trees" and "iron horses"—telegraphy and railroads, respectively—which were closing the distances between the coasts.

The Civil War left tens of thousands disillusioned and displaced and many began to dream of a new beginning in the young West. The migration into territory shared previously by Native Americans and a relatively small number of frontier trappers and explorers caused a rising hostility between the new settlers and the Native American tribes.

As this tension escalated into widespread warfare both Native Americans and the U.S. Army used various signaling systems to communicate over the vast prairies. Plains Indians' smoke signals used different space and time intervals between smoke puffs to convey messages from mesa to valley and from woodland to prairie. These smoke signals could convey messages significantly faster than the couriers of the Army and its cavalry units, which were assigned to patrol the vast lands between the Mississippi and Missouri Rivers, and the Rio Grande, and gave Native American warriors an edge. Because wigwag was limited by terrain and weather and telegraph wires and poles were vulnerable targets, the Army introduced a mechanism called the heliograph, from the Greek *helios*, (sun) and the German *graphein* (to write), which was used by civilians and European military forces alike during the latter part of the 19th century. Mirrors which could be adjusted to reflect the sun's light sent messages over distances of up to 30 miles, and the heliograph was widely used in areas that lacked telegraphy stations.

Such messages did not traverse the distances unnoticed. In fact, many signals were intercepted by resourceful Native American signalers, who interrupted and confused these optical signals with their own mirrors, which were usually made of reflective pieces of silica. Though the western territories were often in dispute, the sun belonged to everyone.

THE FIRST WORLD WAR

World War I was known as the war to end all wars. After years of tension, old antagonisms erupted among Europe's royal families and their military chains of command who were competing to expand their empires in Asia and Africa. At the same time, minority groups within the European bases of these empires were also seeking political and economic autonomy. Each rival nation formed unions with other empires that were either geographically and economically compatible or that provided a balancing force in a region where their mutual enemy was strong. These alliances, the armaments buildups they engendered and the violent independent movement in the Balkans inexorably led to conflagration.

In 1914 the simmering tension came to a head when a Serbian nationalist in Bosnia assassinated the archduke of Austria, Francis Ferdinand. Austria declared war on Serbia on July 28. In subsequent war declarations based largely on the military alliances formed prior to 1914, two opposing sides emerged. The Allies, or the Triple Entente, was made up of

Great Britain, France and Russia, with tiny Serbia and distant Japan. The Central Powers consisted of Germany and Austria-Hungary, who were later joined by Bulgaria and Turkey.

With cryptology still in its fledgling stages, many opportunities for an early Allied victory were lost. The pivotal Battle of Tannenberg was a classic battlefield example of the necessity of communications security. When the war began in August 1914, two huge armies under Czar Nicholas of Russia attacked strategically important East Prussia, which was defended by only one major German army. Although numerous, the Russian forces lacked the necessary communications equipment or encryption methods to protect the many dispatches needed to complete their sweeping plan. Without enough ciphers to conceal messages, Russian wireless operators began sending dispatches in unprotected cleartext. Alert German eavesdroppers heard these exchanges and reported them to their commanders, who planned a brilliant counteroffensive. In a series of engagements through the last days of August, the Kaiser's generals won a crushing victory. Poor crypto-methods and carelessness caused the deaths of some 30,000 Russian soldiers, the capture of nearly 100,000 troops and the loss of vast quantities of matériel.

The inadequacy of the Allies' prewar intelligence-gathering operations had also been revealed after the interception of a radio message from the German warship, *Göben*, in August 1914. The missive was deciphered, but too late to prevent the battle cruiser from shelling Russian ports on the Black Sea, prompting strategically located Turkey to ally itself with Germany and the Austro-Hungarian empire. In the wake of this setback, England's Parliament ordered a redoubling of the military's interception and solution effort. This resulted in the initiation of a cryptanalytic group known officially as Section 25 of the Intelligence Division, but known popularly as Room 40 after their location in the Admiralty Building. Led by the director of naval education, Sir Alfred Ewing, Room 40 soon became the Allies' premier code and cipher branch.

Luck first smiled on the Room 40 group with news from a German light cruiser, the *Magdeburg*, that had run aground in the Baltic Sea in September 1914. Its commanding officer tried to scuttle the ship with explosive charges to prevent capture, but these were only partially successful. When the Russian navy closed in on the scene, they found at least one, and possibly two of the German navy's codebooks in the charthouse and passed this data to the Admiralty, then led by Winston Churchill. The text was not solved immediately, but steady effort by the Room 40 analysts eventually led to the breaking of Germany's naval concealment. Within three weeks the novice codebreakers discovered that the codebook's four-letter codewords had been subjected to superencipherment, a revelation which enabled them to decode a series of earlier radio intercepts.

Room 40 was granted a second measure of good fortune in December when a British trawler recovered a chest that had been cast overboard by a German warship during an October clash at Heligoland Bight near Germany's northwestern coast. This yielded another codebook, which Room 40 added to the *Magdeburg* material.

British cryptanalysis was greatly aided by the development of radio direction finding (DF), which was one of the most significant advances in signals intelligence during this time. While radio broadcasting had become popular on both sides because of its speed and accuracy, any listener on the correct wavelength was able to pick up messages, a fact which Room 40 used to their advantage. DF antennaes honed in on radio signals, enabling monitors to pinpoint the location of the transmittor. The foresight of Winston Churchill and former first sea lord Admiral Sir Arthur Wilson led to a successful use of DF to intercept German naval call signs and communications, which were then provided to Room 40.

In January 1915 the decryption experts in Room 40 broke a vital missive that had been broadcast in the German naval code. They learned that the German vessels that had been shelling England's coastal villages planned to rendezvous on January 16 at Dogger Bank, a shallow part of the North Sea off the northern English coast; the British sent Vice Admiral Sir David Beatty to meet them. In the fierce fighting, the German warships *Derfflinger* and *Seydlitz* were badly damaged and the *Blücher* was destroyed. The Dogger Bank victory gave the young Room 40 staff a much needed boost in the eyes of the military.

Despite such successes, many military commanders still viewed deciphered messages with distrust, which the Battle of Jutland made clear. In May 1916 decryptions of intercepted German radio dispatches combined with the new DF technology helped to pinpoint the Kaiser's High Seas Fleet, and a clash seemed imminent on May 31. England's Grand Fleet left its moorings with her commanders anticipating a decisive confrontation that would gain them control of the ocean, but a series of errors proved to be fatal for the mission.

German Vice Admiral Reinhard Scheer, leader of the High Seas Fleet, had shifted the call sign *DK* for his flagship, *Friedrich der Grosse*. Hoping to hide his movements, Scheer placed the important identifier in Wilhelmshaven, a naval center on Germany's Jade River. Although British radio interceptors were aware of this tactic, they did not alert their superiors to this change.

On the morning of May 31, British Grand Fleet commander Admiral Sir John Jellicoe radioed an inquiry regarding the location of call sign DK. British naval authorities radioed that DK was still in its harbor, which led Jellicoe to believe that the entire German fleet was also in the port. In the early afternoon, when elements of his fleet first unexpectedly met and engaged German warships in the North Sea, Jellicoe's confidence in Admiralty intelligence was shaken. His doubts increased when a decrypted Admiralty report about a German cruiser's position located its position almost exactly beside Jellicoe's own warship. It was later discovered that the cruiser's own navigator had given the wrong position and cryptanalysts had solved the report correctly, but Jellicoe remained disillusioned with these intelligence reports. As the day wore into evening, the naval battle continued with both sides sustaining damages and victory eluding both.

Sometime after nine o'clock in the evening, Vice Admiral Scheer gave orders to break off the conflict and retreat, and the Room 40 staff correctly cryptanalyzed his commands. Doubting the accuracy of these decryptions, however, Jellicoe ignored the Room 40 news about the direction of Scheer's withdrawal. Instead, he heeded a report by the British warship *Southampton* that gave an alternate course for the High Seas Fleet.

Scheer's retreat vindicated the Room 40 cryptanalysts, and Jellicoe missed an opportunity to surprise Scheer's fleet. Despite this series of errors, historians still consider this a victory for Britain as the High Seas Fleet did not come out in force again to engage the Royal Navy during the First World War.

In October 1916 the leadership of Room 40 was passed from Sir Alfred Ewing to Captain William R. Hall, director of naval intelligence, who headed the unit when the most crucial interception of the war occurred. This message came to be known as the Zimmermann telegram. Germany's transatlantic cables had been cut by the English vessel *Telconia* in August 1914, forcing German overseas messages to be sent by two alternate channels. One was the "Swedish Roundabout," a Swedish cable line from Stockholm to Argentina (by a point of British access). In Buenos Aires, the cables were transferred to German hands and forwarded to Washington. The second route was from Berlin on the U.S. diplomatic cables that passed through Copenhagen and then London before crossing the Atlantic. Shortsighted as it now appears, U.S. leaders actually approved the latter route in an attempt to maintain relations with Germany and remain neutral in the conflict.

On January 16, 1917, German Foreign Minister Arthur Zimmermann sent an encoded telegram from Berlin. Most chroniclers agree that the communication was sent by both cable routes. The dispatch then went to Washington to Count Johann von Bernstorff, the German ambassador in the United States, and was forwarded to the Kaiser's ambassador in Mexico City, Heinrich von Eckardt.

Germany was about to engage in unrestricted submarine warfare against neutral shipping. Fearing that this would push an uncommitted United States clearly into England's camp, Zimmermann proposed an alliance with Mexico. In exchange for making war on the United States to distract America from joining the overseas battles, Mexico was promised the regions of Texas, Arizona and New Mexico that it had lost in the Mexican War of 1846–48. The offer also involved "generous financial support" and indicated that Japan might join the scheme to threaten the United States in the Pacific.

The telegram was intercepted by Room 40, who deduced these facts from further intercepted exchanges commenting on the telegram as well as from other German transmissions. The telegram's decryption was made to look like a U.S. intelligence success so the Germans would not suspect the contribution of the British cryptologists, thereby hiding the involvement, and even the existence, of Room 40.

On February 22, 1917, Captain Hall showed the recovered plaintext of the telegram to a U.S. embassy official in England. A top-level decision was made to release the telegram and some limited supporting details to the Associated Press. On March 1, 1917, the exposé shocked the United States. The reasons why Zimmermann admitted his involvement in the scheme remain unclear, but when he did it swept away all suspicion of a conspiracy to dupe the United States into entering the conflict. President Wilson asked for a declaration of war on April 2 and Congress agreed on April 6, 1917. The Room 40 solution seekers were at their zenith of achievement.

Room 40 was not the only cryptanalytic unit operating during World War I. Germany's ally, Austria-Hungary, was also fully engaged in a cryptologic battle. Before the onset of the war, chief of intelligence Colonel Max Ronge had bought a copy of an important Italian military staff encryption, the *cifrario rosso*, or red cipher. When Italy left the Central Powers in 1915, it became an enemy and Austria-Hungarian radio intercept posts began listening eagerly for Italian radio transmissions.

Austrian cryptanalysts became adept at solving the Italians' dispatches even when the keys of the *cifrario rosso* were changed. From the decryption of other codes, cryptanalysis chief Major Andreas Figl and his staff were able to discern Italian defensive tactics, which aided an Austrio-Hungarian drive in the spring of 1916. A few months later, other crypt-analytic successes enable them to halt an Italian offensive along the Isonzo River and inflict heavy losses on Italy.

It was not until the autumn of 1917 and the disaster of Caporetto that the Italians discovered that their dispatches were being read. At Caporetto from October 24 to November 12, the Italian forces lost 40,000 men and 275,000 more were wounded. By comparison, the Austrio-Hungarians suffered only 20,000 killed and wounded. After this bitter defeat, Italy developed better protection using codes enciphered with numerals. Thanks to improved Italian cryptography, the Austro-Hungarian generals were not able to depend on air-wave interceptions for military strategy for the remainder of the war.

Another infamous case involved a woman whose controver-sial spying was exposed through wireless intercepts and crypt-analysis. Margaretha Gertrude Zelle (1876–1917) was a Dutch woman, who as Mata Hari (Eye of the Dawn), had spiced her stage act with exotic costumes and choreography reputedly adapted from Javanese and Indian temple dances. When the war broke out she was recruited by Germany and designated secret agent H-21. In an effort to learn military secrets, Mata Hari carried on several affairs with influential men in Spain and France.

During December 1916, the German naval attaché in Berlin sent radio communiqués to Madrid requesting money and new orders for Agent H-21. England's Room 40 staff soon deciphered these and other radio intercepts along the Madrid-Berlin connection. This information was given to their French allies, who used it to uncover a crucial superenci-pherment. Arrested in France and confronted with evidence of the financial requests made on her behalf, Mata Hari claimed these were payments from her paramours. She was convicted nevertheless and sentenced to death. Though his-torians debate whether her information was of much value to the Central Powers, Zelle was by all accounts a courageous woman. She refused a blindfold and bravely faced a firing squad near Vincennes, France, in October 1917.

In the autumn of 1917, the arrival of the American Expeditionary Force (A.E.F.) troops in Europe, backed by the United States' expansive industrial capacity, changed the course of the war. As these troops burst upon the battlefield, their radio, telephone and courier dispatches urgently needed encryption protection, but as with so many other U.S. war plans, establishment of the Code Compilation Section was delayed until December 1917.

In charge of the Code Compilation Section was Ohio native Howard Barnes. He and his assistants were the first Americans to create two codebooks in the field, *The American Trench Code* and the *Front-Line Code*. Applied for a time as the American First Army entered the front lines, their concealment was easily broken. The superencipherment was weak and deemed too time-consuming for cryptographers to replace amid the pressure of battle.

The Code Compilation Section then produced a much improved group of field codebooks in the summer of 1918. Known as the River Codes, they were named *Potomac, Suwannee, Wabash, Mohawk, Allegheny, Hudson* and *Colorado*. These were one-and two-part codes that contained terms for typical battlefront needs such as ammunition, supplies and weapons. Having learned from British and French experiences with lost codebooks, the A.E.F. set up its codes for quick replacement. When the *Potomac* was captured, the *Suwanee* was rapidly brought into service.

In the ensuing months, codebook production levels were increased to provide protection for the messages of the expanding number of troops that were landing in Europe. A Second Army was formed and provided with its own series of concealments called the Lake Series, which consisted of the *Champlain, Huron, Osage* and *Seneca*. These field codes concealed exchanges within divisions or between divisions and higher command centers and were backed up by special codes such as the *Emergency Code List*.

Despite these massive efforts on the cryptographic front, it was the breaking of a major cipher that helped ensure an Allied victory. The credit for this achievement belongs to the

French Bureau du Chiffre and Georges Painvin, a 29-year-old artillery captain and cryptanalyst. Introduced to ciphers by one Captain Paulier, he demonstrated a quick aptitude and on March 5, 1918, the Germans' new cipher presented him with his greatest challenge.

This cipher method was known as ADFGX and was created by Colonel Fritz Nebel, a contemporary of Painvin and a skillful radio staff officer who recognized the importance of concealing radio messages. Historians generally agree that he and other officers chose the letters *a*, *d*, *g*, *f* and *x* because of concern about possible jumbled airwave messages. The Morse code equivalents for those letters were quite distinct and helped the Germans avoid confusion when transmitting telegrams.

Painvin struggled with the ADFGX and its cipher cousin, the ADFGVX, over the next three months. Exhaustive study of wireless transmission locations, troop movements, ammunition requests and fuel ratio lists finally paid off at a crucial moment.

Kaiser Wilhelm's generals had planned a major offensive, hoping to win the war before the arrival of the American troops tipped the scales. Thanks largely to Painvin's work, the spearhead of the assault was pinpointed between Compiègne and Montdidier, towns 50 miles north of Paris.

The Allies rushed men and materials to the region in preparation for the attack. When the charge came on June 9, the lines of the defenders buckled but did not break. For almost a week the outcome teetered between the violent cannonades and charges of the two sides. By the time the clouds of gunpowder, mustard gas and pulverized soil had settled, the German offensive had been halted. There would still be other terrible exchanges, but the direction of the war had turned.

THE 1920s AND 1930s

In the tenuous peace that followed World War I, many hoped that the need for secrecy and conspiracy was over. Herbert Yardley, however, a cryptologist who had served in the war, continued to push for the maintanence of an American cryptographic unit. In 1917, spurred by concerns about the European conflict, Yardley had persuaded the War Department to set up a cryptology office as a branch of Military Intelligence, known as MI-8. This small but energetic group served well under Yardley's direct guidance. They dealt with such matters as German translations, invisible inks and sabotage investigations, providing crucial information through the final days of the European upheavals.

After the war, wary of a postwar decline in cryptology, Yardley once again applied himself and gained the confidence of Secretary of State Frank Polk. With Polk's help, enough funding was secured to begin a revised version of MI-8 in New York City. Named for the secret mail interception rooms of the late 1600s and the 1700s, the American Black Chamber was the nickname of America's first real cipher bureau. Under the auspices of the War and State Departments and supported by cooperative U.S. telegraph companies, Yardley and his staff began solving communications between embassies, businesses and trade associations.

In a brownstone on East 37th Street near Lexington Avenue, Yardley and his score of part-time staff members were eventually able to achieve surprising successes, despite their often makeshift materials and limited salaries. One of their primary achievements was the solving of several of Japan's most important diplomatic codetexts, a task made all the more difficult by the Japanese ideographic writing. In the summer of 1921 the chamber members uncovered the contents of a Japanese telegram from London to Tokyo and discerned that Tokyo was using an alphabet called *kata kana*, which reportedly used some 70 syllables that had been given roman letter equivalents.

The Chamber crew first broke common syllables and words that included *ari, aritashi, daijin,* and *gyoo,* and by using frequency counts and manipulating combinations of letter groups, the staff found links between a series of two-letter codewords and their hidden equivalents. This phonetic syllabary was placed in code form, with the messages segmented and mixed in order to cover typical letters at the beginnings and ends of words as well as salutations and closings of communications. An example given by Yardley illustrates some of the solved equivalents:

CODETEXT:	RE	UB	BO	AS	FY	OK	RE	OS	OK	BO
PLAINTEXT:	do	i	tsu o		wa	ri	do	ku	ri	tsu

```
(Germany)         (stop)        (independence)
```

The codebreakers reaped substantial rewards for their strenuous efforts. By the time of the important naval disarmament conference in Washington in November 1921, the Black Chamber members were regularly reading Tokyo's as well as others' coded telegrams. Sometimes within hours, U.S. officials knew the negotiating positions of Nippon and other sea powers.

Japanese codes sent prior to the conference revealed that their leaders would accept certain terms regarding ratios of numbers and tonnage weight of ships. If they were pressed to bargain, Japan's conference delegation was to make a series of concessions. Originally they had sought a ship ratio of ten to seven. This meant that they wanted to be able to build seven capital vessels, ships weighing more than 10,000 tons with guns of 8 inches or more, to every ten built by America and England. Every ratio difference of 0.5 was equivalent to 50,000 shipping tons. These numbers were significant because, in addition to their military capability, capital ships were symbols of national prestige.

With Tokyo's secrets in mind, the United States was able to exert extra bargaining clout at the conference. Japan's representatives had to accept a ratio of ten to six. The closing agreement, called the Five-Power Treaty, included Japan, England, America, France and Italy. Thanks largely to Yardley's staff, the disarmament conference ended as an important diplomatic victory for the United States.

Cryptanalysis came to public attention again in the domestic scene when cryptanalysts revealed a national oil scandal in the United States, named Teapot Dome after a U.S. Navy oil reserve located in Wyoming. Rumors had circulated for years that officials in the Republican Warren G. Harding administration, including Veterans Bureau and Justice Department employees, had profited from illegal tampering with government oil bases. In 1924, the Senate launched an official inquiry through the Public Lands Committee. Committee staff persons discovered coded messages, including one sent to Edward Doheny of the Pan American Petroleum Company. Unable to decipher them, the Senate sent them to the U.S. Army Signal Corps and its code section, headed by William Friedman.

Friedman testified that the message to Doheny was sent in a private code belonging to the Pan American Petroleum Company. He also revealed that messages exchanged with *Washington Post* president and Harding's co-conspirator, Edward McClean, were encoded with a method belonging to the Bureau of Investigation of the Department of Justice.

The widely publicized exchanges piqued the curiousity of Americans with tantalizing communications that included codewords like *opaque, hosier, bedraggled* and *chinchilla*. Friedman decoded these telegrams and others by studying the Bureau of Investigation's codebook, which McClean had obtained after being accepted as a special agent of the Justice Department.

Friedman's findings and further Senate inquiries eventually exposed widespread fraud and bribery. Secretary of the Interior Albert Fall had secretly leased the Elk Hills, an oil reserve in California, to Edward Doheny, and Teapot Dome to the oil company of Harry F. Sinclair. Doheny, in turn, had "loaned" $100,000 to Fall. Sinclair had added to the bribery with a herd of cattle for Fall's range, $85,000 in cash and $223,000 worth of bonds.

In 1927 the government won a suit to have the leases canceled. Though a jury acquitted Sinclair, Doheny and Fall of conspiracy to defraud the government, Sinclair was convicted of jury tampering and Fall was convicted of bribery, fined $100,000 and sentenced to a year in prison.

With America's increasingly isolationist policies taking effect, the U.S. military was forced to rely on listening posts to keep abreast of events abroad. The team that operated in America's chain of electronic ears that arced the northern Pacific during this time was known as the On the Roof Gang. In the 1920s, a Navy listening post was placed in the U.S. Consulate in Shanghai, where radio eavesdroppers attempted to intercept Tokyo's diplomatic messages. During 1926 and 1927, the first U.S. Asiatic Fleet intelligence units were established at Guam and Olongapo (later the Cavite Naval Base) in the Philippines.

In 1928 a special classroom was built on top of the Navy Department's building in Washington, where "roofers" were taught by instructors of the navy's cryptologic organization, OP-20-G. One of the primary objectives of the four-month instruction period was to increase the number of radio listeners who could understand *kata kana*, used in the Japanese version of Morse code.

From the late 1920s, Navy listening stations were established at such varied outposts as Bainbridge Island near Seattle; Dutch Harbor, Alaska; and Heeia on Oahu, Hawaii. The *H* from Heeia soon became the basis of the codename for the Pacific Fleet's intelligence unit in Hawaii, Station Hypo. The post at Cavite Naval Base in the Philippines was codenamed Station Cast. Codenamed Station Negat, for the *N* in Navy Department, OP-20-G in Washington was the auditory nerve center of each of these human and electric eardrums.

The On the Roof Gang members who passed the difficult training course went to duty posts overseas. Along with the *kata kana* intercepts and translations used by Navy encryption solvers, the Gang worked with direction-finding and traffic analysis. While stateside citizens struggled with the Depression and hoped that the New Deal would revitalize the economy, the roofers listened for the Japanese dashes and dots that eventually spelled danger on the horizon.

In the late 1920s, just a few years after the remarkable achievement of the Five Power Treaty, the Black Chamber's appropriations were sharply reduced. Changes in management policy also affected the chamber as the two telegraph company giants, Postal and Western Union, began to be less cooperative. Without their full compliance, the Black Chamber's analysts could not work at their full potential. But the telegraph companies were only one part of a larger problem as far as the chamber was concerned.

In the spirit of international good will that pervaded the middle to late 1920s, cryptology as well as spying were considered by many to be unsavory, immoral and even unpatriotic activities. Herbert Hoover's high moral tone in his 1928 election victory turned popular opinion away from cipher interception and analysis. It was President Hoover's secretary of State, Henry L. Stimson, who dealt the chamber its most telling blow.

A well-meaning, idealistic man, Stimson was horrified to learn that the Black Chamber was supported by the State Department. He wanted his branch of the government to be beyond reproach in its dealings with other countries and did not believe that gentlemen and diplomats should be reading communiqués intended for others. With self-righteous indignation, he stopped all funding for the chamber and it soon had to be closed.

Understandably bitter about the demise of his brainchild, Yardley put his recollections together on paper in a book entitled *The American Black Chamber*, which was as surprisingly popular as it was controversial. Yardley claimed that his book was meant to alert free people to the dangers of weak codes and narrow-minded cryptographic policies, but livid critics labeled it "scandalous" and "irresponsible" when it was published in 1931. Introduced to a wide audience through excerpts in *The Saturday Evening Post*, Yardley's book was a financial and critical success. While literary reviewers were pleased with its narrative energy and cloak-and-dagger dramatics, other American cryptologists, including William Friedman, believed that its candid revelations about American codebreaking and ciphersolving harmed U.S. intelligence in both the present and the long term.

Friedman's contention was seemingly proven when overseas observers reported that other nations were changing their coding methods. This was certainly true in Japan, where the book caused a particular sensation. Yardley's account of the decoding of Japan's diplomatic code during the important Washington disarmament conference was seen as a serious breach of honor. Several historians have since argued that these disclosures were responsible for the Imperial Army and Navy's increased caution with their communications when they later initiated war plans in Asia.

As America's brilliant but brief cryptologic efforts came to a virtual standstill, changes were taking place in Japan on the eve of their invasion of Manchuria. The Japanese Tokumu Han, or Special Section, was a decryption unit introduced into the communications department of the Japanese naval general staff in 1925. In 1929, a fully organized *tsushintai* (radio interception unit) was set up in the navy ministry and in 1932, when Japan occupied Shanghai, a radio unit known as the X Facility accompanied the naval landing force. It was this unit that provided the analysts of the Tokumu Han with intercepts from the Nationalist Chinese forces of Chiang Kai-shek. Although Chiang's Kuomintang troops had been trying to counter Japanese expansion into China, Tokyo's superior naval forces and their advanced cryptanalytic capabilities proved to be too strong for the Chinese troops.

The Tokumu Han was also actively trying to break the U.S. State Department's Gray code, a one-part code used by diplomats in Asia. The Japanese cryptanalysts discovered a series of vowels and keywords with the assistance of the Tokyo *Kempetai* (military police), who clandestinely obtained telegram scraps from U.S. Embassy wastebaskets. The solvers finally removed the mask by identifying *nadad* as "stop/period," opening the door to further discoveries that ultimately compromised the State Department's cryptosystem.

The solving of the Gray code led to a military reversal for China. In 1932, X Facility analysts in occupied Shanghai were able to break a Chinese code similar to a digit code that was used to send business telegrams. When its message was compared to a Tokumu Han decryption of a Gray transmission, the combined solutions exposed a planned Chinese airstrike on Japanese troops. Reacting quickly, Japanese carrier-based planes struck the Nationalists' aircraft on the ground at

Ch'ang-sha, destroying most of the Nationalists' planes in one attack.

In the autumn of 1932, the Japanese tanker *Erimo*, which had been shadowing U.S. maneuvers in the Pacific, learned that the U.S. Navy was using a rotor-based encrypting machine known as the Electric Code Machine or ECM. The Japanese had already been studying similar rotary systems and they soon penetrated the cipher generated by this rotor version.

American cryptanalysis of some Japanese exchanges soon showed that the Japanese had broken the ECM's conceal-ments, so the U.S. Navy returned to applying a strip cipher with alphabets on sliding stips of cardboard or paper. Several variations of this basic device were in use at this time. One was the Navy Code Box (NCB), created by Lieutenant Russell Willson in 1917, which was reportedly still in service in some parts of the Navy in 1935. Another strip cipher, designated the CSP 642, was also in use until at least 1941. Considered to be the lowest-echelon cryptographic method, the CSP 642 had 30 alphabet strips with arrangements varied by not using the full amount available every day. For example, on one day 25 strips were used, on another, 28 would be put in place. The Navy's repertoire also included the M-138-A, which was also used by the U.S. Army. Its 100 strips, with as many as 30 in service at a given time, more than tripled the total number of slides available with the CSP 642 and provided thousands more potential alphabet combinations.

Gradually, as U.S. cryptosystems multiplied in number and complexity, the successes of the *Tokumu Han* shrank in number.

THE SECOND WORLD WAR

In the worldwide depression that marked the 1930s, no nation was more aware of the changes in warfare strategies and cipher advances than was the prewar Third Reich. While their French rivals were spending millions of francs on the Maginot Line of forts, Hitler and his generals were preparing *Blitzkrieg*, lightning war. Swift Panzer tank divisions and their supply convoys drove around the flank of the supposedly impregnable Maginot bastions and rendered their armaments useless. While members of the French and Belgian general staff offices were delivering orders by bicycle-riding couriers, the Luftwaffe was using paratroopers to capture key sites and sabotage armaments and communication systems well behind Allied front lines. As the defenders scrambled to regroup with messages sent over downed telephone lines, the Wehrmacht was using a well-coordinated radio system to assist their brutal invasion.

Seven months after the outbreak of World War II, Winston Churchill became Britain's prime minister. Drawing from his experiences in communications intelligence and security during the First World War, Churchill understood the real importance of the early effort to solve German ciphers. In the bleak months when England faced Hitler alone, it was Churchill who used the powers of his office to support the various intelligence staffs. One of these was the fledgling cryptanalysis group at Bletchley Park.

Bletchley Park was the idyllic Victorian-era estate near Bletchley, some 45 miles northwest of London, where the small British cryptanalysis group, the Government Code and Cipher School, moved during the beginning of the war. A number of brilliant individuals with diverse linguistic and mathematical skills gathered there with their support staff members and soon a miniature village arose. Thus began, in often makeshift conditions, the enormous project of breaking German encryptions, with special attention given to the Enigma.

The Enigma was a rotor-based enciphering machine that had the potential to create immense numbers of electrically generated alphabets. As the highest levels of the Reich's chain of command recognized Enigma's potential, they relied more and more heavily on it and its successors to encipher high-security messages. Bletchley Park's mammoth task was code-named Ultra, the name of the English Admirals' code from the days of the Battle of Trafalgar. Ultra was one of the best kept secrets in all of military and espionage history. Facts about the cryptanalysis of Enigma secrets were disclosed in 1967 by Polish historian Wladyslaw Kozaczuk and by French resistance fighter Gustave Bertrand, a participant in the acquisition and transfer of Enigma secrets, in 1973. However, the existence of Ultra was not revealed to the general English-speaking public until 1974 with the publication of Frederick Winterbotham's *The Ultra Secret.*

Ultra was aided by early knowledge of the Enigma device from French and Polish sources. Polish analysts had attained access to an Enigma version sold commercially in the 1920s, and their intelligence agents had photographed and drawn diagrams of an Enigma machine mistakenly shipped to Warsaw in 1929. Scholars at the University of Poznan also solved important mathematical problems regarding Enigma decipherments. In October 1931, a German Ministry of Defense clerk named Hans-Thilo Schmidt, codenamed Asche, or Source D, began selling secret cryptographic documents to the French. Over an eight-year period, deliveries reputedly included an instruction manual for a military Enigma, a codebook of keys, sample encipherments with plaintext equivalents and machine modifications.

In the mid-1930s, Poles at the AVA radio factory near Warsaw and French technicians both built versions of the Enigma that enhanced Allied research. In 1938 Polish efforts with multiple motor-driven rotors led to a machine dubbed *bomba* after a type of Polish ice cream. Named "the Bombe," or the bomb, by the British and French, this cipher machine helped decrypt improved Enigma machine variations.

In 1939 secret meetings in Paris and Warsaw led to a necessary division of effort among the Poles, the French and the British. Bowing to Great Britain's broader scientific and technical base, Poland and France agreed to entrust their own cryptographic contributions to teams of British analysts such as those at Bletchley Park. Chief British cryptanalyst Alfred D. Knox and Cambridge mathematicians Gordon Welchman and Alan Turing, along with other scholars, began work on new attacks against the Enigma system. Construction of an improved Bombe began in 1939 and by late April 1940, the result was a copper-hued cabinet around 8 feet high and 8 feet wide at its base. Its inner works were a maze of wires arranged in electrical circuits that tried to imitate the

Enigma's rotors. The machine was set up in Hut 3, a large structure built on Bletchley's grounds.

With the help of England's intercept stations, recordings were made of messages from enemy military, diplomatic and intelligence organizations. When Enigma transmissions were identified, they were placed on tape and fed into the Bombe. With a noise like a series of rapidly working knitting needles, the apparatus often succeeded in finding the keys enciphering the communications. Working in Hut 6 for army and air force transmissions and Hut 8 for those of the navy, the analysts endeavored to unravel their concealments. The Bombe became the precursor of a series of devices that counteracted Nazi cryptographs throughout World War II and began the cryptographic battle that Churchill later called a "wizard war." It was indeed the wizardry of cryptanalysts that contributed much to the Allies ultimate victory.

In America, the short-sighted isolationist policy that was pervasive after World War I held the country with a grip that directed public and governmental attention away from increasing signs of worldwide danger. But on December 7, 1941, the infamous assault on a relatively unknown place called Pearl Harbor turned America's ideology upside-down.

Much blame has been laid upon various United States cryptographic units for their failure to correctly interpret the events leading up to Pearl Harbor. Questions surround everything from late warnings to mistaken radar sightings, and the answers continue to be speculative at best. In the cryptanalysts' defense, records show that American defensive positions were stretched very thin and that both troops and cryptographic groups lacked the best military technology of the day. The military did, however, overlook the importance of several warning signs that could have alerted them about the impending peril.

One source of such information was Copek, the codename for a highly secure cryptosystem linking U.S. Pacific bases with Washington, D.C., before and during World War II. Its messages were sent by the U.S. Navy's ECM machine, whose electric rotors assured an advanced level of encipherment for its day.

The Copek-associated organizations and their radio network exemplified some of the best and worst aspects of communications security of their time. Copek linked Washington's Negat Station with the U.S. Navy's intelligence centers Hypo, in Hawaii, and Cast, at Cavite in the Philippines. These sites had different responsibilities for intercepting and breaking Japanese cryptomethods: the Hypo group tried to reveal the code of the Japanese admirals, while the Asiatic Fleet intelligence squad at Cast struggled to solve the main Japanese navy operational cryptosystem, JN25.

The Copek system was kept secure partly because it was not used as frequently as other methods. Due to questionable, overcautious security decisions, the Hypo center in Hawaii did not have decipherments of the "J" series of transposition ciphers used by Japan's consulates, despite the fact that one such consulate was located near the important U.S. naval base at Pearl Harbor. Many critics believe that if the base been given access to these ciphers, several warning signs could have been recognized in time to counter the Japanese attack.

In addition, the crucial information filtering through the cryptanalytic chambers from the Japanese cipher machine "Purple" was not disseminated to all distant outposts. Even before their entry into the war, American cryptanalysts had been involved in trying to decipher Japanese ciphers and had broken one of the Japanese machine-made ciphers which was codenamed Red. One of the last missives broken in Red cipher was the announcement of Japan's introduction of a more advanced cipher machine in February 1939, called the Alphabetical Typewriter '97, which the Americans codenamed Purple.

Purple was a complex polyalphabetic machine that could encipher English letters and create substitutions numbering in the hundreds of thousands. This capability presented an immense challenge that at first seemed insurmountable. Nonetheless, America's premier cryptanalyst William Friedman and his associates applied themselves with feverish intensity to the task of unraveling the cipher.

The vast majority of Japanese cipher solutions were achieved by the Signal Intelligence Service (SIS), a branch of the Army led by William Friedman, and by OP-20-G, a section of Navy intelligence directed by Laurence Safford. One of these two groups' greatest achievements was the analog mechanism built to decrypt Tokyo's Purple diplomatic messages.

American analysts were already familiar with specific salutations and closings, thanks to the information about Tokyo's diplomatic codes uncovered by Yardley and the American Black Chamber during the Washington disarmament conference. Military radio stations around the Pacific constantly monitored radiotelegraph transmissions and every possible clue was sought in even seemingly mundane communications. Frequencies and patterns slowly began to emerge.

Blanks were filled in by lucky breaks as Japanese cipher senders made mistakes with the unfamiliar system and repeated dispatches to make corrections. After an extraordinary pencil-and-graph-paper effort to chip away at the shield, one of Friedman's civilian technicians, Leo Rosen, had the breakthrough realization that telephone selector switches could help enhance tests of decryption techniques. This gave rise to a different way of viewing the entire problem of electrical connections and processes and a month later, in August 1940, Friedman and his associates had drawn up a blueprint for their own analog mechanism. The first working device was a maze of wires and clicking relays built by Rosen at a cost of $685. By September 1940, U.S. analysts were using the Purple Analog to decrypt Japan's most secret diplomatic communiqués.

Adopting General Joseph Mauborgne's nickname for cryptanalysts, Friedman called his staff "magicians." As the military began to see the decryptions produced by the Purple analog, the word "magic" became more broadly applied to other cryptographically produced intelligence. Yet as U.S. differences with Japan escalated, the dissemination of the Magic decrypts became a point of divisive and lasting controversy.

Information derived from the Purple Analog was made available to President Roosevelt, some of his cabinet members and to high-ranking military officers. A few of these officers chose not to distribute parts of this intelligence to all outposts. Declassified documents later revealed that two such base commanders were Admiral Husband Kimmel and General Walter Short at Pearl Harbor.

Despite serious lapses in communication between Washington and its Hawaiian outpost, many cryptanalysts made important discoveries on their own. A crucial decipherment was made by American linguist and Japanese specialist Dorothy Edgers. A former resident of Japan, Edgers had 30 years of direct experience with the subtleties of the language. It was this familiarity with the Japanese language, rather than any real cryptographic experience, that had helped her obtain employment as a translator with the Z Section of OP-20-G in Washington.

On Saturday, December 6, 1941, a collection of decrypted, but not translated, Japanese diplomatic messages was placed on her desk. They had been concealed in low-security consular ciphers such as the PA-K2, the general intelligence emphasis being on other communications.

Edgers noticed something of particular interest in one of the decryptions. She mentioned it to her brother and fellow linguistic expert, Fred Woodrough, who advised her to continue with it, which she did until well after normal working hours. While working on the translation, she mentioned the communication to her superior, Lieutenant Commander Alwin Kramer. But the overworked Kramer was concentrating on higher-level messages such as Purple and did not have time to consider the significance of her findings. When Edgers completed her efforts, she left them on the desk of a chief clerk.

Edgers had found the so-called lights message, a cable from Japanese consul Nagao Kita in Hawaii to Tokyo, concerning an agent near Pearl Harbor and light signals from a beach house sent to a Japanese submarine. The information to be exchanged by this system directly concerned the Pacific Fleet at Pearl Harbor, but the translation did not leave Commander

Kramer's desk in time to be understood and appreciated. By the next day, other decryptions had claimed her superiors' attention.

Late on the night of December 6, several Japanese embassy dispatches were picked up by Navy radio stations and sent on to the Navy Department in Washington. By the fateful morning of December 7, a Purple-enciphered Japanese government reply regarding negotiations with the United States had been deciphered. Tokyo had decided to break current negotiations with the United States by 1:00 P.M. that same day.

OP-20-G and SIS cryptanalysts knew this by 7:30 A.M. Washington time. It was not yet dawn in Hawaii. Most U.S. personnel were still sleeping peacefully at Schofield Barracks, Wheeler Field and on Battleship Row. By the time the sun rose in Hawaii that morning, the horizon was dotted with Japanese planes.

Despite later criticisms of misjudgment, inaction and poor communications, the U.S. cryptology staffs had done their work as quickly as the methods and governmental limitations of that time permitted. They broke every important missive necessary to validate reasons for action. After all, no Japanese message specifically stating "Attack Pearl Harbor" ever existed to decipher. Nonetheless, the consequence of the surprise attack was heavy damage to Pearl Harbor and the nearby military air installations. The losses included 2,403 people killed, including 68 civilians; 1,178 injured; 18 warships, including battleships, cruisers and destroyers, put out of action. The U.S. Navy also lost 21 scout bomber planes, 13 fighter planes, 4 dive bombers and 46 patrol planes; and the Army Air Force lost 18 bombers and 59 fighters and suffered heavy damage to airfields and hangars.

These shocking losses at the supposedly impregnable island bastion united the 48 states in a common cause. U.S. leaders could no longer put their faith in an honor code according to which "gentlemen" ambassadors politely overlooked each others' communications.

Severely limited by isolationism and some of her leaders' lack of foresight, America entered a war that was already dominated by the Axis powers. Germany had created warfare technology that made U.S. factories look obsolete by comparison. Spies were infiltrating Allied overseas bases almost as rapidly as their howitzers were demolishing cities and villages.

In addition to the massive task of restructuring the economy and industry in preparation for war, America desperately needed to improve its intelligence capabilities to act decisively in defense of the nation's interests. The strength of America's considerable resources, money and scholarship was poured into revitalizing its cryptology units. After its entry into World War II, the balance of intelligence-gathering and encrypting capabilities shifted in favor of the United States.

NAVAL BATTLES

Naval battles were of critical importance to both the Axis and the Allied powers. The Battle of the Atlantic refers to the struggle to control the sea territories, as Nazi U-boat squadrons, known as wolf packs, wreaked havoc on military and merchant shipping. In addition to their skill and cunning, the U-boat commanders were aided immensely by German cryptanalysis, especially from the German naval high command (Oberkommando der Kriegsmarine), whose experts managed to solve top-secret Royal Navy cryptomethods early in the war.

Probably one of the greatest sources of information for the German forces on both land and sea was the Beobachtungs-Dienst, or B-Dienst. Translated as the "Observation Service," B-Dienst was the name given by the commander of the German navy, Admiral Karl Dönitz, to its small but very capable cryptanalysis group. In March and April 1940, as Hitler was formulating plans for the invasion of Norway and the seizure of its rich ore supplies, the B-Dienst broke British codes and revealed a Royal Navy plan to mine the harbor of Narvik, Norway. This knowledge helped the Nazis to arrange a decoy attack, fool the British and successfully land invasion troops on Norwegian soil.

The B-Dienst also broke many of the British merchant vessels' codes that were broadcast over the BAMS (Broadcasting for

Allied Merchant Ships) code network. In the second full year of the war, the B-Dienst was deciphering communiqués that exposed the positions of entire convoys. Orders that were meant to warn supply ship captains away from dangerous areas were instead used by the Germans to direct wolf packs to their prey. Throughout the spring of 1941, Allied losses increased to an alarming level, with an average of one ship being destroyed every 16 hours. By March 1943, the B-Dienst was even reading an English broadcast called the "U-Boat Situation Report" and these warnings to vessels actually helped the U-boats counteract the convoys' defenses. The wolf packs were at the peak of their success: B-Dienst analysts were able to single out convoys at different ports and determine their courses, travel times and speeds. From March 16 to March 19, with the loss of only one submarine, Dönitz's squadrons sank 21 vessels totaling 141,000 tons. For some time, this also enabled the German naval high command to keep the small German surface fleet away from a battle against potentially superior forces and avoid entrapment of its submarines.

After much loss of life and naval tonnage, the Allies developed a multilevel plan to combat the wolf packs. In addition to better organized convoys, increased destroyer escorts, aerial surveillance defense and sonar, the Allies applied better cryptography to ensure communications security. German naval ciphers, such as Hydra and Triton, were solved through Bombe/Ultra intelligence with the aid of a naval Enigma and documents retrieved from captured U-boat 110 in May 1941. These cryptanalytic successes, along with the British decryptions of the German Enigma ciphers, gave the Allies access to vital facts such as the U-boats' refueling points and allowed them to attack when the U-boats were most vulnerable. These factors played a role in eventually turning the tide of the Battle of the Atlantic.

As Japanese conquests in Asia began to threaten U.S. interests in the region, special attention was focused on the Japanese fleet's activities. The On the Roof Gang and American cryptanalytic units focused their skills on intercepting and solving the Japanese fleet encryptions.

Before the outbreak of the war, Allied listening posts and cryptanalysts had concentrated on the Imperial armed forces' cryptosystems and the Japanese government's diplomatic concealments. After several successes in breaking these cryptomethods, American analysts turned their attention to the supply vessels that provided food, medical supplies and weaponry to the Japanese forces overseas.

The Japanese supply ships' code was called the Maru code, after the Japanese suffix attached to ships' names, and used a four-number code called the S Code. After a five-month effort to break this system, a solution was finally accomplished in early 1943 by the Station Hypo intelligence team in Hawaii. Information provided by the decryptions enabled the Pacific Fleet intelligence center to track Japanese convoy routes from their position reports, made on a precise basis at regular time intervals. By mid-1943 the unmasked communications were providing a wealth of information, including timetables, routes and destinations of the convoys. With this knowledge, along with improved submarine tactics and better torpedoes, the rate of sinkings increased substantially. As of January 1944, U.S. submarines were destroying Japanese merchant ships at a rate of almost 330,000 tons a month. By 1945 over 8.5 million tons had been relegated to the deep.

The Imperial Navy's fleet cryptosystem was codenamed JN25, as it was the 25th Japanese naval code studied by the Americans. The JN25 continued to plague cryptanalysts as Japan advanced further into the Pacific. Finally, in 1940, cryptanalyst Agnes Driscoll of OP-20-G succeeded in breaking into the shield. The solving of parts of the JN25 proved to be a critical turning point for the Allies in the Pacific.

With the crucial information derived from radio transmissions covered by JN25, the U.S. Navy planned the Doolittle raid, the famous bombing raid on Japan led by Colonel James H. Doolittle. In early 1942 Allied morale in the Pacific was sorely in need of a boost. American forces at Pearl Harbor had been devastated. The Philippines was under heavy attack and the combined U.S. and Philippine defense forces were being forced down the Bataan Peninsula. The Japanese army and navy had also made large gains in a wide area of Southeast Asia.

To demonstrate American military strength in the Pacific region, the Joint Chiefs of Staff and the Pacific Fleet commander Chester Nimitz agreed on a bold plan. A two-carrier group led by Admiral William Halsey, with the codename Task Force Mike, sailed toward the Japanese islands in April 1942 with a secret mission. Aboard the carrier *Hornet* were U.S. Army B-25 medium bombers, while the carrier *Enterprise* held fighter planes to protect the ships.

Relying on the information provided by more JN25 breaks, Admirals Nimitz and Halsey felt fairly certain about the locations of Tokyo's naval and air squadrons. They also knew about some of Japan's immediate intentions, all of which gave them the freedom to risk the bold attack. On April 18, against great odds, Colonel Doolittle lifted his bomber off the *Hornet*'s lurching deck and led 16 B-25s into history. Though relatively little damage was done in Japan, the supposedly impenetrable skies over its cities had been breached. Allied confidence lifted as banner headlines declared, "Doolittle Dooed It."

The difficult and sometimes confused efforts to drive wedges into the Imperial Navy's cryptomethods now started to pay some much-needed dividends. By mid-April 1942 cross-checked clues were strongly indicating that the Imperial Navy was preparing an assault codenamed RZP on Port Moresby in New Guinea, which was an important fortress in its own right and a vital part of the larger Australian defense plan.

In this unfamiliar region Allied troops relied heavily on reports from the Islands Coastwatching Service (ICS), the Australian intelligence services and their cryptologists trained to guard Australia's long shoreline. Founded in 1919, the ICS was made of up of fishermen, planters, farmers and local civic officials, among others, who used radios to coordinate their reports. During World War II the ICS became a part of Australian naval intelligence and the Allied Intelligence Bureau, which directed and supplied the espionage and intelligence-gathering missions of combined Australian, U.S., British,

Dutch and other resistance forces in the Southwest Pacific. These operations were coordinated with U.S. General Douglas MacArthur's Far East Command.

Many coastwatchers lived a moment-to-moment existence as they hid on islands that had been captured by the Japanese. Their radios connected a region that spread from Port Moresby to the Solomon Islands. Using concealments such as the Playfair cipher, they supplied vital facts about Imperial Navy locations, troop carriers, convoys and enemy landing sites, and their information contributed to important successes in the Solomons (Guadalcanal) and in New Guinea.

At the time of threat to Port Moresby, U.S. naval forces were divided, with two aircraft carriers participating in the Doolittle raid against Japan. Nimitz took a calculated risk and ordered the carriers *Yorktown* and *Lexington* to disrupt the invasion plan.

The two ships and their accompanying Task Force 17, commanded by Rear Admiral Frank Fletcher, met the Imperial Navy on May 7, 1942, in the Coral Sea just northeast of Australia. American flyers sunk the light carrier *Shoho*, an escort for troop transports. Japan's naval cryptosystem JN25 (including the newer JN25b) radio transmissions were detected at more than one point during the battle, but in the midst of the conflict, Rear Admiral Fletcher was not always certain of the reliability of the decryptions.

The battle between the U.S. and Japanese fleets continued until the next day in what would later be known as the Battle of the Coral Sea. It was the first naval engagement conducted solely with aircraft; the opposing ships never even sighted each other. The Japanese carrier *Shokaku* was damaged and some of their support vessels were sunk, but the U.S. Pacific Fleet was hard-hit with the loss of the carrier *Lexington* and the damaging of the *Yorktown*.

Fletcher was later criticized for not making better use of the tactical radio intelligence at his disposal, and some historians believe he let a decisive victory slip from his grasp. He did, however, succeed in thwarting the planned Japanese assault on Port Moresby as well as delivering a considerable blow to Japanese morale.

In June 1942 the U.S. and Japanese navies fought another pivotal naval battle, known as the Battle of Midway. The confrontation itself proved to be one of the most amazing reversals in the annals of military history and validated the importance of signals intelligence for the U.S. armed forces.

As American and Japanese ships were fighting in the Coral Sea, the Imperial General Headquarters in Tokyo issued Navy Order No. 18, setting in motion Operation MI, the invasion of Midway Island and the occupation of selected points in the western Aleutian Islands. The Japanese strategy was to destroy the U.S. Navy in a decisive engagement and force the United States to a peace conference to end the Pacific conflict.

Cryptanalysts reading radio transmissions encrypted in JN25 discovered the designating letters *AF* among parts of the revealed messages. This finding was extremely significant because the Japanese had been giving the coordinates of their geographic locations in a code called the Chi-he system. Previous clues had been pieced together to interpret some of these letter pairs and *AK* was believed to be Pearl Harbor, while *AG* was Johnston Island, located 500 miles to the southwest of the Hawaiian Islands. There was a strong indication that *AF* was Midway. Commander Joseph Rochefort and the Station Hypo team in Hawaii began piecing the puzzle together in May, when they found the decrypted words *koryaku butai* (invasion force) followed by the designator *AF*. The Japanese Second Fleet command had also sent messages to its air units about advancing landing-base equipment and ground crews to *AF*.

To further verify this hypothesis, Rochefort created an elaborate scheme with fleet intelligence officer Edwin Layton and Lieutenant Commander Jasper Holmes. The Midway headquarters was instructed to transmit a plain-language message that its freshwater distillation plant was incapacitated, in the hope that this cleartext would be heard by Japanese listening posts. Within days, the Japanese response was overheard and decrypted. The Japanese transmission of this false report was linked with the designator AF, proving that it was indeed being used as Midway's location code.

Forewarned about the attack, Admiral Nimitz and his staff reacted with an imaginative plan. Task Force 16, commanded by Rear Admiral Raymond Spruance, was made up of the carriers *Enterprise* and *Hornet*, supported by five heavy cruisers, one light cruiser and nine destroyers. On May 28, this convoy sailed for a location named Point Luck, 350 miles northeast of Midway Island. At the same time, a radio deception plan was initiated, involving the seaplane tender *Tangier* at Efate in the New Hebrides and the heavy cruiser *Salt Lake City* in the Coral Sea. These ships sent radio signals in patterns similar to carrier task forces conducting typical communications with scout and fighter planes. Through such tactics, Admiral Nimitz hoped that the Japanese traffic analysts would believe that U.S. carriers and their accompanying fighter planes were far from the Midway region. While this radio fog screen continued, Task Force 17 steamed toward Point Luck. Commanded by Rear Admiral Fletcher, it consisted of the repaired carrier *Yorktown* that had been damaged at the Battle of the Coral Sea, two heavy cruisers and six destroyers.

The Imperial Navy squadrons opposing them were massive but spread out from the Marianas Islands to the Aleutians. According to the Japanese invasion plans, planes from two carriers of the Northern Area Force were to attack Dutch Harbor in Unalaska, followed by the occupation of the western Aleutian Islands of Attu and Kiska. The next day, planes from the four carriers of Admiral Chuichi Nagumo's Kido Butai (mobile strike force), the *Akagi*, *Hiryui*, *Kaga* and *Soryu*, would bomb the island of Midway. The following day, the Second Fleet transports would send troops ashore to occupy the strategic island while 300 miles to the west, Admiral Isoroko Yamamoto, on the combined fleet flagship *Yamato*, awaited the expected Pacific Fleet counterattack from Hawaii with two other battleships, a small carrier and support vessels.

Before the battle began, the Japanese made a long-expected change in the JN25 naval operational cryptosystem and by May 28 they had switched to a new version that the Allies could not immediately decipher. But the change had come too late to hide the plans for Midway. The American fleet intercepted the Japanese force and the resulting battle, which took place from June 3 to June 6, 1942, was a stunning defeat for Japan. The United States lost 347 personnel, 147 aircraft, 1 carrier and 1 support vessel. Japan's losses totalled 2,500 personnel, 322 aircraft, 4 carriers and 1 cruiser.

Beyond the tonnage and men lost, the Imperial Navy suffered a severe psychological setback. Its vaunted position as ruler of the Pacific was swept away in those early June days, never to be regained.

ESPIONAGE AND COUNTERESPIONAGE

The war on land continued inside occupied Europe, played out underground in resistance movements or in interception chambers where cryptanalysts struggled to predict their enemy's next move. As well as important contributions from brave members of the Resistance throughout Western Europe, much information about the Germans' plans came from the Soviet resistance group known as the Rote Kapelle.

After the Soviets signed a nonaggression pact with the Third Reich in 1939, they maintained full-fledged espionage networks, such as the Rote Kapelle, to watch the Nazis. German for "Red Orchestra," the Rote Kapelle was begun by an organization of skilled Soviet cryptographers with the first hints of the threat from Germany. Its name came from the Soviet radio transmitters, called "music boxes," that broadcast from inside the Third Reich and other European countries, whose operators were known as "pianists." Despite the information provided by the Rote Kapelle, Stalin failed to heed numerous enciphered warnings about the Germans' impending attack.

With its front operations, forged papers, false identities and other tools of espionage, the network remained in a holding pattern until June 22, 1941, when Wehrmacht and Luftwaffe forces roared across the Soviet Union's borders, launching Operation Barbarossa. On cue, the orchestra quickly began to transmit a symphony of data about the Nazis, including battlefield plans and information on aircraft production, political decisions, energy development and consumption.

The transfer of these secrets was protected by ciphers based on the Nihilist cipher and the straddling checkerboard, and used in conjunction with texts such as Honoré de Balzac's *La Femme de Trente Ans* and Guy de Teramond's *Le Miracle du Professeur Wolmar*. In general, Soviet cryptosystems at this time were superior to those of Germany. While Hitler's foreign office was compromised by spies and traitors, Soviet ambassadors protected their vital diplomatic transmissions with one-time pads, a type of unbreakable cipher.

Nevertheless, beginning about December 1941, high frequency direction finding (HFDF) carried out by the Funkabwehr (radio counterespionage force) succeeded in locating some of the Rote Kapelle's important broadcast sites and by December 1942 the orchestra had been effectively silenced. Some of the compromised sites and transmitters were used for a time by German intelligence in a *Funkspiel*, or "radio game" that involved the use of an enemy's radio set or network to send false facts. German agents intended to divide the Soviets and the Allies by indicating that the latter were conducting secret peace negotiations with the Nazis and even providing them with weapons and supplies.

Fortunately for the Allies, surviving Rote Kapelle members managed to alert their superiors to this ruse. Through a series of clever maneuvers involving faked clues about intelligence sources, nonexistent agents and half-truths about real events to keep up appearances, Soviet espionage officials forced the Nazis to disclose valuable information in order to maintain what they thought was a successful ruse. By the time the Berlin authorities realized that the tables had been turned, much damage had already been done.

A second important espionage operation was run by a Dr. Richard Sorge. Using the cover of journalism, Sorge had operated a spy ring for Russia in Shanghai, China, from 1929 to 1931. As a member of the Nazi party and a friend of high officials, he had easy access to Japan in the 1930s. Working for a top German newspaper, Sorge made many important contacts in Tokyo's government and military and gathered vital information about the two most dangerous enemies of the Soviet Union.

Sorge's personal radio broadcaster was Max Clausen, who had had previous experience in transmissions during World War I. They used a straddling checkerboard cipher similar to that of the Rote Kapelle to encipher their messages. For an added numerical key, Sorge and Clausen applied digits from innocuous texts like trade statistics logs. As the texts were publicly available, they were not as unbreakable as a series of random, one-time use digits, but they were strong enough to foil German counterespionage eavesdroppers.

Among his many discoveries, Sorge learned two crucial facts. One was not heeded by Stalin and nearly led to complete disaster for the Soviets. The second discovery was noted and acted upon at a pivotal time.

Sorge had determined that Hitler was intending to attack the Soviet Union and had obtained a nearly accurate date. However, Clausen's radioed alert to Moscow was dismissed and the Nazi juggernaut rolled on June 22, 1941. Despite this blow, Sorge kept plying his connections for information and his persistence finally paid off. Moscow had stationed a considerable number of troops in Russia's eastern regions to defend it against Japanese threats, but one of Sorge's key Japanese contacts confirmed that Japan would be making advances into the Pacific instead.

Based on Sorge's information, Red Army commanders were able to divert tens of thousands of seasoned troops and tons of equipment westward in time to check the Wehrmacht and Luftwaffe siege of Moscow. These new forces and extremely bad weather united to halt the Nazi onslaught and prevent the capture of the city.

Sorge was not so fortunate. A series of informants outside and inside his ring led to his and Clausen's arrest in October 1941. Under severe interrogation, Clausen gave up their encryption secrets which were used against them in their subsequent trial. Clausen was sentenced to life imprisonment and Sorge was hanged in November 1944.

A third very successful Soviet espionage operation came to be known as the Lucy Ring from the codename of its most prominent spy, Rudolph Rössler. His codename was derived from his residence in Lucerne, Switzerland, the neutral territory which gave the Russian agents more leeway than they had in German occupied nations. The founder of the ring was Alexander Rado, codenamed Dora, a Hungarian Communist.

Rössler's sources of top-secret information included a group of World War I friends who had since 1918 moved high up in the Wehrmacht hierarchy. Like Rössler, they were anti-Nazi and provided key military information as the war began between Germany and the Soviet Union. Using numbers derived from a book of Swiss trade data, the Lucy Ring radioed facts concealed with the straddling checkerboard system. The information was immensely important as it concerned Germany's order of battle—military strength, organization and location—during Hitler's invasion of Russia. These vital details enabled Red Army commanders to plan particular battle strategies and overall tactics until October 1943, when Swiss police finally located and broke up the ring after much pressure from the Funkabwehr.

The Rote Kapelle, Sorge and Lucy groups did much to counterbalance the Third Reich's early military successes, but the Germans also had their share of espionage agents. In 1942 a group of German spies involved in what was known as the Kondor mission sought to obtain information about the British forces opposing General Erwin Rommel, "the Desert Fox," in the North African desert. The two key members of this mission were John Eppler and Peter Monkaster. The former was born of German parents in Egypt and had acted as an Abwehr agent provocateur attempting to inflame long-smoldering Arab hatred of Great Britain. Peter Monkaster was a long-time German resident in East Africa working as an oil mechanic.

In May the two made their way through British and Egyptian checkpoints with the help of forged papers and their ability to speak English. Once in Cairo, they gathered a group of supporters, including staunch Arab nationalists, a pro-Nazi priest and an exotic dancer named Hekmeth Fahmy. A primary attraction at the Kit Kat Cabaret, Mademoiselle Fahmy was one of Egypt's most popular *danseuses du ventre*, belly dancers. She was also an agent for the Free Officers' Movement, a group of anti-British Egyptian army officers who sought independence for Egypt and were eager to undermine the British cause. Hekmeth was already actively involved in liaisons with a Major Smith of the British Government Headquarters in Cairo. Monkaster and Eppler rented a houseboat on the Nile, in which they installed one of their radio transceivers. During the Major's trysts on Mademoiselle Fahmy's nearby houseboat, they studied his documents and obtained facts about British troop dispositions and plans. The important details were encoded with a book code based on *Rebecca*, the 1938 novel by Daphne du Maurier. The concealed messages were broadcast to fellow agents in one of two locations, Rommel's *Horch* (wireless intelligence) Company or the Wehrmacht's radio outpost in Athens. There, with the aid of a copy of the novel, the information was decoded and sent to the appropriate parties.

Fortunately for the British Eighth Army, carelessness and a courageous counterespionage agent foiled the Germans' plans. While scouting for information, Eppler met a young woman named Yvette, paid her for her company and asked to see her again. She agreed, but not because of his charm or his pound notes. She was a member of the Jewish Agency, which was working with Britain's secret intelligence service, MI-6. Finding Eppler's behavior and plentiful money suspicious, she visited his houseboat again and made a valuable discovery. While Eppler and Monkaster slept amid empty bottles and full ashtrays, Yvette found a copy of *Rebecca* and some notepaper covered with groups of letters and grids. Before she could leave the houseboat and reach her contacts, she was arrested by local police who were watching suspicious activity in the area. She did, however, get an opportunity to reveal what she had learned.

Britain's field security in Cairo closed in on the Kondor pair, who refused to relinquish their secrets despite thorough interrogation. With Yvette's information regarding the book code, British cryptanalysts were able to solve the *Rebecca* code.

Impersonators replaced the jailed Kondor team and began sending false information to Rommel. These faked reports were believed and had an important effect by confounding German actions in the region.

In 1941, the third member of the Axis, Italy, succeeded in compromising the U.S. Black Code, which was used to conceal the communiqués of U.S. ambassadors and military attachés sent around the world. Although the Italian army's security and intelligence group, the Servizio Informazione Militare (SIM), contained a cryptanalysis department that had succeeded in breaking ambassadors' and generals' dispatches, it was espionage, not cryptanalysis, that handed the SIM the important Black Code. When the Black's secrets were exposed in September of that year, it posed a serious security breach for the United States.

The agent was a messenger assigned to the U.S. military attaché in Rome. In August 1941, he had reportedly enabled the Italians to carry out a black bag operation by either making an impression of an office key or providing agents with a safe's combination. This gave an SIM team the means of entering the U.S. Embassy office, opening the safe and photographing the Black's contents.

The SIM reportedly gave some copies of decryptions to the German Abwehr, a counterespionage group. It is unclear whether or not the German codebreakers had already solved the Black Code on their own initiative, but in either case the Reich's cryptanalysts were soon reading United States attachés dispatches, which included vital facts about the Allies' military and economic situations.

The Axis powers achieved another coup in Egypt in 1942, when important Allied military secrets were inadvertently divulged by the U.S. military attaché in Cairo. The attaché, Frank B. Fellers, gave too-frequent and detailed reports about British actions in North Africa. German eavesdroppers snared the reports concealed by the Black Code, and after solving and re-encrypting them in a German code, they were dispatched to Rommel.

The Desert Fox's Afrika Corps had been on the defensive in the latter part of 1941. By early 1942, thanks to the information contained in the intercepts, he counterattacked and sent the British reeling back some 300 miles over dearly won land. He continued to be successful in outwitting the British as Fellers's ongoing reports provided important military data.

It was not until mid-June of 1942 that a war prisoner told the British about the interceptions. The British had also solved the U.S. Black Code and began studying Fellers's dispatches to check their contents. They realized that his attention to detail had been providing the Germans with their apparent omniscience in North Africa and warned the American authorities. Fellers was quietly removed in July. His replacement in Cairo was equipped with the M-138 cipher device with mixed alphabet strips, rendering new communiqués secure against German analysts.

Without his source of secret intelligence, Rommel's decisions for allocating his limited tank forces and fuel suppies became much more difficult. When the British Eighth Army launched a devastating surprise attack at El Alamein, Egypt, on October 23, 1942, they sent the Desert Fox into a final flight from North Africa.

With the additional information provided by spies on both sides, the "war of the wizards" continued as cryptanalytic teams struggled to break each other's ciphers. One of the advancements at England's Bletchley Park cryptanalysis center was a machine called Heath Robinson, an electronic match for Germany's early *Geheimschreiber*, or secret writing machine. Designed in 1943, the Heath Robinson, which had advanced teleprinter aspects of the Baudot code and the Murray code, succeeded in breaking some of Hitler's directives and German foreign office exchanges.

Among many such inventions, possibly one of the most remarkable achievements was the creation of the Colossus, a predecessor of the modern computer about the size of three large wardrobes, that was used to break an improved ten-rotor *Geheimschreiber*. Established at Bletchley Park in December 1943, Colossus's photoelectric cells, radio tubes, high-speed drive system and extra tape-reading mechanisms helped the cryptanalysts check thousands of cipher characters a second. In 1944–45, Colossus and a Mark II successor provided data about the most secret radio transmissions within the hierarchy of the Third Reich. These included dispatches ranging from overall defensive plans to specific troop movements and provided the Allies with their most telling descriptions of Axis intentions.

THE END OF THE WAR

As the war dragged on, the highest Allied command secretly met and planned D-Day, an attack of unprecedent scope. Codenamed Overlord, the Allied invasion of Nazi-held Europe was the most massive invasion in history. The successes of cryptology played a critical role in planning this feat: British Ultra decryptions of the Enigma and later cryptosystem machines revealed vital plans from many levels of the German command; the BRUSA Agreement, signed in May 1943, included careful plans for data exchanges, security procedures and a new lexicon of codenames and codewords, among other arrangements between the United Kingdom and the United States; the United States military shared Magic decipherments of Purple-covered Japanese diplomatic exchanges between Berlin and Tokyo with Britain, enabling the Allies to learn about German plans for counterattacks and fortifications that were transmitted to the Japanese.

Resistance groups in occupied Europe also intercepted written messages, radio transmissions and Nazi communications. As this work required agents to work in close proximity to the Wehrmacht camps, the Resistance used passwords, open code signals and other cryptologic means to convey messages across battle lines, arrange rendezvous and protect their identities. The Allies practiced many types of COMSEC (communications security) measures, from the "Loose lips sink ships" campaign, enciphered ship-to-shore and air-to-ground transmissions and codewords to maintain high-level secrecy. Innumerable jargon codes in the form of single sentences, poems, personal messages and even weather forecasts filled the airwaves over the English Channel as D-Day drew closer. The

Resistance groups awaited these orders to initiate their own assignments, which included sabotage, seizures of strategic sites and ambushes of Wehrmacht patrols or reinforcements.

Fortitude was the codename for the tactical operations of cover and deception intended to hide the actual timing of the D-day invasion and the location of its staging area in Britain. The assault, codenamed Neptune, was made up of operations codenamed Fortitude North, which was directed at Scandinavia, Fortitude South, which targeted Belgium and the Channel coastline of France, and Fortitude South II, which took place in the Channel areas after the D-Day invasion.

The elaborate Fortitude plans were determined at a conference, codenamed Rattle, held in Largs, Scotland, beginning in June 1943. Top-level strategists chose western Europe as the site of the crucial Allied landing and formulated the broad strategy of aerial, electronic, espionage, camouflage and other deceptions that would keep the German command on the defensive.

Fortitude North was intended to keep some 27 Nazi divisions in Norway, Denmark and Finland under threat of imminent attack. An operation codenamed Skye created a mythical British Fourth Army of 350,000 men supposedly assembled in Scotland, along with a nonexistent 15th Corps from the United States and equally imaginary Russian troops.

Fortitude South invented a First United States Army Group seemingly poised to strike at the Pas de Calais in France. The idea was to pin down the powerful German 15th Army and keep it away from a direct defense of Normandy. The deceptions surrounding this ruse were codenamed Quicksilver.

These fictitious maneuvers involved the transmission of codes and ciphers that were intended to be intercepted by German intelligence. Several were accomplished by radio tricks wherein the Allies made intentional cryptographic errors designed to give misleading details about the imaginary armies and other forces. Radio messages from southern England were conveyed to a mock headquarters in Dover, leading German eavesdroppers to conclude that a large force was poised in Dover, another site opposite the Pas de Calais.

These misconceptions were further enhanced by squadrons of dummy vessels purposely gathered in the ports of southeastern England, as wireless deception and reports by double agents created a sense of real activity around the nonexistent naval forces. Encrypted messages were sent to false air bases containing plywood bombers and fighters. The contrived radio reports and agents' "intelligence" was verified by the Luftwaffe when their aerial reconnaissance sighted these harmless air forces.

The Axis policy of sharing facts about military preparations and activities also provided crucial details thanks to Magic intercepts and solutions. Japan's ambassador in Berlin, General Hiroshi Oshima, was a virtual font of secret information. Like other Japanese embassy and consular officials at distant sites, Oshima sent his reports to Tokyo via radio. The messages were enciphered by the Alphabetical Typewriter '97, which U.S. analysts had already solved. Among his many detailed encrypted radiograms, Oshima reportedly mentioned a number of facts about Germany's defenses along the Westwall (the European coast) and the Siegfried Line (the German border). The U.S. decryptions became Magic information that greatly benefited the Normandy invasion and subsequent military decisions on the European continent.

On June 1, 1944, the British Broadcasting Corporation broadcast a jargon code to the French resistance, based on phrases from the poem "Chanson d'Automne" by Paul Verlaine. The first line, translated as "The long sobs of the violins of autumn," announced that the mission was imminent. On June 5, a phrase from the second stanza, "Wound my heart with a monotonous languor," indicated that the invasion would begin in 48 hours at 00:00 hours. These codes signaled the Maquis and other Resistance groups to begin to carry out their plans. The success of these and many other operations with sabotage squads, airborne troops, naval commandos and Resistance units combined to make the invasion of June 6, 1944, a monumental achievement which turned the tide of the war in Europe.

On the Normandy beachhead itself, the U.S. 849th SIS provided important solutions to Nazi communiqués. This field unit solved a Wehrmacht dispatch that gave General Omar Bradley's 12th Army time to counter a strong assault on a lightly defended position. The 849th also aided Bradley with intelligence important to battlefield tactics, as the Allies gained ground hedgerow by hedgerow through the French countryside.

No matter how diligent or skilled an interception/cryptanalysis unit is, it is at a loss if the enemy does not send communiqués, or is extremely careful when doing so. Such was the case with Hitler's last desperate gamble, a winter counteroffensive that came to be called the Battle of the Bulge.

Many military historians have examined the manpower, armaments and supply buildup that led to the surprise assault by Panzer troops and Luftwaffe units on December 16, 1944. Some have concluded that the Allied unpreparedness was due to a failure in intelligence gathering. As once-classified records have become public, they have confirmed that there were indeed intercepts that showed troop movement. Friendly spies and informers did conduct human intelligence (HUMINT) observations and picked up gossip about groups of trucks or fuel collections in certain areas. During the autumn of 1944 a few deserters also divulged some clues about something "big" in the future, and aerial reconnaisance showed pattern of activity on bridges, roads and railroads that had not already been thoroughly wrecked by bombing.

But the Allied commanders could not be certain whether this troop movement and concentration was a real buildup to offensive action or simply deployments for a different defensive strategy. They did not believe that the Germans could, or would, mount an attack through areas of hilly, thickly wooded land, and so left such areas more thinly guarded along the lengthy combat front.

One such lightly defended region was the Ardennes forest in southern Belgium and northeastern France, which area became the very point of the offensive's spearhead. Mistaken preconceptions about German intentions and capabilities had led to near calamity.

After the loss of many Allied troops, the Wehrmacht's surge that caused the bulge was halted. At Bastogne, a radio intelligence team helped turn back the armored waves. The SIS crew studied the Germans' mobile radio exchanges and picked up a number of valuable clues, including one dispatch that allowed General George Patton to heavily batter a formidable paratroop division.

With the passing of the winter of 1944–45, the Allies were again on the march to breach the Siegfried Line and enter the Nazi's heartland. The overview intelligence nets of Ultra and Magic, as well as the combat cryptanalytic units, listened and solved the desperate, then dying, orders of the crumbled Third Reich.

The war in Europe was over, but the future of the Pacific still seemed uncertain. In April 1943 an important symbolic victory had occurred when Japanese Admiral Isoroku Yamamoto, commander-in-chief of the Japanese combined fleet and architect of the attack on Pearl Harbor, was ambushed and killed.

At 59 years of age, Yamamoto was considered a top strategist and a bold leader. The Japanese forces had suffered numerous setbacks in the Solomon Islands, and he planned to travel by plane to the region for a morale-boosting, one-day inspection tour. In preparation, a detailed description of his itinerary was radioed to the area's Japanese commanders with the news concealed by the then-current edition of the JN25 and an Imperial Army code.

On April 13, U.S. listening posts snared the dispatches and solved them. The very punctual admiral was scheduled to fly within the range of U.S. aircraft located at Henderson Field at Guadalcanal. The analysis revealed such a specific schedule that military leaders were able to plan a precise interception of Yamamoto's plane. There was a potential danger that success would cause the Japanese to suspect ther messages were being read, but that concern was outweighed by the desire to eliminate such a prominent leader and simultaneously avenge Pearl Harbor.

On April 18, as Yamamoto's party of two bombers and six escorting fighters flew near the coast of the Bougainville Island, they were attacked by eighteen U.S. Army P-38 fighters. Yamamoto was shot down and cryptanalysis was credited with a success equal to an important military victory.

Largely supported by the U.S. Navy, troops from nations including Australia, Great Britain, France, New Zealand and the Netherlands struggled to push back the Japanese presence in the Pacific. As the war in the Pacific dragged on, a weapon was being developed in America that many hoped would never have to be used—the atomic bomb.

During the development of the nuclear warheads in America, codes and ciphers were used to help protect the top-secret research. The massive project's codename was the Manhattan Engineer District, also known as the Manhattan Project. Code names were also given to individuals, such as General Leslie Groves (Relief), Niels Bohr (Nicholas Baker), and Enrico Fermi (Henry Farmer), and to research centers, including Los Alamos, New Mexico (Site Y) and Oak Ridge, Tennessee (K-25). The bomb itself was referred to as S-1, the Gadget and the Thing. The U.S. Signal Corps also used its ciphers to protect telegraphic communications concerning the bomb. Personalized jargon codes and self-styled double-talk protected private conversations, and some telephone calls were concealed by encryptions such as the Quadratic code.

The first atomic device was successfully detonated in the desert of Alamogordo, New Mexico, on July 16, 1945, during a test codenamed Trinity. The news was transferred by jargon code to the new president, Harry S. Truman and Secretary of War Henry L. Stimson in Potsdam, Germany, who were attending a conference on the future of Europe and the ongoing war with Japan.

As the conflict continued through the summer, plans were drawn up for a direct invason of Japan, codenamed Centerboard. Many military leaders feared the possibility of huge losses of Allied personnel if the war in the Pacific should continue. Furthermore, the Japanese would not accept an unconditional surrender that would threaten the continuance of the Emperor's throne. Based on these and many other factors, President Truman and U.S. military commanders made the controversial decision to drop atomic bombs on Japanese cities.

On August 5 the first atomic bomb was dropped on Hiroshima, followed three days later by a detonation over Nagasaki. In the wake of these attacks, Japan formally surrendered on September 2, 1945, and World War II had finally ended.

The end of World War II ushered in the beginning of the nuclear age. While the United States seemed to have a monopoly on this new and awesome power, in spite of all the security precautions, atomic secrets were being stolen even as they were being created. An extremely clever Soviet spy network had spread through the United States, Canada and England to reach all levels of research and development, and within four years of the first atom bomb detonations, the Soviet Union had its own nuclear weapon.

THE POSTWAR ERA: SPIES AND SECURITY

Adolf Hitler's thousand-year Reich had lasted for twelve horror-filled years. With its downfall, the world looked forward to a time of peace. Yet even as American and Soviet forces were meeting at Torgau on Germany's Elbe River in April 1945, the seemingly solid foundations of the victorious alliance were cracking.

The Yalta Conference in the Crimea had established that Nazi-occupied nations were to be freed, borders re-created and new economic policies begun, but ideological differences, varied national aims and misinterpretations of this agreement led to a breakup of the wartime East-West pact. Eleven months after the meeting on the Elbe, on March 5, 1946, Winston Churchill made a speech at Westminster College in Missouri, announcing that "from Stettin in the Baltic to Trieste in the Adriatic, an iron curtain has descended across the continent."

An iron curtain in the form of Soviet armies, secret police and political organizers had indeed fallen with brutal force along the borders of Eastern Europe. Within a year of expelling the Nazi invaders, much of Eastern Europe faced an equally menacing aggression from Moscow. The Yalta agreement had been broken, and in place of free elections and local autonomy were Red Army tanks, propaganda machines and the stifling of political opposition.

In the next four years the United States responded with everything from humanitarian relief for Soviet-blockaded Berlin to the massive economic recovery program called the Marshall Plan and the military alliance of NATO. Despite continued international pressure, the Soviets refused to leave the occupied countries. Using radio networks and one-time pad ciphers to contact each other, the Red Army and spy networks that had fought the Nazis remained entrenched in Eastern Europe. In many cities their troops took from the Wehrmacht, cells of Communists and their sympathizers were either in position or were placed to act on behalf of the Kremlin planners.

These spy networks had stretched into the heart of U.S. security itself. Some five months before Churchill's speech, the western Allies had received a first-hand shock concerning Soviet treachery through the revelations of a former Communist loyalist.

Igor Gouzenko had been a cipher clerk at the Soviet embassy in Ottawa, Canada, until he defected on September 5, 1945. He had been primarily responsible for encrypting the official transmissions of Moscow's Ambassador Georgy Zarubin and a military attaché, Colonel Nikolai Zabotkin, who led the Canadan networks of the GRU, the Soviets' military intelligence organization. Gouzenko's documentation of these men's activities, drawn from their telegrams and some pages from Zabotkin's diary, gave conclusive proof of security penetrations in Great Britain, Canada and the United States. Of most concern to national security agencies was the evidence that atom bomb research was a top Soviet spy target. As a result of Gouzenko's testimony, several Canadians were convicted of espionage and imprisoned and the treason of British atomic research scientist Alan N. May was exposed. After Gouzenko's defection, he, his wife and son went into an early version of a witness protection program in Canada.

The FBI and U.S. military security groups had been aware that the Soviets had active spy rings around the world, but during the Second World War the FBI's primary focus had been counteracting Axis espionage. The Soviet Union had been an anti-Axis ally and Stalin was even called "Uncle Joe" in some circles.

U.S. military intelligence did maintain surveillance of potential Soviet espionage, despite being greatly understaffed. During World War II's national security emergency, cable and telegraph companies had made copies of other nations' messages available to the government. Encrypted Soviet telegrams, including those of Soviet diplomats and trade organizations, had been gathered intermittently since 1939. On February 1, 1943, an official effort was begun by U.S. Army intelligence to decrypt these exchanges.

The demands of war greatly expanded the SIS and divided its activities, but interest in the Soviets' transmissions continued. At a new intelligence site in Virginia called Arlington Hall, study of the cables showed that five cryptographic systems had been used by five different groups communicating with Moscow. In order based on the volume of messages gathered for each, the groups were: trade representatives; diplomats, including embassy and consular business; the Sovet secret police or KGB; army GRU; and Navy GRU. Analysis of the traffics' concealment covers indicated the concealment was a one-time system. Soviet representatives at the above sites and in Moscow were superenciphering a numeric codes with the random digits of one-time pads.

The process seemed impenetrable until October 1943 when U.S. Army analysis revealed that the pads' numbers had not remained completely random. It seemed quite likely that a major blunder had occurred during the production of the pads at the Soviets' cryptographic materials' center. Numerals had been duplicated on more than one series of pads and this security breach enabled the American solution seekers to compare dispatches and seek repeating patterns in the message digits.

U.S. cryptanalysts labored continuously to locate these additive groups, peel them and solve the codenumbers and codewords beneath them. As some breaks began to occur and a growing list of spies' covernames and references to clandestine activities became apparent, Army G-2 intelligence officials contacted the FBI.

Along with the Gouzenko disclosures, the FBI had been given a series of leads about the Soviets' growing operations. In 1945 they closely questioned ex-Communist journalist Whittaker Chambers, whose 1939 warnings about Moscow's espionage apparatus had gone unheeded by a State Department official. In 1942 an earlier FBI interview had proved fruitless due to Chambers's excessive immunity demands and mixture of true and deceitful answers. In November 1945, many of Chambers's accusations were corroberated by supportive informer Elizabeth Bentley.

An experienced spy courier and agent handler, Bentley was a Soviet defector who had provided the FBI with a number of name and place connections. Both made their charges public before the House Committee on Un-American Activities, better known as HUAC, in July and August 1948. Their testimony caused a sensation.

In October 1948 FBI Special Agent Robert Lamphere became the full-time liason between U.S. Army intelligence and the FBI. He met with a brilliant analyst named Meredith Gardner at Arlington Hall. Partial solutions provided clues that Lamphere applied to espionage investigations that he and others conducted through the latter 1940s and early 1950s. News of the investigations and arrests filled the headlines. The resulting convictions included those of Judith Coplon, a Justice Department employee, who passed documents to her contact, Valentin Gubitchev of the Soviets' UN mission; Klaus Fuchs, a British scientist, who divulged several atomic bomb secrets to Soviet agents; Harry Gold, an American chemist, who conducted industrial espionage and acted as a courier of atomic bomb secrets; and the Rosenberg ring, made up of atomic bomb spies Julius and Ethel Rosenberg, David Greenglass and Morton Sobell.

The message-solving project continued under various code-names such as Jade, Bride and VENONA. The British and other NATO allies shared facts from their own counterespionage projects in exchange for additional information from the United States. In spite of much effort and computer-enhanced cryptanalysis, various parts of the transmissions remained unsolved and later investigations never achieved the same success as the 1948–53 inquiries. While VENONA was formally concluded on October 1, 1980, its full magnitude was not officially announced to the general public until July 1995 during a series of declassifications. Many of its findings are still being examined.

With the stark evidence of espionage in its highest organizations, the Cold War became a daily concern for America's policy makers. Many domestic and international security agencies developed to keep track of potential animosities. The mutually agreeable results of the BRUSA agreement led to meetings with Australia and Canada and in 1947 a number of nations signed a top-secret pact named UKUSA, the acronym for the United Kingdom–United States Security Agreement with the United States, Great Britain, Canada, Australia and New Zealand. Each nation agreed to share signals intelligence information within the sphere of influence where it was best able to gather intelligence. The United States covered certain Chinese frequencies from its listening posts in Japan, Taiwan and Korea. England had priority over other Chinese channels from its position in Hong Kong. Australian eavesdroppers covered regions across the South Pacific as well as sections of the Indian Ocean. Canada and the United States shared monitoring stations in North America and the vital northern polar region. Other NATO countries, as well as South Korea and Japan, later joined as signatories designated as third parties with restricted access. Though now broader in scope, the specific intent of the pact continues to be the maintenance of a highly secret signals intelligence network.

U.S. defense and foreign policy strategies had been based largely on the sole possession of the atom bomb and advanced scientific infrastructure. This monopoly was shattered in 1949 when Russia successfully detonated her own atomic device. With the help of stolen research provided by American traitors, the Soviet Union had done what had seemed impossible for a war-torn country to accomplish. The long-held view of Russia as a snowbound, backward behemoth had to be sharply re-evaluated; the Kremlin was given due credit as a formidable foe.

In the wake of these revelations a new U.S. security organization, the Armed Forces Security Agency (AFSA), was created in 1949 to coordinate the armed forces intelligence and communications tasks. Although the AFSA was given control of strategic intelligence and communications duties, the individual services still maintained tactical control of communications intelligence responsibilities in potential combat areas. Additionally, the U.S. Communications Intelligence Board was created to coordinate data transfers among the AFSA, the FBI and the State Department. What at first seemed to be an effective streamlining process soon encountered its own series of complications. Some historians blame the AFSA arrangement and its continued interservice and civilian security service rivalries for contributing to the U.S. intelligence failure to recognize the danger signs of the Korean War.

The admitted intelligence problems during this war prompted President Truman in 1952 to replace AFSA with the National Security Agency (NSA), which achieved much better coordination of intelligence gathering and security in the ensuing years. Still the center of American cryptology, the top-secret NSA develops concealment methods for various U.S. government departments and analyzes the codes and ciphers of other nations. Located in Fort Meade, Maryland, the NSA also creates and maintains a number of signals intelligence-gathering systems around the globe, from outer space to the oceans' depths.

A global intelligence network has evolved to supply the NSA with data. This network combines human intelligence (HUMINT) provided by the Central Intelligence Agency (CIA), whch was founded in 1947 as the successor to the World War II Office of Strategic Services, the first global U.S.

spy ring; military intelligence agents; land-based listening posts; and sea and airborne platforms. Information gathered from these sources is sent to various NSA departments according to its type, level of encryption complexity and region. Although the NSA analysts do not claim high rates of success for high-level encryptions, they maintain their vigil, hoping for a miscue of repeated ciphertext, an accidental transmission in cleartext and new computer enhancements for quicker electronic analysis. Defectors with encryption secrets are welcomed, though warily.

One such defector was Russian Reino Hayhanen. Carefully prepared in the late 1940s for an espionage assignment, Hayhanen was given the cover identity of Eugene Maki, an actual American who had disappeared in Finland. As Maki he was able to apply for and receive a passport to the United States. After arriving in October 1952 in New York City, Hayhanen sent for his new wife. They lodged in different parts of the New York area before purchasing a home near Peekskill. But Hayhanen was not seeking Hudson Valley scenery; he wanted a fitting site for a radio transmitter. His main cipher was a very unwieldy pencil-and-paper cryptomethod called VIC, after his codename, that combined a substitution table and two transposition tables. While awkward and time-consuming, it did manage to fool American intelligence.

Upon his arrival Hayhanen began drop site contact with Soviet deep-cover agent Rudolf Ivanovich Abel, whom Hayhanen had been sent to help as a courier and communications lieutenant. Abel was an experienced operative who was equipped with top-level training and encryption systems. Beginning in Canada in 1948 with aliases such as Andrew Kayotis and Emil Goldfus, Abel eventually made his way to New York City. Posing as a retired photofinisher and artist, among other covers, he kept a radio receiver in his Latham Hotel room, where he tape-recorded communications from Moscow and later decrypted them.

Hayhanen and Abel began their unseen contacts with a series of prearranged drops of microfilmed messages placed, for instance, in a space behind a loose brick under a Central Park bridge. Communications were also hidden in hollowed bolts and flashlight batteries.

It is unclear when problems in the Hayhanen-Abel relationship began. Some historians believe that a rift occurred when a newspaper boy discovered a hollow nickel and turned its contents, microfilmed numbers wrapped in tissue paper, over to the FBI. Other sources state that after the pair had finally met, Hayhanen's careless work habits and weakness for liquor finally alienated Abel. In 1957 Abel ordered Hayhanen to return to Moscow for a "rest." Hayhanen traveled as far as Paris, where he defected to the U.S. Embassy and divulged his association with Abel as well as facts about their cryptosystems.

Abel was arrested in his Latham Hotel room in June 1957. A one-time pad was found hidden in a hollow block of wood, along with other incriminating material. He was held as an illegal alien and convicted of espionage, attempting to obtain defense secrets and failure to register as a foreign agent. Sentenced to 45 years' imprisonment, he was exchanged for U-2 pilot Francis Gary Powers in 1962.

Hayhanen also exposed the treachery of U.S. Army sergeant Roy Rhodes, stationed at the U.S. Embassy in Moscow, who in 1951 had been blackmailed and bribed into providing cryptographic information to the Soviets. Rhodes was court-martialed and sentenced to five years' imprisonment.

Hayhanen's last years are shrouded in mystery. After being given political asylum in the United States, he is variously reported as having succumbed either to an automobile accident or to alcohol.

England also had its share of high-level infiltration, but the counterespionage results were less spectacular due to security breaches in British intelligence by Cambridge University graduates such as Anthony Blunt, Guy Burgess, Donald Maclean and Harold Kim Philby, and other Soviet agents who hampered investigations. In 1959 Britain's Government Communications Headquarters made a discovery that renewed interest in the World War II Soviet transmission in England. GCHQ learned that the Swedish Signals Intelligence Service had intercepted some GRU radio exchanges between agents in Great Britain and their headquarters in the Soviet Union. The Swedish authorities were persuaded to circumvent their standard neutrality policy and provide the radio data for analysis. Giving this project the codename HASP, GCHQ analysts began to attack the GRU cryptosystems.[1]

Following advances in computer-enhanced cipher solving, the first real breaks in HASP came in 1972, revealing that dozens of British journalists, technicians and scientists had provided information to the Russians during the war. Among them was a Cambridge cinema expert and freelance journalist named Ivor Montague, who passed facts relating to the Labour Party and other political groups. Even more jarring was the treason of J. B. S. Haldane, a respected biochemist who worked with Admiralty submarine-related experiments.

The GRU decryptions finally proved the existence of a long-operating spy ring that included the deep-cover agent Sonja, also known as Sonia, the atomic bomb spy Klaus Fuchs prior to his arrest, and links with Alexander Foote, a Soviet operative in Switzerland and a multiple defector. Most stunning were the realizations that many members of MI-5, the British security and counterintelligence agency, were criminally negligent and the entire system was riddled by Soviet moles.

THE RACE FOR TECHNOLOGY

During the 1950s and early 1960s security agencies on both sides of the Cold War struggled to outflank each other on diplomatic tactics, but by the 1970s the war of the wizards had come to have a decidedly different strategy. Advances in science and industry and the development of jet planes, nuclear submarines, intercontinental ballistic missiles and computers was making technology the focus of concern. America's scientists were making quantum leaps in so many areas that the Soviets and others began to fear a "facts gap." A new era of industrial spying evolved and gave birth to a new group of technological agents.

[1] As with the VENONA declassifications, information about HASP is still being declassified.

Technology thieves placed highest priority upon computer technology. America's industrial centers were given the special target emphasis that the KGB had once applied to the atomic Manhattan Project. One of the primary examples of technological treason occurred in the mid-1970s with the case of "the Falcon and the Snowman."

The first Rhyolite satellite had been put into place in March 1973 as a signals intelligence orbiter eavesdropping on Soviet microwave signals and telemetry data. The usually secrecy-minded Soviets had not believed that either transmission method could be intercepted. For about four years, the United States had a very valuable advantage.

By the middle of 1977, however, a spy named Christopher Boyce had informed the Soviets about their vulnerability. The Falcon (Christopher Boyce) and the Snowman (Andrew Lee) were young friends from California who shared a mutual disillusionment with U.S. foreign and domestic affairs in the late 1960s. Aided by the lax security of an aerospace company, TRW, Lee and Boyce were able to pass vital information to Soviet agents operating in southern California and Mexico City. Though his work in TRW's communications center, Boyce provided Lee with CIA satellite reconnaissance data and top-secret cipher lists. Lee was responsible for transporting these documents and selling them to the Soviets.

Their operation continued unsuspected for a number of years. Finally, Lee's impatience in making a Russian embassy contact led to his detention by Mexico City police. During rough questioning, he divulged information about Boyce that led to their arrest on charges of espionage. After a trial that kept the nation captivated, Lee was convicted for life and Boyce was given a forty-year sentence.

The damage to U.S. intelligence was irreparable. Based on the information they provided to Moscow's agents, the Soviets began protecting their launches with much better encrypted telemetry signals, severely compromising U.S. surveillance of Russian ballistic missile tests.

In 1985 the American public was again shocked when newspaper headlines proclaimed the FBI arrest of retired U.S. Navy warrant officer John Walker. Investigations revealed a complex pattern of deception involving this family of Navy men. In trials during 1985 and 1986 John Walker, his son Michael, John's brother Arthur and family friend Jerry Whitworth were directly linked to massive long-term military espionage. In John's case the spying had continued for as long as two decades on naval bases from Virginia to California. Although he had brazenly carried top-secret material in a duffel bag, it took the efforts of his troubled ex-wife Barbara and daughter Laura to expose him.

As a Navy radio operator, John Walker used a KL-47, the mainstay of naval communications, an electronic rotor machine that was a more advanced version of the German Enigma machine. The KGB had desired the KL-47's solution so avidly that they had placed their best scientists on the mission.

Soviet experts succeeded in providing Walker with a palm-sized, battery-operated device called a continuity tester, which revealed the wiring pattern of the KL-47's rotors. Soviet scientists used this knowledge to re-create the circuitry, and then they applied computers to search the millions of possible encryption variations. When Walker also provided real cipher/code key lists after his tours of duty, the Navy's protection methods were completely undermined.

Aided by his lieutenant commander brother, his son who was stationed on the aircraft carrier *Nimitz* and Whitworth, a California cryptology expert, Walker's treason was catastrophic for American security. Among the compromised systems were top-secret strategies for submarine and surface fleets; submarine base locations and equipment capabilities; the SOSUS (Sound Surveillance System), the Navy's sonar network, which uses hydrophones to track Russian submarines; underwater map-making technology; results from Polaris, Tomahawk and Harpoon missile tests; and computer data from space-sea surveillance and ways to penetrate the computers.

From 1984 through the spring of 1987, top-secret cipher machine data was stolen by high-tech "jewel thieves" in Moscow. In 1987 news reports carried a series of accounts implicating some U.S. Marine embassy and consulate guards in treasonous activities committed after having been seduced by female Soviet agents.

During the Cold War, Jewels was the codename for the supposedly closely guarded cipher machines in Moscow used by U.S. diplomats and cipher clerks from the NSA and CIA. The equipment was located on the ninth floor of the U.S. Embassy in a specially secured chamber called the Communications Programs Unit (CPU). Within this chamber was another room where the cipher machines encrypted and decrypted transmissions between the embassy and CIA headquarters in Langley, Virginia. In another part of the CPU, different machines transmitted data gathered by the NSA eavesdropping operations or exchanged dispatches between the State Department and its ambassadors in Moscow.

Many precautions had been taken with the CPU's physical surroundings, such as its walls, electrical wiring, power and air supplies to ensure complete security. The cipher machines themselves were state-of-the-art NSA designs with keynumbers in magnetic strips that were changed every 24 hours. The messages were also double encrypted through two cipher machines. These security processes gave an aura of invincibility to the CPU and the Jewels.

Despite these safeguards, the Soviets had managed to clandestinely alter the "clean" power line to the cipher machines, so that the electronic security filters had been bypassed and signals were being diverted from the Jewels. Circuit boards and silicon chips had been replaced in the printers for the cipher machines, allowing KGB operatives to pick up and record signals from the unencrypted side of the communications circuits and printer processes. With the plaintext in hand, they could then proceed to make comparisons with encrypted messages and reproduce the crucial cipher keys. The top-secret encryptive equipment in Moscow and Leningrad had been turned into a huge listening device.

The result of these leaks was a security nightmare for the United States, with CIA agents revealed and expelled, NSA eavesdropping practices revealed, U.S. negotiating positions compromised and the arrest and execution of Russian informants for the CIA.

Many headlines and some controversial investigations followed. While some of the other Marine guards in both Moscow and Leningrad were court-martialed, demoted or discharged, only Marine Sergeant Clayton Lonetree, a Native American Winnebago, was sentenced and received a prison term of 25 years at Fort Leavenworth, Kansas. No Marine officers or State Department or other high security officials faced such consequences, despite information that later suggested that the high-level officials had been negligent.

The NSA had been given some important counterespionage information regarding security breaches by a defector who went to French authorities in 1984. The DGSE, the French intelligence agency, also informed its NSA contacts that the French embassy in Moscow had found bugging devices in its typewriters. In a counterspy plan codenamed Gunman, the NSA had transferred typewriters from the Moscow embassy and the Leningrad consulate back to the United States. They discovered that the technically advanced bugging mechanisms picked up the movements of the print balls of the IBM Selectric typewriters and intermittently transmitted encrypted signals to antennas, such as one hidden in an embassy chimney. The antennas then transferred the data to KGB agents in a nearby apartment house. Officials had also been aware that Soviet agents had planted listening devices in other parts of the old U.S. embassy and a new embassy built next to it in the mid-1980s, yet this knowledge had not led to full U.S. security reviews in either Moscow or Leningrad. Only in August and September of 1987 were the CPUs from both sites shipped back to the United States. They were found to be thoroughly compromised and had to be replaced.

In the wake of these revelations, the FBI conducted a detailed investigation called Power Curve. Completed in June 1988, it concluded that many more individuals than the named Marines had committed security violations or had taken part in prohibited dealings, but neither the CIA nor the NSA had wanted to disclose high-level breaches. The full truth may never be known.

While the Marine guards scandal was evolving, another of the many strange ironies in espionage history took place. The year 1985 was known as "the year of the spy" because of cases such as the Walker family and other traitors including Clyde Lee Conrad (U.S. Army), Richard Miller (FBI), the CIA's Edward L. Howard and Ronald W. Pelton (NSA), who sold secrets about NSA interceptions of the Soviets' communications.

As these cases were breaking, and U.S. counterespionage officials were scrambling for explanations, a high-level KGB operative named Vitaly Yurchenko defected in August 1985. In charge of important KGB operations in the United States and Canada, Yurchenko seemed like a gleam of light in the gloom, a much-needed counterspy victory after several perfidious losses. Amid comments made during FBI and CIA questioning sessions, Yurchenko raised concern about moles inside the CIA. He established his authenticity by exposing Pelton and Howard, who had sold secrets regarding CIA operations in Moscow to the KGB. While Pelton was arrested and sentenced to a life imprisonment term, Howard was apparently tipped off and escaped FBI capture. He later resurfaced in the Soviet Union, having been granted asylum.

For a time Howard's treachery seemed to be the explanation for a series of CIA setbacks that had occurred in the Soviet Union, including exposed operations and the arrests and executions of Soviet citizens who had aided the United States. In November 1985 Yurchenko stunned U.S. counterintelligence by defecting back to the USSR. Questions swirled anew about everything inYurchenko's testimony and its mixture of apparent truths and suspected lies.

When the Marine scandal broke in 1987, the exposed Jewels' secrets appeared to add to the puzzle pieces. Many experts believed that the CIA's losses of agents and information sources had been due to Howards' revelations and the Jewels' thefts.

In the last week of February 1994, the CIA was jolted by the verification of its long-held mole fears. Someone inside the agency had been directly divulging facts about the CIA's operations for years. He was Aldrich Ames, a longtime employee whose father, Carleton, had also been a CIA counterspy official in the 1950s. Fatefully, Aldrich had been one of Yurchenko's debriefers in 1985.

In contact with many Russians and their associates as a spy recruiter for the United States, Ames had reportedly become a turncoat in the early 1980s. He applied cryptographic techniques to conceal his schemes, some as simple as in "meet at B" (for Bogota, Columbia), while others involved a series of chalk-marked mailboxes with codenames like "north" and "smile." The marks signaled various brief answers ranging from "travel on" or "travel off" to news that more detailed information had been retrieved from a hidden drop site.

A joint CIA-FBI counterespionage task force was formed in 1991 to improve security. After some inexplicable delays, they began to suspect the Ameses and gathered evidence against them. The antispy team tapped phones, studied bank accounts, sought typewriter and computer ribbons in the couple's trash and probably applied a new detecting mechanism that picked up and recorded the electromagnetic waves emanating from their computer screen. The combined evidence was damning.

Subsequent to his arrest, in return for leniency for his accused wife Rosario, Ames confessed the details of his treason and was sentenced to life in prison. For $1.5 million he had given up the codenames and missions of as many as ten Soviet citizens who were providing the United States with important intelligence. It was later confirmed that they had been executed. Experts are still adding up the full extent of the resulting damage.

Only two months after the Ameses were arrested, CIA veteran Harold J. Nicholson began a brief but also damaging espionage career for Moscow. Nicholson was deputy chief of the CIA's station in Kuala Lumpur, Malaysia, and his alimony payments, the expenses accumulated by his girlfriend and the debts from three college tuitions had put him in a deep financial bind.

From latter April 1994 until his arrest in late November 1996, Nicholson had accepted $120,000. In return, he had identified agents that he had earlier trained at the CIA's instruction site near Williamsburg, Virginia. He also downloaded CIA computer data, photographed other documents and fingered American businesspersons in Russia who helped the CIA.

His sudden spending sprees and failure to pass three lie-detector tests led to CIA-FBI surveillance and arrest. Contributing to his life prison sentence were state files still on his computer hard drive and cryptic postcards to his spymasters. They contained suspicious phrases like "the snow should be fine," and were signed with his covernames, such as Nevil R. Strachey.

More recently, a traitor in the FBI's ranks had become an on-off-and-on-again operative for Moscow. At his trial in the spring of 1997, it was learned that Earl Pitts had been involved in espionage since early 1990. Described as a man with a strangely troubled perfectionist psychology, Pitts had become enraged at the "FBI's bureaucracy" and decided to make himself a self-styled Soviet spy. After meeting his agent handler in the New York Public Library, Pitts entered a confused world of half-truths. He gave his spymaster real facts about Russian diplomats under FBI surveillance, but, suffering from a guilty conscience, he soon contrived other data to maintain his value as a source. For a time when he left New York City and returned to Washington, D.C., he disassociated himself from his espionage past and was considered "dormant" by his Soviet contacts.

It was a turned Russian diplomat who aided U.S. counterespionage and exposed Pitts. The FBI arranged a sting operation, pretending to be Russian agents again interested in Pitts. In exchange for $15,000 Pitts provided the sting team with parts of a telephone scrambling/encrypting mechanism. His attempt to expose this security system and the $224,000 he had been paid by the Russians for other real and faked data were important evidence that led to a 27-year prison sentence.

THE NEW CRYPTOLOGY:
PUBLIC AND PRIVATE SECURITY ISSUES

While codes and ciphers have remained primary tools for spies, cryptology has also become more familiar to the general public. From post–World War II revelations about Magic and Ultra to news exposés about security breaches by hackers and white-collar criminals, many have become more aware of cryptological issues.

In the 1950s and 1960s, organizations like the NSA reacted to the increasing needs for computer and communications protection. Publically accessible cryptography was required for general civilian uses such as banking transactions or health facts in databases. In 1973 and 1974 the National Bureau of Standards sought encryption systems that would be publicly known but made secure with private keys. The name for this national information security procedure began to be called the Data Encryption Standard (DES).

From a small group of possibilities, one method came to the fore. Developed by Horse Feistel of IBM, the process was based on computer-enhanced transposition. It was learned that the binary computer digits enabled the transposition of numbers to be far more involved and thus more secure than the shuffling of letter positions in previous standard transpositions. At IBM this encryption process had begun as a "demonstration cipher."

Known first as "Demon" and then "Lucifer" (presumably from "demonstration" and a near-spelling of "cipher"), the DES is a complicated encrypting procedure built upon groups of 64 plaintext bits. After meetings with the NSA, the bit totals were altered to 56 plus 8 for transmission accuracy checks known as parity bits. In 1975 this system was presented as the Federal Information Processing Standard in the *Federal Register*.

Many debates arose among private industry and communications cryptologists about the quality of the DES and its real security. A number of people questioned the NSA's involvement, claiming that the U.S. government had weakened the process to enable government experts to make secret decryptions. After a number of meetings, articles and debates, the same 56-bit key (plus the parity bits) was printed again in January 1977 as *Federal Information Processing Standards Publication 46*. Along with federal government applications, the DES was also promoted for better business security. While rival companies and other nations have made claims from time to time that the DES has been broken and that the NSA routinely reads encrypted electronic mail or commercial deals, as of 1998 no verified case exists of its being solved inside or outside the NSA's headquarters.

As the DES was being completed, a new type of cryptography reached the headlines in 1976. Known as two-key or public key cryptography, the process was developed at Stanford University in California by Professor Martin Hellman and research students Whitfield Diffie and Ralph Merkle. Their collaboration presented the public key as a solution to the problems inherent with the distribution of individual secret keys. Their system had a primary basis of two keys. One was publicly available in a published listing or directory, while the other was intended to be private. It was believed computationally impractical, if not impossible, to learn the private decrypting key by possessing the public encrypting key.

For a time the Hellman-Diffie-Merkle process appeared unbreakable, but in 1982 important basics of its security system were broken by a trio of mathematicians at the Massachusetts Institute of Technology, Leonard Adleman, Ronald Rivist and Adi Shamir. Soon this group had developed a competitive two-key procedure based upon prime numbers. Their public key version is called RSA, based on its inventors' initials. Due to its many computations RSA remains slower to implement than the DES and serves primarily for security in networks where there are many communicants and key exchange is a problem.

When DES and the two-key methods were first reported, they were the subjects of much controversy, conjecture and misunderstanding even among experts. Computers began to be applied to organize and file increasingly more personal financial, insurance and medical data. Increasingly dependent on phones, faxes and satellite systems for their profit margins, neither businesses nor individuals wanted to confront glitches of any type, be they hackers or error-prone transmissions.

Matters of security, long the domain of the FBI or military intelligence, became very personal. Once only conscious of Social Security, automobile license, bank deposit box or employment-linked numbers, people began to see their lives be digitized on a daily basis. Some of the many alphanumeric designations include: door and gate opener codes, voicemail digits, health plan numerals, medical records codes, phone credit cards, automated teller PIN numbers, beeper, cell phone and frequent flier numerals as well as customer ID digits from on-line purchases.

While many of these letter and digit combinations are not intentionally secret encryptions, many of them involve aspects of personal safety and financial well-being. The necessity of cataloguing and verifying these codes has brought the individual into a community of interchanged data. With a Social Security number or a few employment facts, hackers can tap into databases that provide general information such as addresses and listed and even unlisted phone numbers. With such basics, impersonators can use their victims' identities to make unauthorized purchases, gain medical records or commit certain types of fraud. The news is filled with reports of small-scale scams as well as larger financial institution swindles.

People are increasingly seeking privacy and security through cryptographic methods. One such system for e-mail was called Pretty Good Privacy (PGP). Developed by a computer expert named Phil Zimmerman in 1991, PGP was the bane of many government officials who have had valid reasons for concerns about high-level secrecy in public communications.

Using court-approved phone taps and eavesdropping devices, the FBI, military counterspies and similar groups have had past successes against suspected criminals. While allowing an individual's right to privacy, many fear that if people or groups can communicate freely on networks without fear of surveillance, the consequences could be disastrous for public safety.

In the early 1990s security specialists began to promote procedures that implemented a new technology generally known as "key escrow." It involved sending and receiving equipment, including computers and phones, that electronically and mutually chose algorithms from millions of available keys to encrypt conversations or data exchanges. The keys were to have been held by two secure agencies chosen by the U.S. Attorney General. To obtain the keys for investigative cryptanalysis, law enforcement or anti-spy groups would have to obtain court-approved permission. One particular chip developed for this access capability and which received substantial publicity called the Clipper Chip. Although the proposal seemed promising, key escrow was confronted with substantial opposition and did not obtain public approval.

While other security processes are currently under review, the issue of private rights versus national security needs will continue to be a crucial debate of the new cryptology and the new millenium.

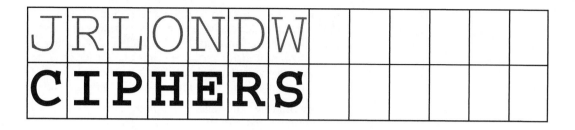

CIPHERS

One of the major categories of cryptology, ciphers are a method of concealment in which the primary unit is the letter. The word *cipher* derives from the Arabic *sifr*, meaning "nothing." Ciphers were extremely important to the advanced Arab civilization that flourished in the seventh century as well as to the Greeks and Romans. After the fall of the Roman empire, ciphers largely fell into disuse. In the Middle Ages, ciphers were again applied in limited forms by Catholic Church officials and later gained popularity as an alternative to codes and nomenclators.

Cipher making boomed during the mid-19th century with the advent of telegraphy. Codebooks were especially vulnerable to capture during battle, and producing new books to replace them was a difficult and laborious process. The capture and solution of many codes during World War I enhanced military interest in ciphers as ciphers offered a high degree of protection if their keys were changed regularly.

By the 20th century ciphers had begun to supersede codes as a military concealment method. Their manipulation of independent letters and syllables made them easily adaptable to modern electronics, such as digital equation and tabulating equipment. The invention of machines that could rapidly and electronically shuffle letters to create polyalphabetic concealments offered hitherto unimagined cipher complexity.

While codes substitute entire words, ciphers work with discrete letters and can be applied to communications in a variety of ways. Their two main functions are *substitution* and *transposition*, both of which involve types of letter transformations. Transposition shifts the original text, causing the normal order of the letters to be disarranged. Substitution replaces the message letters with numbers, symbols or other letters. A cipher's letter unit can also be made up of a pair, known as a *bigram* or *digraph*, or sometimes larger letter groups called *polygrams*. The substitutes that are used to replace the normal alphabet are known as the *cipher alphabet*.

When an unconcealed message, or *plaintext*, is covered with a cipher, it is enciphered and the result is called the *ciphertext*, *encryption* or *cryptogram*. After the planned recipient deciphers the ciphertext, the revealed message is also called the *plaintext*. The term "decipher" was once also used to describe an enemy's attempts to solve a message, but "cryptanalysis" has been preferred since William Friedman coined the term in 1920. A dispatch that is sent with no protection is known as *cleartext*, or "in the clear."

Ciphers are often grouped in five-letter sets as a letter check against accidental repetitions and for convenience during transmission. This grouping was standardized when international telegraph conferences in the 1870s and 1880s set rates for sending letter and number groups. Five digits or letters became a standard for fees in different communications systems. Knowing that five characters were expected, it became easier for an addressee to check where the groups should begin and end and thereby establish orderly transmissions patterns.

Ciphers also require a *key*, which determines the operations of a given cipher. A *keyword* might signify the pattern of letters in a cipher alphabet, while a *keynumber* could specify the order of the letters in a transposition. In the following example, the first and third letters of the keyword *here* set the pattern of the alphabet.

PLAIN ALPHABET:

h	i	j	k	l	m	n	o	p
q	r	s	t	u	v	w	x	y	z	a	b	c
d	e	f	g					

CIPHER ALPHABET:	r	s	t	u	v	w	x	y	z
a	b	c	d	e	f	g	h	i	j	k	l	m
n	o	p	q						

In the alphabet, *h* and *r* are 10 letters apart. This directs the resulting arrangements in a cyclical shift of 10 letters through the standard 26 letters. Using this particular alphabet, the word *message* becomes *wocckqo*.

If the keynumber 34512 is given for a transposition cipher, it would indicate that the third letter of the original word is transposed to become the first, the fourth letter becomes the second and so forth. If this keynumber is applied to the plain-text word, *leave*, the encryption would read *avele*.

The key can be arranged prior to sending the cipher or it can be conveyed within the encryption. It should be varied for security purposes as a frequently used word or phrase can be recognized by an experienced cryptanalyst. It should not, however, be altered without notification as it determines the procedure to be followed, the structure of the created cipher and the pattern of decipherment by the recipient.

Various devices give ciphers additional protection. Cryptographers often make use of *nulls*, meaningless letters inserted into a cipher alphabet or matrix. Nulls complicate the decryption efforts of unintended third parties by disrupting anticipated sentence patterns, word lengths and syllable groups. Encrypted words or digits can also be superenciphered. In this process the encryption is itself enciphered by transposition or substitution, providing an extra layer of cryptographic protection.

New advances in mathematics and computer science have expanded cipher applications to new heights. Still, it is the basic relationship between ciphers, keys, transposition and substitution that forms the base of the millions of computer-generated concealment possibilities.

TRANSPOSITION CIPHERS

Transposition is a method of encipherment whereby the letters of a message are rearranged without actually replacing a word's letters with other letters, numbers or symbols. For example, the word *watchful*, when transposed, can be *afcwluth* or *flwhuatc*. The letters change their positions relative to each other, but retain the same identities. A common example of a transposition technique are anagrams, which became popular in the 17th century when they were developed by such men as Galileo and Johannes Kepler. Anagrams take the letters of a word, and by rearranging them, create new words. For example, the word *are* becomes *era*, *bat* becomes *tab*, *procedures* is *reproduces* and *percussion* can become *supersonic*, among a surprising number of examples.

The history of transposition is vague, but may have begun when an ancient scribe chose to alter the characters of his king's name for security or ritual purposes. Written documents show that both the Spartans and the Arabs knew about forms of transposition by the fifth century C.E. The Spartans developed the *skytale*, one of the earliest forms of transposition and the first known cryptographic device. According to records left by Greek historian Thucydides, the skytale consisted of a cylinder of wood or, in some cases, an officer's baton around which a strip of cloth, leather or papyrus was tightly wrapped. The plaintext was written on this strip down the length of the staff and meaningless nulls were placed around the missive.

While some efforts were made to create more than one sensible phrase, these attempts proved more time consuming than protective. When the papyrus or leather was unwound, the loosened strip of letters would appear to be a jumble to an intercepting third party.

A courier would then carry the dispatch to another camp, where the communication was wrapped around another staff of the same length and thickness. Once the parchment was wound closely again, the proper letter order would be realigned and the message revealed. Particularly effective dur-

A skytale

ing campaigns, the skytale did not require an elaborate code to memorize or possess and was relatively easy to encrypt and decrypt.

In Europe during the Middle Ages, some friars or monks dabbled in letter shifts and reversed phrases. By the time of the Renaissance, cryptographers in Italy were practicing transposition and developing new applications of it and for substitution, its concealment partner. The Neapolitan scholar Giovanni Porta discussed transposition as a primary concealment method in his classic *De Furtivis Literarum Notis*, published in 1563. By then, however, the nomenclator was fast becoming the principal masking system and transposition techniqueswere forgotten.

Transposition was revived in the United States during the Civil War, when a form called "route" was used by Northern forces to conceal telegraph messages from Confederate wiretappers and cryptanalysts. Although today transposition is rarely used as the sole concealment method, it is applied in combination with other crytographic methods to give an encipherment additional protection.

The security in transposition ciphers relies upon a wide range of possible permutations, which can deter potential interceptors. An example of a permutation which affords a basic level of protection is the blocking of a message into groups of eight letters and spaces. The first and last and the fourth and fifth letters or characters within each group are then transposed. The symbols • and . representing spaces and punctuation should also be transposed:

PLAINTEXT: hold • 800 • shares • of • acmo • industries. • nate • smith. •

PERMUTATION: 0ol • d80h • shraes • • f • camooindsutrie. n • ate • smtih. •

Transposition is as effective in concealing low-security missives as is substitution; however, transposition ciphers have a limited degree of security against experienced cryptanalysts as the original letters continue to exist within the cryptogram. This method is not as secure as more advanced forms of substitution, wherein letters or other characters have their identities changed.

Reverse Transposition

A very basic type of transposition is one wherein the plaintext is written in reverse. For example, take the message "meet her at nine tonight." In ciphertext it becomes:

CIPHERTEXT: `thginotenintarehteem`

Dividing the string into groups of five for transmission, the cryptogram becomes:

`thgin   oteni   ntare   hteem`

Another possibility is to reverse the words alone to form the reverse cryptogram:

`teem   reh ta  enin   thginot`

This cryptogram can then be sent in a five-letter group:

`teemr   ehtae   ninth   ginot`

REVERSE VARIATIONS

Using the first reverse example above, each of the five-letter groups can then be altered by exchanging the first letter for the third and the second for the fourth:

FIRST CIPHERTEXT: `thgin   oteni   ntare   hteem`

SECOND CIPHERTEXT: `githn   enoti   arnte   eehtm`

If this entire sequence was again written backwards the new encryption would read:

`mthee   etnra   itone   nhtig`

Geometric Ciphers

A geometric cipher is a transposition cipher in which the elements of the plaintext are entered in a geometric design, known as a matrix, an array or a grid. By moving in a planned direction, up one column of letters, down another, or diagonally, a different order of the letters results. A key defines the size and direction of the route for the addressee. Codenames for certain persons and places, meaningless nulls and diagonal alignments can also help to conceal the message.

Simple geometric forms involve very basic ways to rearrange letters. For example, the words *meet her at nine tonight* can be placed in a rectangle as shown:

m	i
e	n
e	e
t	t
h	o
e	n
r	i
a	g
t	h
n	t

The ciphertext is drawn from the horizontal pairs to create the cryptogram:

mi   en   ee   tt   ho   en   ri   ag   th   nt

These can be transmitted as groups of five letters each, forming the ciphertext:

miene   tttho   enria   gthnt

ROUTE TRANSPOSITIONS

Route transpositions are so named because they have a direction or pattern of placement and shifting by which the cryptographer arranges the letters. The plaintext is placed in the cells of these configurations in one prearranged pattern and transcribed in another to produce the ciphertext. Route ciphers are associated with a kind of geometric design, and while rectangles appear most frequently, triangles, circles and even trapezoids, a plane figure with four sides, two of which are parallel, have proved useful to cryptographers.

To ensure successful decryption of a transposition cipher, the communicators must prearrange four details of the cipher. First, the size and shape of the geometric pattern must be pre-determined; second, the starting place and the direction of the message inscription; third, the starting place and direction of the message transcription; and finally, the group lengths for the final ciphertext.

For example, the missive "he is the allies' courier" could be written with one half of the letters on top of the second half and the apostrophe deleted. Punctuation marks should always be eliminated from the ciphertext as they are an easy giveaway. For example, an apostrophe is usually followed by an s or a t, which gives a cryptanalyst a first chink into the cipher's armor.

The plaintext is laid out as follows:

h	e	i	s	t	h	e	a	l	l
i	e	s	c	o	u	r	i	e	r

This across-the-page arrangement is already a form of rectangle, though not strictly a closed design such as a grid or matrix. Letters can be transcribed from this array in more than one pattern. A simple technique is to transcribe the letters in groups of five in a top then bottom sequence. The first group of five is taken from the top row, the second group from the bottom row, the third from the top and fourth from the bottom, making the slightly concealed cipher:

heist   iesco   heall   urier

The cipher can be given more protection by taking out alternating pairs of letters from the top and bottom:

```
he   ie   is   sc   th   ou   e   ri   ll   er
```

which can then be placed in five-character sets and sent as:

```
heiei   sscth   ouear   iller
```

Deciphering tips: You must know the original pattern of arrangement. If the arrangement is groups of five on top and then below, place the first five letters, *heist*, in the top row and the second group of five, *iesco*, beneath them. Continue for the third series of five, *heall*, above and the fourth, *urier*, below.

If the missive was concealed by alternating pairs, divide the scrambled letters into pairs and begin the above-below process: *he* above and *ie* below; *i* combined with *s* (from the second five-letter group) as a pair above *sc*; *t* and *h* as a pair above *ou*; *ea* above *ri* and *ll* paired with *er* beneath them. This creates the two lines of letters reading:

```
heistheall
```

```
iescourier
```

Reading line-by-line from left to right, recover the message.

Rail Fence

A very simple form of route ciphers is the rail fence, named for its fencelike appearance, which is the result of aligning rows of letters, then shifting them. The rail fence was a popular method in the early decades of cryptography. It faded with the rise of more complex systems such as nomenclators in the Middle Ages and codebooks in the 15th and 16th centuries. It regained some of its popularity during the American Civil War, when it was used for concealments of military messages as well as by Union and Confederate spies.

A depth-two rail fence has two rows and an *n*-column array, with the number of columns determined by the length of the given message. The plaintext is written in columns along a cryptic route, that runs, in this example, from the top to the bottom of each column in succession. This means that every other letter is written slightly lower. For example, the message "lamp signal her cottage at midnight" would be encoded as follows:

RAIL FENCE:

```
l  m  s  g  a  h  r  o  t  g  a  m  d  i  h
 a  p  i  n  l  e  c  t  a  e  t  i  n  g  t
```

The ciphertext can be made by writing the letters left to right by groups of five, beginning with the top row:

CIPHERTEXT: lmsga hrotg amdih apinl ectae tingt

Deciphering tips: Divide the letters into two equal groups with a vertical line, leaving 15 letters on each side:

lmsga hrotg amdih | apinl ectae tingt

Beginning on the left with *l*, read each letter in the left half of the ciphertext followed immediately by the corresponding letter of the right, in this case *a*, then back to *m*, followed by *p* on the right side, continually alternating until the plaintext is recovered.

Variations on the Rail Fence

Taking the enciphered line used above, *lmsga hrotg amdih apinl ectae tingt*, letter variations can be achieved by writing them with one row from left to right and the other in reverse, or by reversing both of them:

A. FORWARD/BACKWARD:

```
l  m  s  g  a  n  r  o  t  g  a  m  a  i  n
 t  g  n  i  t  e  a  t  c  e  l  n  i  p  a
```

CIPHERTEXT: lmsga hrotg amdih tgnit eatce lnipa

B. REVERSED:

```
n   ı   a   m   a   g   ɔ   o   r   n   a   g   s   m   ı
  t   g   n   i   t   e   a   t   c   e   l   n   i   p   a
```

CIPHERTEXT: hidma   gtorh   agsml   tgnit   eatce lnipa

A three-level version of the rail fence yields a different combination of five-letter groups for the same plaintext:

```
ı     s     a     r     ɔ     a     ɑ     n
  a   p   i   n   l   e   c   t   a   e   t   i   n   g   t
    m     g     h     o     g     m     i
```

CIPHERTEXT: lsart   adhap   inlec   taeti   ngtmg hogmi

Deciphering tips: For the backward/forward style, you must know beforehand which letter clusters or series of letter clusters are written backward. Once they are reset in proper order, place the dividing line halfway between the letters. Follow the alternating rail fence procedure described above to decipher the message. For the reversed form, divide the full sequence of letters in half. Place the first 15 letters, *l* back through *h*, on the left side of the dividing line and reverse them to read *l m s g . . . a*. Place the second group of 15, *t* through *a*, to the right of the dividing line to read in regular order *a p i n l . . . t*. Follow the rail fence deciphering procedure to decipher the whole message.

To decipher the three-level rail fence, place the first eight letters of the cryptogram so that they fill the top row of the "fence" (*l s a r t a d h*); the next 15 letters form the middle *row* (*a p i n l e c t a e t i n g t*); and the final seven make up the base level (*m g h o g m i*) in an inverted pyramid. Read in a zigzag pattern, from top to bottom to top, to recover the message.

Horizontal Routes

The application of more-specific geometric patterns gives more positioning variety. In the horizontal route, the plaintext characters can be placed in a rectangle or square depending on the number of letters and the size of the required matrix. The missive "secret pact today" can be placed in a three-by-five matrix:

s	e	c	r	e
t	p	a	c	t
t	o	d	a	y

These rows can then be transcribed in reverse to form a thin cover, (A), or can be transcribed in columns (B):

A.   erces   tcapt   yadot

B.   sttep   ocadr   caety

Other versions vary the horizontal inscriptions with every other letter being placed in a cell:

s	c	e	p	c
t	d	y		

The new grid is formed by placing all the first group of letters in consecutive squares from left to right until completed. The second series of letters (those previously skipped) are placed next in line to form the grid:

s	c	e	p	c
t	d	y	e	r
t	a	t	o	a

Even a straight transfer of these letters to ciphertext is more varied, creating *scepc tdyer tatoa*. A simple reversal adds further concealment:

```
cpecs   reydt   aotat
```

Deciphering tips: You must know if the arrangement is a straight transfer or a simple reversal. If it is a straight transfer, place the letters in the array from the top left to bottom right and read them from left to right: *scepc tdyer tatoa*.

If it is a simple reversal, *cpecs reydt aotat*, place the letters of each word in the right hand column beginning with *c* in the upper right hand corner and continuing with *pecs* to the left to fill the first line. Next, place the letter *r* in the right column's middle cell and use the following letters, *eydt*, to fill in the cells in the middle row. The letter *a* begins in the bottom right corner with *otat* moving left across the row.

In both the straight transfer and the simple reversal placement sequence, you are restoring the grid containing every other letter. Place the first line in every other cell of a new matrix. Fill in the blanks with the remaining letters. With all the cells filled, read the plaintext from left to right.

Alternate Horizontals

Alternate horizontal routes begin their changes from the moment of inscription. The dispatch "contact their headquarters at once" is placed in a five-by-six rectangle in rows of alternating direction.

c	o	n	t	a
e	h	t	t	c
i	r	h	e	a
r	a	u	q	d
t	e	r	s	a
e	c	n	o	t

When drawn out of this array, the original message can be further concealed by reversing every other row, beginning with the first, to produce the encryption:

atnoc   ehttc   aehri   rauqd   asret   ecnot

Deciphering tips: Beginning with the first letter group, place the first letter, *a*, in the cell in the upper right corner of the matrix. Cross the top row from right to left with the following letters, *t n o c*. In the first cell of the second row place the first character of the second group, *e*. Place the following letters, *h t t c*, sequentially from left to right. The same type of alternating series follows; place the third character cluster (*aehri*) from right to left beginning with *a* on the right; place the fourth group in cells from left to right (*r* through *d*); the fifth from right to left (*a* across to *t*); and the sixth group, *e* through *t*, from left to right.

This process arranges the letters as shown in the sender's array. Read the words as alternate horizontals by going from left to right, then right to left to recover the plaintext.

Vertical Routes

Vertical routes accomplish a similar level of concealment by placing the plaintext in the cells in a standard top-to-bottom direction:

c	t	h	a	a
o	t	e	r	t
n	h	a	t	o
t	e	d	e	n
a	i	q	r	c
c	r	u	s	e

Notice now that there is a difference in concealment possibilities between the standard horizontal and this vertical version. The basic horizontal required changes in its inscriptions and/or transcription to change the English left-to-right reading pattern. The alternate horizontal form achieved this by rearranging the letters' placement from right to left, but by its very nature the vertical route creates scrambled outtake characters:

`cthaa    otert    nhato    teden    aiqrc    cruse`

Any added character placement exchange or transcription variation adds to the strength of the concealment. The following example is a simple reversal of the outtake groups:

`aahtc    treto    otahn    nedet    crqia    esurc`

Deciphering tips: For a standard vertical route, take the first of the five-character groups and place it in the top row of the matrix. Place each succeeding letter cluster below in order. Starting with the left column, read down each column in turn to see the plaintext words.

If the groups have been transcribed in reverse, you must be informed of this change. Use the prescribed pattern of changed lines and reset these characters in their regular order before placing them in their matrix rows.

Alternate Verticals

This version has aspects of an extended rail fence. The example, "contact their headquarters at once," is placed in the matrix in alternate verticals, or a down-and-up direction.

c	r	h	s	a
o	i	e	r	t
n	e	a	e	o
t	h	d	t	n
a	t	q	r	c
c	t	u	a	e

Removal from the matrix by standard left-to-right gives this cryptogram:

crhsa   oiert   neaeo   thdtn   atqrc   ctuae

The groups can also be reversed from right to left:

ashrc   treio   oeaen   ntdht   crqta   eautc

Or they can be transcribed in an alternating left-to-right, right-to-left pattern:

crhsa   treio   neaeo   ntdht   atqrc   eautc

Deciphering tips: Knowing the vertical array and the directions for each given message, place the letters in the array accordingly. For example, directions for standard alternate vertical would instruct the recipient to begin in the upper left corner and fill the cells down the left column, up the second, down the third and so on. A planned direction shift such as a reversal could be called "alternate A." Alternating rows could be designated "alternate B." These alterations must be reversed before the cryptogram is placed back into the array.

Diagonal Routes

A change in the typical horizontal or vertical placement occurs with diagonal routes. For the communiqué "formula sent to you by courier stop answer today," the plaintext is arranged in diagonals from bottom left to top right.

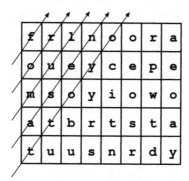

A standard horizontal transcription then reads:

```
frlno   oraou   eycep   emsoy   iowoa   tbrts   tatuu
snrdy
```

An alternate vertical provides:

```
fomat   utsur   leobu   sryyn   ocitn   rsoeo   rpwtd
yaoea
```

The diagonal inscription permits even more transcription variety than do horizontal or vertical routes.

Deciphering tips: You must know the size of the matrix, the diagonal placement and the transcription direction. If the transcription is standard horizontal, place the same letters in the same cells, beginning at the top left.

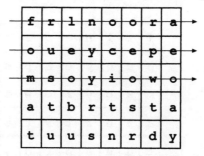

Read in diagonals from top left to bottom right to reveal the message.

If the transcription is alternate vertical, the last cryptic group sent would be *yaoea*. Knowing the directional pattern, insert the first letter group, *yaoea*, beginning in the very bottom right corner of the design with *y* and going up the column to *a* at its top. Place the second group, *rpwtd*, in the next column running downward. Reverse the alternating verticals until *tamof* is traced up the far left column.

f	r	l	n	o	o	r	a
o	u	e	y	c	e	p	e
m	s	o	y	i	o	w	o
a	t	b	r	t	s	t	a
t	u	u	s	n	r	d	y

Read the letters diagonally from the top left to reveal the plaintext.

Alternate Diagonals

Using the previous diagonal array, the cipher maker can bring out the letters as alternate diagonals:

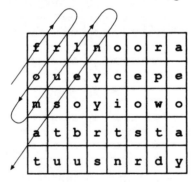

Beginning in the upper left-hand corner, this route elicits:

```
fromu   inesa  ttoyo  ocybu  urier  apots  nsweo
trday
```

Deciphering tips: Knowing that this is an alternate diagonal transposition, place the last letter, *y*, in the lower right corner. The next-to-last letter (in this case *a*) is placed directly above *y* in its column. This begins a reversal of the alternating process wherein the recipient traces from *a* to *d*, moving in an alternate diagonal from *d* to *r* t *o*.

f	r	l	n	o	o	r	a
o	u	e	y	c	e	p	e
m	s	o	y	i	o	w	o
a	t	b	r	t	s	t	a
t	u	u	s	n	r	d	y

The reverse process continues up to *e* then down to *w*, *s* and *n*. When the reversal is concluded and *l*, *u*, *m*, *o*, *r* and *f* are placed in their cells, you have the message set as a standard alternate diagonal for decryption. Trace the message in alternate diagonals until you have read the entire message.

Clockwise Routes

Also known as spirals, clockwise routes begin at either a corner or a central section of the given geometric design. In this example, the truth about a computer network's apparent protection measures can be sent with a standard rectangle that is given a "spiral" twist.

t	h	e	s	e
c	u	r	i	t
y	i	s	a	f
a	c	a	d	e

The message "the security is a facade" is wound clockwise to form:

`these   tfeda   caycu   riasi`

The turns are made ever tighter toward the center. On the first turn to the right *e,t, f* and *e* are used, and the second turn on that side can only use *i* and *a* before concluding with *s* and *i*.

The clockwise spiral can also originate with a central letter such as *r* and continue through *i a s i*, up and around through *u h e s e*, down to *t f e d a*, then on to *c a y c t*. The clockwise turns are maintained in as continuous a circle as possible from the inside out.

Deciphering tips: The spiral patterns are all important here, so you must prearrange either a specific corner or a particular central letter as a starting point. For a regular wind you would be told to begin in the upper left corner of the planned array. Begin with the first letter of the first cryptogram group, *t*, and move from left to right across the first row until its cells are filled. Place the first letter of the second cryptic group, *t*, directly beneath the last letter of the first row (in this case *e*) and the letters following *t*, namely *f e d a* are placed down the far right column and then angled to the left across the base row until the word ends at *a*.

t	h	e	s	e
				t
				f
		a	d	e

Fill the last row with the third cryptic group, *caycu*, moving up the far left column until you reach the top row. At this point, turn to the right on the second row until the third group of letters end.

t	h	e	s	e
c	u	r	i	t
y	i	s	a	f
a	c	a	d	e

Fill the remaining open blocks with the last group, *riasi*, turning inwardly until the message is complete. Read in standard left-to-right to reveal the message.

Counterclockwise Routes

A counterclockwise turn is also possible, starting from a corner and going backwards:

t	h	e	s	e
c	u	r	i	t
y	i	s	a	f
a	c	a	d	e

Starting at the lower left corner, the route follows the letters:

acade ftese htcyi sairu

Or beginning at the center and going out counterclockwise from *r*:

ruisa iseht cyaca defte

Deciphering tips: Knowing that this is a counterclockwise spiral sequence, begin at the lower far left corner. Place the

letters of the first cryptic group across the base row, beginning with *a* and concluding at *e*. Turn up the far right column of the array with the second series of letters and wind
them to the top of the column and across the first row, moving to the left. As can be seen in the original encryption-
forming grid, the position of the original letters is restored.
Again, read from left to right to see the message.

Columnar Transposition

In a columnar cipher, a key determines the order in which
the letters are taken from the columns. Inscription in a rectangle follows a horizontal pattern from left to right and top
to bottom. In the following example the letters will be
aligned as ciphertext in four-letter groups.

The key is numerical and is formed by a simple process. Suppose the keyword is *diplomat.* Its letters are simply numbered
according to their position in the standard English alphabet
(1 to 26), so that $d = 4$, $i = 9$, $p = 16$, $l = 12$, $o = 15$, $m = 13$,
$a = 1$, $t = 20$. This key word and its corresponding numbers
are placed above the plaintext message, "microfilm will be
arriving by courier," which is itself placed in a rectangle:

```
d   i   p   l   o   m   a   t

4   9  16  12  15  13   1  20
```

m	i	c	r	o	f	i	l
m	w	i	l	l	b	e	a
r	r	i	v	i	n	g	b
y	c	o	u	r	i	e	r

The first alphabet letter in the keyword, *a* (1), and the column beneath it with the letters *iege* becomes the first line of
ciphertext. The keyword's next letter in alphabetical order, *d*
(4), has the column letters, *mmry.* This procedure is followed
in alphabetical order to complete the ciphered message:

`iege mmry iwrc rlvu fbni olir ciio labr`

Deciphering tips: Knowing the keyword (*diplomat*), prepare a rectangle or other preset geometric figure, with the necessary number of spaces for letters based on the length of the key and the message. Write the keyword above them and number them according to their position in the alphabet. Take the first group of ciphertext and place it in the column underneath the lowest number, in this case 1. Take the second group and place it the column of the next lowest number, column 4. Continue this procedure until you have filled in all the spaces.

d	i	p	l	o	m	a	t
4	9	16	12	15	13	1	20

m	i		r			i	
m	w		l			e	
r	r		v			g	
y	c		u			e	

By the time the first four columns are filled in, the original facts about the arrival of the microfilm are beginning to appear. Read the letters in a normal left-to-right manner in order to recover the full message.

Other alternatives are to encipher in reverse or in alternate forward and reverse. Both the sender and the recipient must agree upon such changes before the message is sent.

INTERRUPTED COLUMNAR

This method has blank cells placed at preset points in the columns. The empty cells introduce variety to the lengths of the segments. The keyword *verify* guides the order of the transmission "she is the only person who is trusted." The letters are placed horizontally and around the gaps:

v	e	r	i	f	y
5	1	4	3	2	6

v	e	r	i	f	y
s	h		e		i
s		t	h		e
o	n		l	y	p
	e	r	s		o
n		w	h	o	
	i	s	t	r	
u	s	t		e	d

The letters are withdrawn from the matrix according to the alphabetical order of the keyword's letters. The first letter is *e* (1) and the letter group beneath it is *hneis*. This is the first ciphertext group. Beneath *f* (2) there are only four letters, *yore*. For a fifth letter, the cryptographer goes to the third column (below *i*) and takes the *e* to make the cryptic group *yoree*. The remaining letters under *i* form *hlsht*. Beneath *r* (4) is *trwst*; below *v* (5) is *ssonu*; and under *y* (6) is *iepod*. The fully drawn out ciphertext becomes:

hneis   yoree   hlsht   trwst   ssonu   iepod

Deciphering tips: You must have the matching matrix and the specific arrangement of the cells to be left blank, for example 3 and 5 of row 1, 2 and 5 of row 2 and so on. Using the keyword *verify*, place it on top of the matrix and number each letter from 1 to 6 according to alphabetical order. Begin with the first cryptic group, *hneis*, and place it in the column beneath *e* (1). Then place *yoree* in the column below *f* (2). Knowing the pattern of blanks, you can see that the second *e* of *yoree* will not fit. Move it over into the first cell beneath *i* (3). The placement of the succeeding cryptic groups in the column concludes with *ieopd* below *y* (6). Reading in standard left-to-right fashion, reveal the missive.

DIAGONAL COLUMNAR

This style combines a mixture of columnar and diagonal route influences. The keyword is *hamiltonsville* and it governs the communiqué "website sabotage at their first location probable stop take all precautions today." Set up the matrix as shown with repeating letters in the keyword numbered sequentially. For this example, the two letter *i*'s are out of sequence, *i* (5) then *i* (4), to aid the alignment of diagonals on the left and right sides:

h	a	m	i	l	t	o	n	s	v	i	l	l	e
3	1	9	5	6	13	11	10	12	14	4	7	8	2

w	e	b	s	i	t	e	s	a	b	o	t	a	g
e	a	t	t	h	e	i	r	f	i	r	s	t	l
o	c	a	t	i	o	n	p	r	o	b	a	b	l
e	s	t	o	p	t	a	k	e	a	l	l	p	r
e	c	a	u	t	i	o	n	s	t	o	d	a	y

The cryptogram is formed by taking a diagonal line from each of the first four key numbers and alternating them across the matrix. In this example, the diagonal patterns begin under column *a* (1) on the upper left of the matrix and cross the array in a downward diagonal from *e* to include *ttpi*. The next diagonal begins at column *e* (2) with the letter *g*. Following *g* on a diagonal down to the left is *talt*. The third diagonal is *h* (3) with *waaot* on an angle to the right and column *i* (4) has *oirko* angling to the left.

n	a	m	i	l	t	o	n	s	v	i	l	l	e
3	1	9	5	6	13	11	10	12	14	4	7	8	2
w	e									o			g
	a	t						i			t		
		a	t				r			a			
			o	p			k		l				
				t	i	o		t					

These diagonals form the first four ciphertext groups, *ettpi gtalt waaot oirko*. The fifth group begins the process of vertically transcribing the remaining letters from the now altered array. For example, the first column under *a* (1) is used to begin the fifth ciphertext group with *csc* and, from the column beneath *e* (2), the letters *ll* to form *cscll*. The remaining letters below *e* (2), *ry*, join *eoe* under *h* (3) to become the cryptic group *ryeoe*. This process continues by combining letters in groups of five from the columns in their numerical number. The ciphertext then becomes:

```
ettpi  gtalt  waaot  oirko  cscll  ryeoe  erbos
tuihi  tslda  bpapt  asrpn  einaa  feste  otboa
```

Deciphering tips: Knowing the keyword and the pattern of the four diagonals, fill in the cells of the diagonal's characters. Fill in the first column beneath *a* (1) with the fifth cryptic group *cscll*. Since the diagonals have filled two of this column's cells with *e* and *a*, only the letters *csc* are needed to complete the column. The letters *ll* go to the column under *e* (2). This column is filled with *ry*, the first two letters of the sixth cryptic group *ryeoe*. Place the remaining letters *eoe*, beneath *h* (3). It is one letter short, and that letter is the *e* taken from *erbos*.

h	a	m	i	l	t	o	n	s	v	i	l	l	e
3	1	9	5	6	13	11	10	12	14	4	7	8	2

w	e									o			g
e	a	t							i			t	l
o	c	a	t					r			a		l
e	s	t	o	p			k			l	l		r
e	c			t	i	o			t				y

Continue filling in each successive column around the diagonals' already filled cells from each letter group until the matrix is complete. Read from left to right to see the plaintext words taking shape.

DOUBLE COLUMNAR

This method combines more than one transposition style and involves two arrays with different keywords. The keyword's letters are numbered according to their position in the alphabet: $a = 1$, $h = 8$, $m = 13$ and so on. The two keywords are *secure* and *locale*, and the missive is "the key list site is guarded at night." Using the first keyword, set up the matrix as shown:

s	e	c	u	r	e
5	1	3	6	4	2

t	h	e	k	e	y
l	i	s	t	s	i
t	e	i	s	g	u
a	r	d	e	d	a
t	n	i	g	h	t

The ciphertext letters are withdrawn by columns according to alphabetical order. Thus *e* is the first letter and below *e* (1) is the column containing *hiern*. The other letters are brought out accordingly to form the cryptogram:

```
hiern   yiuat   esidi   esgdh   tltat   ktseg
```

These letters are inscribed in the rows of the second array beginning with *hiern*. Because the keyword *locale* has six letters, indicating a column width of six letters, the letter *y* from the second cryptic group is placed in the sixth cell of the first row. This procedure continues from left to right until the cells are filled.

```
l   o   c   a   l   e

4   6   2   1   5   3
```

h	i	e	r	n	y
i	u	a	t	e	s
i	d	i	e	s	g
d	h	t	l	t	a
t	k	t	s	e	g

Removed by columns again according to the keyword's numerical order, the concluding encryption is:

```
rtels   eaitt   ysgag   hiidt   neste   iudhk
```

Deciphering tips: Align the two rectangles and keywords side by side with empty cells. Beginning with the keyword *locale*, place the cryptic groups in its matrix's columns according to the alphabetical order rule. Beneath *a* (1) is *rtels*; beneath *c* (2) is *eaiit* and so on, completing one stage of the process with *iudhk* under column 6.

Take out the letters in this matrix by standard horizontal rows beginning with *hierny* from row 1. Place these letters in the first matrix according to the alphabetical order of the first keyword, *secure*. The letter *y* carries over to the first cell beneath *e* (2) and that column is filled by the letters *iuat* from the group *iuates*. The letters *es* carry over to column 3 and are filled by *idi* of *idiesg*.

s	e	c	u	r	e
5	1	3	6	4	2
h	e				y
i	s				i
e	i				u
r	d				a
n	i				t

The rows from the second matrix continue to fill the columns of the first matrix until column *u* (6) is complete with *ktseg*. Reading normally from left to right, you can see the message appear.

Four Corner Transposition

Special figures have been incorporated to shuffle letters. One such method is known as "four corner." In this example, enough crossed-line figures are drawn to accommodate the total number of letters in the message. The plaintext is placed around the figures in a preset manner, in this case clockwise. The order "proceed to Philadelphia stop examine their key exchange" can be arranged as follows:

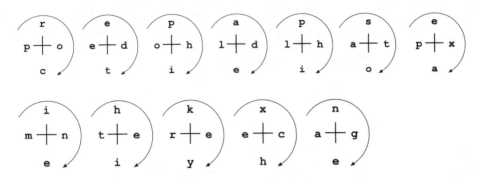

The letters are taken out by standard horizontal means. The transcription sequence takes five from the top row, *repap*, ten from the middle, *poedo hldlh*, and five, *ctiei*, from the lower row of the first five figures. The process continues in the same way through the next series of five figures and 20 letters.

The closing letters are drawn from those remaining in the concluding figures (two groups) going from top to bottom. This makes the cryptic group *xneca* and the partial *ghe*. To give five letters to the final encryption group, the nulls *x* and *z* are added.

CIPHERTEXT: repap   poedo   hldlh   ctiei   seihk
                atpxm   utere   oaeiy   xneca   ghexz

Deciphering tips: Knowing the total number of figures and the use of the nulls, place the encrypted letters across the top points of the first five designs, the next 10 across the middle points, and the next five at each base point:

r   e   p   a   p

p + o   e + d   o + h   1 + d   1 + h

c   t   i   e   i

The next group of five figures are filled in the same way:

■ 160

As you know that the final two designs will hold eight letters, place them accordingly. Delete the two nulls, *x* and *z*. Read each figure in a clockwise direction to reveal the plaintext.

Triangle and Trapezoid Transpositions

Both triangles and trapezoids can help make transpositions. The plaintext is placed in the figures according to the number of words and the shape of the figure, which can be expanded or reduced to suit the missive.

The phrase is: "peace can be achieved when it is built upon vigilance and strength." The keyword, *diplomatics*, is numbered sequentially in alphabetical order. For the first figure, a triangle, the plaintext is placed horizontally from the apex to the base. Because of the width of the base, the keyword is repeated and its repeating characters are numbered sequentially.

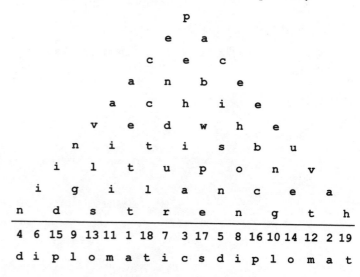

The columns' letters are removed starting at the top of each column, even if there is only one letter per column. Above *a* (1) are the letters *aett*. To complete a five-character group, the fifth letter, *a*, is taken from above *a* (2). The second cryptic group is formed from the letters on top of *c* (3)—*pehia*. The next encryption is formed from *d* (4), the letter *n*, and *d* (5), *cisn*, to make *ncisn*. Following them are *i* (6) and *i* (7) which form *iendu*. The *r* of *i* (7) is combined with *ehon* of *i* (8). This procedure continues until all the letters are used. The final ciphertext is:

```
aetta   pehia   ncisn   iendu   rehon   ngeng
aiivt   vlsue   i       debc    abwpe   cctlh
```

Deciphering tips: Deciphering is accomplished with the same-sized triangle and keyword governing the spacing pattern. You must also know if the words should be returned in a base-to-top or a reverse order. Begin with *a* (1), so in the column above it, the first cryptogram group *aeet* is placed from top to bottom to fill the column's cells. The remaining *a* is placed in the only cell of the column above *a* (2). Place the next five cryptic characters *pehia* above column *c* (3) starting at the top with *p*. Above *d* (4) the single cell holds *n*. The next four letters, *cisn*, are placed in *d* (5)'s cells from top to base. This process continues until the cells are filled.

```
                                   p


                            e     c

                    a

                            h     i

                    e

                            i     s

                    t

                            a     n              a

   n                t

   4                1       3     5              2

   d                a       c     d              a
```

Begin at the top of the triangle and read the consecutive rows beneath it to recover the plaintext.

With a trapezoid, the preliminary arrangement is slightly different. The angled columns require a different letter and keyword alignment, giving the trapezoid distinct variations.

The same keyword, *diplomatics*, and message, "peace can be achieved when it is built upon vigilance and strength" (plus one null, *z*), gives the following design:

				p	e	a	c	e			
			c	a	n	b	e	a			
		c	h	i	e	v	e	d			
	w	h	e	n	i	t	i	s			
b	u	i	l	t	u	p	o	n			
v	i	g	i	l	a	n	c	e	a		
n	d	s	t	r	e	n	g	t	h	z	
3	4	9	6	8	7	1	11	5	2	10	
d	i	p	1	o	m	a	t	i	c	s	

The diagonals are begun at the base from *a* (1) with *nnpid* being traced on a diagonal upward to the right edge of the design. To form a five-character group *c* (2) and its letters ha are combined with *d* (3)'s *nvb* to make *hanvb*. The remaining characters of *d* (3), *wccp*, are joined with *i* (4)'s *d* to form *wccpd*. Then *i* (4) contributes *iuhha* for one cryptic cluster and leaves an *e* to add to *i* (5)'s ten and *l* (6)'s *ti* to form *etenti*. Then *l* (6)'s *lnebc* makes its own cryptic group. This process of moving up the columns and combining characters continues until *s* (10) and *t* (11) combine to make *zgcos*.

CIPHERTEXT: nnpid hanvb wccpd iuhha etenti lnebc eaute arlti veesg ieina zgcos

Deciphering tips: You must have the trapezoid, the keyword and the transcription arrangement. Beginning with *a* (1) place the group *nnpid* in the diagonal formed from *a* (1) to the upper right. For *c* (2) with only two cells for *h* and *a*, the remaining three letters *nvb* are placed in *d* (3). The *wccp* series fills *d* (3). Shift *d* to begin *i* (4). Next *iuhha* fills most of the diagonal.

```
                    p      e

                 c     a

              c     h                          d

           w     h                          i

        b     u                          p

     v     i                       n                   a

  n     d                       n                   h
  ─────────────────────────────────────────────────────────
  3     4     9     6     8     7     1     11    5     2     10

  d     i     p     l     o     m     a     t     i     c     s
```

The number of cells in the diagonals govern whether the letter groups can fill the cells or must be carried over to the next diagonal. Read from left to right to see the plaintext.

Word Transposition

Methods have also been developed to transpose entire words. An example of a military dispatch whose order is directed by a keyword *friday* is:

f	r	i	d	a	y
3	5	4	2	1	6
move	to	area	one	at	sunrise
attack	enemy	deployed	south	of	fortress
stop	seize	all	weapons	and	supplies

The alphabetical order of the keyword's letters directs the placement of the words listed beneath them. For example, *a* (1) has *at of* and *and* below it. The word *friday* is placed first in the dispatch to indicate that it is the keyword, followed by the words *at of and*. Column *d* (2) aligns the words *one south weapons* and these are placed next in the ciphertext. Continue in this way from *f* (4) through to *y* (6). When the full ciphertext is formed, it appears as follows:

CIPHERTEXT: friday at of and one south weapons move attack stop area deployed all to enemy seize sunrise fortress supplies

Deciphering tips: The word *friday* is included as it defines other facts for the addressee. The number of message terms (18) governs the depth of the words, which is three (18 divided by the six letters of the keyword) beneath each digit. Number each keyword letter in alphabetical order and place the words beneath the numbers in numerical order. The first three words of the message, *at of and*, are placed below *a* (1); the next three under *d* (2); and the next trio beneath *i* (4) as follows:

f	r	i	d	a	y
3	5	4	2	1	6
move		area	one	at	
attack		deployed	south	of	
stop		all	weapons	and	

When the columns are filled, the dispatch becomes clear.

Word transposition is not a strong concealment to an experienced third party. During the American Civil War the Union forces expanded lists of routes that included a series of 12 ciphers, some of which consisted of dozens of pages for the routes and accompanying codenames. Each of these routes was made up of a keyword, codenames for important names and nulls to complete lines and to confuse enemy interceptors. The keyword determined the size and direction of the route. Following is an example of a Civil War word transposition:

KEYWORD: watch

CODENAMES:

brass (gunboats)

bronze (Vicksburg)

copper (bombardment)

gold (General Grant)

iron (Mississippi)

lead (1:00 p.m.-message time)

metal (New Orleans)

nickel (wired)

ore (telegram)

pewter (five)

silver (Admiral Farragut)

steel (General Sherman)

NULLS: air, earth, fire, ice, water

Message (with codenames and nulls) ready for transmission:

gold has nickel a ore to metal and will have
silver send pewter brass up the iron to bronze
for copper hold your sector steel lead air earth
ice fire

For the clarity of this explanation, the codenames used here are all generally associated with metals and the nulls are variations of the basic elements. In larger, actual routes, the words would have been varied for greater secrecy since similar words could give enemy analysts a clue with which to break the cipher.

The order of the alphabet letters in watch is *a*-1, *c*-2, *h*-3, *t*-4, and *w*-5, and its length is five letters, which becomes the width of the route. By prearrangement between the sender and the recipient, this route will go down column 1, up 2, down 3, up 4, and down 5.

Below is the completed route with keyword for directions:

```
watch gold to silver the copper lead air hold
iron send metal has nickel and pewter to your
earth ice sector bronze brass will a ore have
up for steel fire
```

Deciphering tips: Using the method described above, use the keyword to work out the number of columns and the number of words per column. For this message, 30 words divided by five keyletters means that there are six words in each column. You also know that the plaintext will run down the first column, up the second column, down the third and so on. Place the first group of six words below *a* (1) and take the next six words up the column of *c* (2). The following groups of six alternate down *h* (3), up *t* (4) and down *w* (5). Using the codenames from the code list, set up a key letter and numeral arrangement as follows:

A	C	H	T	W
1 (keynumber)	2 (keynumber)	3 (keynumber)	4 (keynumber)	5 (keynumber)
gold (General Grant)	has	nickel (wired)	a	ore (telegram)
to	metal (New Orleans)	and	will	have
silver (Admiral Farragut)	send	pewter (five)	brass (gunboats)	up
the	iron (Mississippi)	to	bronze (Vicksburg)	for
copper (bombardment)	hold	your	sector	steel (General Sherman)
lead (1:00 PM)	air (null)	earth (null)	ice (null)	fire (null)

With the codenames decoded and reading from the first words, General Grant, horizontally across the rows, recover the plaintext:

```
General grant has wired a telegram to New
Orleans and will have Admiral Farragut send five
gunboats up the Mississippi to Vicksburg for
bombardment hold your sector General Sherman
1:00 p.m. null null null null
```

SUBSTITUTION CIPHERS

Substitution is a method of cryptography in which the plaintext is replaced by letters, numbers or other symbols while its original word or letter order remains unchanged.

For example, the word *security* could be replaced by *rdbtqhsx*, which is simply comprised of the letters directly preceding each letter of the plaintext in the English alphabet. It could also be replaced by the digits 19, 5, 3, 21, 18, 9, 20 and 25, which reflect the numerical position of each letter in the alphabet, or by the symbols #&*$-/=+. Each of these examples is a substitution cipher since individual letters are replaced. For added security, substitutions are sometimes combined with transposition so that the letters of the plaintext are not only replaced, but also shuffled. Besides ciphers, substitution is also used to form codes, in which whole words are replaced at once, as opposed to the letter-for-letter substitution used to create ciphers.

Monoalphabetic Substitution

Monoalphabetic substitution is a cipher system in which a single cipher alphabet is used to conceal the letters of a plaintext communication. By the early 1400s in Europe, monoalphabetic ciphers had become the most frequently used concealment method. The early nomenclators of the Italian city-states used monoalphabetic letter substitutions to spell out unencoded words. Below is an example of a single-alphabet cipher:

PLAIN ALPHABET:	a	b	c	d	e	f	g	h	i
j	k	l	m	n	o	p	q	r	s	t	u	v
w	x	y	z					

CIPHER ALPHABET:	b	c	d	e	f	g	h	i	j
k	l	m	n	o	p	q	r	s	t	u	v	w
x	y	z	a					

The equivalent letters remain an alphabet even if their order is random.

PLAIN ALPHABET:	a	b	c	.	.	.	m	.	.	.	z

CIPHER ALPHABET:	w	l	x	.	.	.	q	.	.	.	h

A monoalphabet can also use symbol cryptography or numerals, whereby special symbols or numbers can serve as equivalents. In addition to letter-for letter equivalents, a single alphabet can also supply several substitutes for a letter. For example, the plaintext *v* could be represented by a number such as 7, or by a series of numerals such as 21, 28, 36 and 46. Such replacements are called *homophones*. Although they are a plural set in order to increase concealment complexity, they replace only *v*, not any other alphabet characters. Although there are multiple concealment options, this is still considered a monoalphabetic substitution as there is only one cipher alphabet in use.

SHIFT CIPHERS

One of the earliest and simplest forms of letter substitution is the Caesar substitution. This cipher bears the name of Gaius Julius Caesar, who sent encrypted messages during his successful military campaigns in Gaul, which is now modern France, Belgium and parts of the Netherlands, Germany, Switzerland and Italy.

In his text *Gallic Wars*, Caesar described using Greek letters to mask his Latin communiqués. He also used rearrangements of the plaintext alphabet. The Caesar cipher is a cyclical shift substitution, meaning that the cipher alphabet is shifted a predetermined number of steps along the plaintext alphabet. In this particular form, the plaintext letters are substituted for letters three steps along a natural alphabet progression. Using the modern 26-letter English alphabet, the method looks like this:

PLAIN ALPHABET:

a	b	c	d	e	f	g	h	i
j	k	l	m	n	o	p	q	r	s	t	u	v
w	x	y	z

CIPHER ALPHABET:

d	e	f	g	h	i	k	l	m
n	o	p	q	r	s	t	u	v	w	x	y	z
z	a	b	c

The message "Gaul is ours" becomes *jdxo lv rxuv*.

Deciphering tips: As you know that the cipher alphabet is a three place shift, write the plaintext alphabet with the cipher alphabet, moved forward three places, below. Compare the missive's ciphertext to the key to decipher the message.

Shift Cipher with Keyword

A variation on the Caesar cipher uses a prearranged keyword to shuffle the alphabet. The keyword should not contain any repeating letters. For example, if the keyword *cipher* was used, the alphabet would be set up with the keyword in front and the remaining alphabet following sequentially:

PLAIN ALPHABET:

a	b	c	d	e	f	g	h	i
j	k	l	m	n	o	p	q	r	s	t	u	v
w	x	y	z

CIPHER ALPHABET:

c	i	p	h	e	r	a	b	d
f	g	j	k	l	m	n	o	q	s	t	u	v
w	x	y	z

In this example, the last eight letters of the ciphertext are identical to the plaintext, because the keyword does not contain a letter beyond *r*. If the keyword contained the letter *z*, for example, the entire cipher alphabet would be shifted.

Keywords can be changed regularly to make different cipher alphabets and confuse potential interceptors.

Deciphering tips: Set up the cipher alphabet with the keyword and the plaintext alphabet as shown above. Using the equivalents, substitute the ciphertext for its plaintext.

KEYPHRASE CIPHERS

A keyphrase cipher is one in which a phrase forms the key. When Edgar Allan Poe was editor of *Graham's Magazine* in Philadelphia, he asked his readers to challenge him by sending him cryptograms. He described a method purportedly used by the Duchess of Berry, whose secret correspondence to a group of antimonarchists in Paris was written in cipher. The system used the keyphrase *le gouvernement provisoire* as a substitution. In a 24-letter alphabet (with x and z removed), it can be arranged as follows:

PLAIN ALPHABET:	a	b	c	d	e	f	g	h	i
j	k	l	m	n	o	p	q	r	s	t	u	v
w	y								

KEYPHRASE:	l	e	g	o	u	v	e	r	n	e
m	e	n	t	p	r	o	v	i	s	o	i	r
e									

The keyphrase is not a letter-for-letter substitution, but rather a polyphonic one whereby a ciphertext letter can replace more than one plaintext letter. In this example, cipher *e* conceals plaintext *b*, *g*, *j*, *l* and *y*. Cipher *o* masks plaintext *d*, *q* and *u*, while cipher *r* hides plaintext *h*, *p* and *w*. As these examples illustrate, there would probably have been some ambiguity, even for the intended recipient, which suggests that careful instructions may have accompanied the keyphrase.

A type of keyphrase was also used by the Carbonari of Italy and France for their secret messages. Founded in 1811, this antimonarchical revolutionary group was named for the charcoal burners among whom they met and from whom they acquired some of their jargon.

Their method was similar to the one described by Poe, but repeated letters were each given separate identities (and thus separate cipher varieties) with numbers. For example, the first time that the letter *a* appears in the keyphrase it remains *a*. The next time it appears it becomes a^1. The Italian phrase that translates as "Good evening friends of liberty" could have concealed a dangerous vow, "The crown will be ours," as follows:

PLAIN ALPHABET:

a	b	c	d	e	f	g	h	i
k	l	m	n	o	p	q	r	s	t	u	v	w
x	y	z

KEYPHRASE WITH NUMBERS:

b	u	o	n	a	s	e
r	a^1	a^2	m	i	c	i^1	d	e^1	l	l^1	a^3	l^2
i^2	b^1	e^2	r^1	t	$à^4$

PLAINTEXT:

t	h	e	c	r	o	w	n	w	i
l	l	b	e	o	u	r	s

CIPHERTEXT:

l^2	r	a	o	l^1	d	e^2	i^1	e^2	a^1
i	i	u	a	d	i^2	l^1	a^3

Deciphering tips: Using the keyphrase, set up the alphabet as shown above. Make sure that all repeating letters are numbered consecutively. Find the first ciphertext letter, l^2, in the keyphrase-with-numbers line. This corresponds to *t* in the plaintext alphabet. Using this same method, ciphertext *r* is plaintext *h*. Continue until the message is deciphered.

RANDOM SUBSTITUTION CIPHERS

Random substitutions are ciphers in which the cipher alphabet does not follow a simple plan. Plaintext letters have a completely random ciphertext equivalent. The example below shows a mixed cipher alphabet beneath a plaintext one:

PLAIN ALPHABET:	a	b	c	d	e	f	g	h	i
j	k	l	m	n	o	p	q	r	s	t	u	v
w	x	y	z						

CIPHER ALPHABET:	i	x	e	q	y	o	h	f	v
z	a	u	d	b	l	s	g	k	m	p	w	c
r	j	t	n						

The cryptographer enciphers a message by replacing the letters of the plaintext with their cipher equivalents. The warning "outpost in danger" becomes:

PLAINTEXT:	o	u	t	p	o	s	t	i	n	d
a	n	g	e	r						

CIPHERTEXT:	l	w	p	s	l	m	p	v	b	q
i	b	h	y	k						

Random substitution ciphers can also consist of numbers, letters, symbols or any combination thereof. While slightly more difficult to break than shift ciphers, random substitutions are unwieldy, as their decipherment requires the memorization of the ciphertext equivalents, or the carrying of a key which could get lost or destroyed.

Almost infinite versions of random substitution ciphers can be created. It merely requires that every letter in the plaintext alphabet have its cipher equivalent.

Charlemagne's Cipher

Charlemagne, king of the Franks and founder of the Holy Roman Empire in 800 C.E., used this series of symbols as alphabet replacements to correspond with his generals.

Charlemagne's cipher

Zodiac Cipher

The zodiac was very important to the Medieval alchemists who dabbled in a number of experiments, including the quest for the mythical philosopher's stone, which they believed could turn base metals into gold. Because alchemists were often accused of practicing sorcery, they frequently tried to conceal their writing with crude symbol substitutions.

The following chart illustrates how zodiac signs and letters could be matched.

Zodiac cipher

⊙	a	sun	♌	n	lion
♃	b	Jupiter	♍	o	virgo
♄	c	Saturn	♎	p	balance (libra)
♆	d	Neptune	♏	q	scorpion
♅	e	Uranus	♐	r	sagittarius
⊕	f	earth	♑	s	capricorn
♀	g	Venus	♓	t	fishes
♂	h	Mars	♈	u	ram
☿	i	Mercury	♒	v	aquarius
☽	j	moon		w	
♉	k	taurus		x	
♊	l	twins		y	
♋	m	cancer		z	

Mary Queen of Scots Cipher

Symbol cryptography played a major part in the downfall of Mary Queen of Scots. Queen of Scotland from 1542 through 1567, she became the victim of a series of devious intrigues that included cipher messages. Mary Stuart rivaled Elizabeth I for the latter's English throne in the 1580s. This put Mary's own life in growing jeopardy and she used concealed writing to protect her correspondence.

Her missives were enciphered using a nomenclator, a system consisting of a codelike list of words and names, and a separate cipher alphabet, shown below.

Mary Queen of Scots' cipher

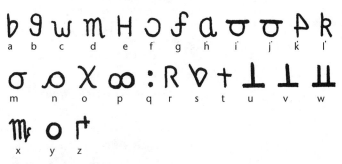

Her name could be enciphered as shown below:

"Mary Stuart"

Chevalier de Rohan's Cipher

An example of a random substitution cipher using the Roman alphabet is that of the chevalier de Rohan. Imprisoned during the 1670s for conspiring to betray his fort Quilleboeuf to the Dutch, his plea depended on the fate of his accomplice, Trouaumont, who lay dying in the same prison. If his friend confessed, Rohan would be convicted of treason and executed. Unsure of his friend's loyalty, Rohan spent a torturous few days in his cell.

Shortly before his trial, Rohan received a bundle of clothing in which he found a note attached to a shirtsleeve. The note contained the following ciphered phrase:

```
mg      eulhxcclgu      ghj     yxuj    lm      ct
ulgc    alj
```

Rohan spent the night struggling to decipher the message, but dawn arrived with the cipher still intact. Quickly broken by the prosecutor's harsh questions, he gave a full confession and the judges ordered him to meet his executioner.

Rohan could have saved himself if he had been able to discern this basic substitution cipher which could be analyzed by repetition. The note he received said in French:

```
Le prisonnier est mort; il n'a rien dit.
```

```
The prisoner is dead; he said nothing.
```

Rose Greenhow's Cipher

American Civil War spy Rose Greenhow applied steganography and cryptography to help the South. She hid secret missives in articles of friends' clothing and concealed documents in objects carried by fellow Secessionists.

She was provided with an encryption method and communication instructions by her spymaster, Colonel Thomas Jordan. Different theories exist about Jordan's sources for the encryption, as well as whether or not Greenhow was the only person to receive the cipher. With q and z omitted, the actual symbol-to-alphabet letter equivalents have been reconstructed as shown:

Rose Greenhow's cipher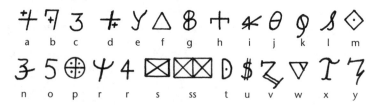

Edgar Allen Poe's "Gold-Bug" Cipher

Poe had already succeeded as an author when he turned his attention to codes and ciphers, and had also established a somewhat dubious name for himself as a cryptanalyst. The story that secured Poe's cryptanalytic fame was "The Gold-Bug." This story was first published in *Graham's Magazine* in 1843, and though it contained flaws in cryptanalysis and concealment methods, it caused a sensation. Within a few years it was adapted as a play and published in a book.

The gold-bug was a fictitious gold beetle found by the protagonist, William Legrand, who was living on Sullivan's Island near Charleston, South Carolina. He made a drawing of the bug for his friend, the story's narrator. It so happened that Legrand drew the bug on a piece of paper that contained a secret message written in invisible ink. Legrand used a flame's heat to make the ink visible and found this random substitution cipher:

The Goldbug cipher

```
5 3 ‡ ‡ † 3 0 5 ) ) 6 * ; 4 8 2 6 ) 4 ‡ .
) 4 ‡ ) ; 8 0 6 * ; 4 8 † 8 ¶ 6 0 ) ) 8 5
; I ‡ ( ; : ‡ * 8 † 8 3 ( 8 8 ) 5 * † ; 4 6
( ; 8 8 * 9 6 * ? ; 8 ) * ‡ ( ; 4 8 5 ) ; 5
* † 2 : * ‡ ( ; 4 9 5 6 * 2 ( 5 * - 4 ) 8 ¶
8 * ; 4 0 6 9 2 8 5 ) ; ) 6 † 8 ) 4 ‡ ‡ ; I
( ‡ 9 ; 4 8 0 8 I ; 8 : 8 ‡ I ; 4 8 † 8 5
; 4 ) 4 8 5 † 5 2 8 8 0 6 * 8 I ( ‡ 9 ; 4 8
; ( 8 8 ; 4 ( ‡ ? 3 4 ; 4 8 ) 4 ‡ ; I 6 I
; : I 8 8 ; ‡ ? ;
```

This simple cipher was easily been broken by an analysis of repeating symbols. Using the principles of repetition, Legrand analyzed the missive to reveal this message:

```
"A good glass in the bishop's hostel in the
devil's seat-forty-one degrees and thirteen
minutes northeast and by north-main branch
seventh limb east side-shoot from the left eye
of the death's head-a bee-line from the tree
through the shot fifty feet out."
```

Although critics correctly pointed out mistakes in Poe's application of cryptanalysis, readers were unperturbed, and the story sparked a new popular interest in cryptography.

Sherlock Holmes "Adventure of the Dancing Men" Cipher

Arthur Conan Doyle and his detective, Sherlock Holmes, used a random substitution cipher in the "Adventures of the Dancing Men." This story involved an English squire's wife who was receiving anonymous messages. Apparently children's scribbles, these odd communications were placed on surfaces around the squire's home and written with chalk.

Adventure of the Dancing Men cipher

These stick men and their varied positions were a symbol cipher known only to the squire's wife, Elsie Cubitt, and Abe Slaney, a criminal from Chicago. Slaney's engagement to Elsie had been ruined by his criminal leanings. After she had traveled to England, he had re-entered her life.

Holmes realized that the symbols were a monoalphabetic substitution and deciphered them. The message shown above read "am here abe slaney come elsie." The squire, however, confronted Slaney and died by gunfire before Holmes could intervene.

Although Holmes was unable to prevent this tragedy, his keen brain did not fail him. Having broken the cipher, he knew the meaning of each figure and the fact that a flag indicated the completion of a word. Holmes sent a request to Slaney in the same method, making it appear to have been sent by Elsie.

Here is Holmes's contrived message. There is one unfamiliar figure, whose plaintext can be guessed by the context:

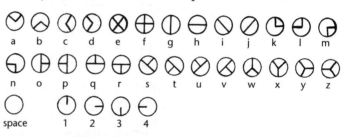

The message read: "Come here a_ once." The upside-down man represented the letter *t*.

The Shadow's Ciphers

In the 1930s a popular superhero was a mysterious man named the Shadow. A Shadow novelette, the *Chain of Death*, contained what was called a "directional code," although it actually functioned more like a cipher than a code:

The Shadow's Chain of Death cipher

The extra symbols at the base of the alphabet make this code different. The lines inside each circle function like an arrow and tell the recipient how to hold the paper. Symbol 1 means hold the sheet normally, top and bottom positioned as usual, and read the message in a regular left-to-right manner. Symbol 2 indicates a turn 90° to the right, and symbol 3 directs the reader to invert the paper. Symbol 4 signifies a turn 90° to the left. There is also a version "B" in which symbol 2 and 4 represent a 90° turn to the left and right, respectively.

These extra symbols can appear before or in the middle of a line of text. These turns make the cipher difficult to analyze by principles of frequency.

The following example reveals the Shadow's true identity:

The Shadow's real identity: Lamont Cranston

As directed by holding the page in the regular straightup way, the substituted symbols spell the Shadow's best-known identity, the crimefighter Lamont Cranston.

Later in *Chain of Death*, the Shadow encountered another type of encryption based upon paired symbols and spaces.

Chain of Death cipher

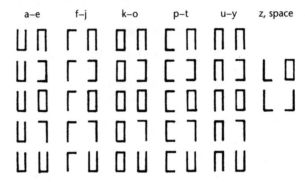

Each alphabet letter is represented by a symbol pair. To begin a communication, it is best to have a starting half symbol such as ⌐. To it is joined the left half symbol of the first message letter. After a space, the right half symbol of the first message letter is set in place. The left half of the second message character is connected to the previous half symbol, then a space, the right symbol, and so forth. The following example spells out the word *code*:

The divisions created by the split figures discourage an interceptor from easily matching figures and suspected alphabet characters. Odd symbols can also be added to serve as nulls.

Deciphering tips: Divide each figure in half and delete the first null figure. Grouping the figures in pairs, use the list of symbols above to recover the plaintext.

GEOMETRIC CIPHERS

As with transposition, types of geometric patterns have been applied to substitution. Around 1640, John Wilkins, a young chaplain in England, proposed a cipher in his *Mercury, or the Secret and Swift Messenger* wherein dots, lines and zigzags were substituted for plaintext characters.

The key was the series of known spaces between the message's alphabet letters, which were either sequential or shuffled. The key was placed along the top or side of a piece of parchment. The given plaintext was then spelled with a dot aligned with each keyletter, with every succeeding dot placed further right. The marks could be left as dots, or connected as lines, triangles or graphlike zigzags. The key could later be removed to ensure secrecy. The recipient possessed an identical key and after aligning it at a prearranged point, he or she read off the ciphertext.

The example below spells out "I love you as you love me."

A zigzag cipher

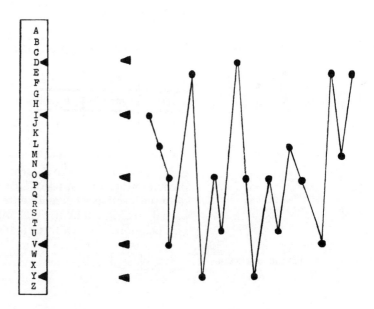

Another type of geometric cipher is the pigpen cipher, which uses square or rectangle designs divided into boxes, or cells, of equal size. Alphabet letters are placed in an open-sided matrix and are varied with combinations of crossed lines and dots. Pigpen ciphers permit an orderly alignment of letter position and identifying mechanisms for both the sender and the receiver of a concealed message.

The Rosicrucian brotherhood and the Freemasons are two groups who used pigpen ciphers and some of Europe's best cryptanalysts were never able to break their simple ciphers. As interceptors were unable to determine whether the symbols represented letters, many believed that the cipher was a type of hieroglyph or rune. This factor was one of the greatest barriers for cipher solvers, as this was a simple cipher which could be solved by analysis of repetition.

The success of the cipher lay in its simplicity, since it did not require a hard-to-remember key or a codeword.

Rosicrucian Cipher

In the example below, which uses the English alphabet, the cipher was arranged with a matrix of nine segments:

The Rosicrucian matrix

abc	def	ghi
jkl	mno	pqr
stu	vwx	yz

On a simple grid, the positions of the letters follow the standard alphabet from A to Z. Other patterns could be used if predetermined by both sender and recipient. In this standard example, each letter is represented by the shape of the box or partial box formed by the grid and by a dot indicating the letter's position. The first three letters are:

Rosicrucian letters a–c

The *a*'s dot is in the first position, or left side; *b* is second in the middle; and *c* is third on the right. The letters *d*, *e* and *f* are differentiated by the shape of the grid's lines, as follows:

Letters d–f

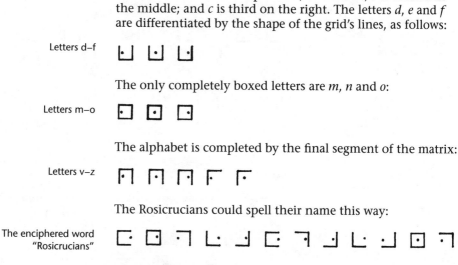

The only completely boxed letters are *m*, *n* and *o*:

Letters m–o

The alphabet is completed by the final segment of the matrix:

Letters v–z

The Rosicrucians could spell their name this way:

The enciphered word
"Rosicrucians"

Eventually, as some individuals left the order and broke their vows of secrecy, this cipher was revealed.

Freemason's Cipher

In the 18th century the Freemasons created a similar crypto-system to protect their business from public scrutiny. As divulged by former members, the Masons used a method of secret writing quite similar to the Rosicrucians' cipher. The matrix and other angles contain dots that replace letters, as shown in this example:

The Freemasons' cipher

As a result of security breaches, most U.S. Masonic grand lodges no longer create written ciphers, preferring to use vocal instructions in private ceremonies.

Using the Rosicrucians' system as a guideline, the following term can easily be deciphered.

It is the first level on which a newly accepted member joins, the Blue Lodge.

Deciphering tips: To reverse a pigpen cipher, draw the alphabet within the appropriate grid. Compare the position of the dots in the cipher to the letters in your grid to decipher the message.

ALPHANUMERIC CIPHERS

Substitution ciphers can also use numbers. In a simple alphanumeric cipher, standard alphabetic positions might be reversed: $A = 1$ through $Z = 26$ could become $A = 26$ through $Z = 1$.

PLAIN ALPHABET:	a	b	c	d	e	f	g	.	.	.
m	n	o	.	.	.	t	u	v	w	x	y	z

CIPHER ALPHABET:	26	25	24	23	22	21	20	.	.	.
14	13	12	.	.	.	7	6	5	4	3	2	1

In the mid-to-late 1800s, "agony columns,"or personal advertisements, were especially popular in newspapers. They acquired their nickname from the number of supposedly private lovers' communications contained in them. To avoid being caught in adulterous affairs or to circumvent parental restrictions, the smitten but often separated pairs corresponded in this disguised way. Because intermediaries could be disloyal and notes were susceptible to interception, the paramours sought the relative convenience and accessibility of the daily news. Correspondents used systems ranging from simple substitutions of alphabet letters (such as *a* for *b* and *m* for *o*) to more involved styles where numbers were equated with one or more letters.

Using the above cipher, a column might have contained these numbers:

18	14	6	8	7	14	22	22	7	2	12	6	7
12	13	18	20	19	7							

As can be seen by this one example, the repeating numbers virtually give away this simple cipher. Those with even rudimentary skills in cryptanalysis were able to decode familiar phrases, salutations and even rendezvous points, if mentioned often enough. The above plea, "I must meet you tonight," was not at all secure when read by cryptogram fanciers.

Franklin-Dumas Cipher

Benjamin Franklin and his supporter Charles William Dumas used a variation of the alphanumeric cipher to conceal reports of his efforts on behalf of the rebelling American colonies. In his communications he used codenames such as *Concordia* for himself and *le grand facteur* for the duc de la Vaugoyon, the French ambassador at The Hague. He also developed a cipher which historian RalphWeber credits as "the best, most reliable and least confusing cipher used in correspondence between the Continental Congress and its foreign agents."

This cryptosystem was based on the consecutive numbering of every letter and punctuation mark in a French passage, of which a portion is shown here, followed by its translation and a sample of the cipher. According to Weber, these lines and the others in the passage, with 682 symbols, probably originated in an essay provided, but not penned, by Dumas.

Voulez-voux sentir la différence? Jettez les yeux sur le continent septentrional de l'Amérique. Dans les résolutions vigoureuses de ces braves colons vous reconnoitrez la voix de la vraie liberté aux prises avec l'oppression.

Do you want to feel the difference? Glance at the continent of North America. In the vigorous resolutions of those worthy colonists you will recognize the voice of true liberty at grips with oppression.

Each letter in the paragraph is numbered consecutively and the numbers are then used as the ciphertext. For example, using the short section shown below to encrypt the word *love*, the ciphertext could be 4 2 8 5, 4 2 1 5, 4 9 8 5 or 4 9 1 5.

1	v
2	o
3	u
4	l
5	e
6	z
7	= (division or hyphen)
8	v
9	o
10	u
11	s

This cipher provided a good amount of concealment since it included different numbers for often-used letters. For example, *e* had 128 different numbers, *a* had 50 and *o* had 44. A simple analysis of repetition would not be sufficient to break the cipher.

Any sample of text can be used in a similar manner. The text must contain all of the letters of the alphabet unless the corresponders come to an agreement that, for example, *i* and *j* will be interchangeable. The longer the passage, the more cipher possibilities it offers and the more coverage it gives against potential interceptors.

Deciphering tips: Using the agreed upon passage, number the letters consecutively. Compare the ciphertext with the numbered passage, finding each number's corresponding letter to decrypt the missive.

"The Man from Scotland Yard" *Cipher*

In *The Man from Scotland Yard*, the Shadow dealt with an alphanumeric encipherment called "double letter." The system was based on a series of equivalents as follows:

CHART 1

8	12	7	2	6	4	1	11	3	10	5	9	13
a	b	c	d	e	f	g	h	i	j	k	l	m

n	o	p	q	r	s	t	u	v	w	x	y	z
20	22	17	21	24	26	18	16	23	25	15	19	14

The letters *a* to *m* are represented by digits from 1 to 13, and *n* to *z* are equated with 14 to 26. This apparently standard equivalency is altered when the system equates either *a* or *n* with 8 or 20, *h* or *u* with 11 or 16, *m* or *z* with 13 or 14 and so forth. To avoid confusion, a keynumber designates whether a letter will belong above or below the numeral 13, the standard English alphabet midpoint. Then number combinations can be made of all the digits from 1 to 13 with the digits that represent the given letter. For example, *a* can be replaced by 5-8, 7-8, 12-8, 3-20, 5-20 or 11-20. All the numerals from 1 to 13 can be joined with *a*'s 8 or 20 base numbers. In comparison, *n* can be replaced by 8 or 20 plus all the numerals from 14 to 26. Thus 14-8, 17-8, 22-20, and 25-20 can be substituted for *n* (8 or 20 plus the numbers above 13).

With the standard arrangement, in which below 13 means *a* to *n* and above 13 means *n* to *z*, the individual letters of the word *crime* could be replaced by any of these examples among others:

PLAINTEXT:	c	r	i	m	e
KEYNUMBER:	(7-17)	(6-24)	(3-23)	(13-14)	(6-24)
	2-7	14-6	5-3	1-13	2-6
	3-17	18-24	6-23	3-14	4-24
	4-7	16-6	8-3	4-13	7-6
	6-17	19-24	9-23	5-14	9-24
	9-7	21-6	11-3	8-13	11-6
	13-17	12-24	25-23	10-14	13-24

The letters *e* and *r* demonstrate the idea of the keynumber. *E* is below 13 and *r* is above. Their mutual digits, 6 and 24, are matched with numerals for 1 to 13 for *e* and 14 to 26 for *r*. For messages of more than one word, preplanned double digits (preferably outside the number scale 1–26) are used to make spaces. This helps avoid transmission confusion for the intended exchangers, as in the following example, the phrase *crime clue*:

PLAINTEXT:	c	r	i	m	e	(space)	c	l	u	e
CIPHERTEXT:	2-7	14-6	5-3	1-13	2-6	27-27	3-17	4-9	14-11	4-24

The letter/number choices do have a sequential pattern and, if guessed, this could be an important break-in angle for a third party. The previous chart could be altered with different letters located above and below the 13 mid-point.

CHART 2

1–13

a	c	e	f	j	m	p	q	s	u	v	x	z
8	12	7	2	6	4	1	11	3	10	5	9	13

14–26

b	d	g	h	i	k	l	n	o	r	t	w	v
20	22	17	21	24	26	18	16	23	25	15	19	14

The words *crime clue* can now be replaced by other combinations such as:

PLAINTEXT:

c   r   i   m   e   (space)   c   l   u   e

CIPHERTEXT:

(12–22) (10–25) (6–24) (4–26) (7–17) (12–22) (1–18) (10–25) (7–17)

1–12   14–25   17–24   11–4   13–7   28–2   2–22   25–18   13–10   6–17

If the cryptographer chose to conceal these digits further, he or she could equate them with alphabet letters. Even standard alphabet equivalents (for example, a = 1, z = 26) would give added security. Because this ciphertext is alphabetic rather than numeric, the communicants would have to preplan double z's if they wished to distinguish them from nulls.

PLAINTEXT:	c	r	i	m	e	c	l	u	e
	1–12	14–25	17–24	11–4	13–7	2–22	25–18	13–10	6–17
	al	ny	qx	kd	mg	bv	yr	mj	fq

CIPHERTEXT: alnyq xkdmg bvyrm jfqzz

The concluding double z's are nulls to complete the five-character group.

Deciphering tips: You know that the letters from *a* to *m* are equated with the numbers from 1 to 13 and have the key numbers 8 and 20. The letter *n*, which is one numeral past 13, is represented by all the numbers from 14 to 26 along with the keynumbers 8 or 20. This the process used earlier for Chart 1.

Use chart 2 for alphabetic replacements. When you receive a communiqué beginning with *a*, you know that it is represented by the numeral 1 in this system. The letter *l* is 12 and the first number combination is 1-12. Check your second chart to find the letter that is in the 1-to-13 row and beneath 12. The letter is *c*. By this same process locate *n* at the 14th alphabet position. The letter *y* is number 25. The duo *n-y* is equivalent to the chart combination 14-25. The letter in the 14-to-26 row and right above 25 is *r*. The next letter *q* is the 17th alphabet character, which puts the letter in the 14-26 row. The letter *x* is the 24th alphabet character. Directly above 24 in the chart is *i*. The letter *k* is 11th in the alphabet and *d* is 4th. The number 11 is in the 1-to-13 group in the upper row where 4 is located. Beneath 4 is the letter *m*. The letters in *m-g* are in the alphabet's 13 and 7 positions. Since 13 is in the 1-to-13 group, it is in the upper row. Beneath 7 is the letter *e*. The word *clue* is recovered the same way.

MULTILITERAL FORMS

Many monoalphabetic forms are uniliteral, meaning that the cipher equivalents are single letters. Multiliteral versions are those whose encryption elements consist of two or more letters. They are typically letter or numeral pairs developed from bilateral matrices with two-letter or two-digit replacements for the plaintext.

Polybius Checkerboard

Polybius was a Greek historian and cryptographer of the second century B.C.E. His signaling system (see Chapter 3, "Signals") led to a basic form of substituting digits for plaintext letters. Each of the letters had a pair of numerical coordinates derived from the numbers of the row and column in which it was located. Using the English alphabet and combining *i* and *j* in a single cell, a five-by-five square could be arranged as below:

	1	2	3	4	5
1	a	b	c	d	e
2	f	g	h	ij	k
3	l	m	n	o	p
4	q	r	s	t	u
5	v	w	x	y	z

In cryptology this arrangement came to be known as the Polybius checkerboard. It provided a transferral of letters to numbers, a pattern that was relatively easy to recall, and a limited amount of figures (numbers and letters) with which to work.

This system can be used to convert the word *cover* to numbers. For example, the letter *c* is found at the junction of row 1 and column 3 making the ciphertext 13.

LETTER	ROW	COLUMN	CIPHERTEXT
c	1	3	13
o	3	4	34
v	5	1	51
e	1	5	15
r	4	2	42

The Polybius checkerboard has been used to turn letters into digits, generate substitution ciphers, and divide the coordinates into parts (fractionating ciphers), among other matrix variations.

Deciphering tips: Set up an identical checkerboard to locate the letters. Use the first number to locate the row and the second to locate the column. The converging points reveal the plaintext letters.

Elizabeth Van Lew's Cipher

During the Civil War, both Union and Confederate spies sent encrypted missives using simple ciphers. A Richmond native, pro-Union Elizabeth Van Lew sent messages via her agents regarding escaped Federal prisoners and Confederate troop strength, while masking the intelligence with an alphanumeric cipher based upon the Polybius checkerboard:

Elizabeth van Lew's cipher

Using the same technique as the Polybius checkerboard, letters and numbers were covered by designator digits. The report "9 wagons here" could be concealed as:

9	w	a	g	o	n	s	h	e	r	e
56	34	53	14	43	63	24	62	11	61	11

Variations on this square could easily be achieved by shuffling the positions of the the numbers and letters. The only requirement is that both sender and recipient possess the same key in order to decrypt the missive.

Deciphering tips: Using the same square for decryption, take the first numeral of each pair (the number 5 in the first letter pair) and trace it across its row. Then take the second numeral of the pair, 6, and trace up its column to find the point where they intersect, which is the digit 9. This process is followed for each ciphertext letter.

Knock Cipher

An audible variation on the Polybius checkerboard was developed by Russian prisoners over the centuries. This system involves sequences of taps representing the rows and columns of a checkerboard of alphabet letters. One tap followed by two indicates row 1, column 2, or the letter *b*. A pause represents the division between rows and columns. If a prisoner had early news of a jailer's pending arrival, he could send the message "guard coming" as follows:

g   knock knock–knock knock
u   knock knock knock knock knock–knock
a   knock–knock
r   knock knock knock knock–knock knock knock
d   knock–knock knock knock knock
c   knock–knock knock knock
o   knock knock knock–knock knock knock knock
    knock
m   knock knock knock–knock knock knock
i   knock knock–knock knock knock knock
n   knock knock knock–knock knock knock knock
g   knock knock–knock knock

Sometimes a square of six rows and six columns was used to include the 35 Cyrillic letters of the old Russian alphabet; at other times a five-by-six matrix was used, with some of these letters deleted.

Deciphering tips: Note the frequency of knocks and pauses. The first set of knocks represents the row number, and second set the number of columns. Compare it to the knock checkerboard to decipher the message.

MULTILITERAL VARIATIONS

The following examples are variations on the multiliteral forms described above. In this form, the standard five-by-five grid dimension becomes three-by-ten. Symbols complete the array.

	1	2	3	4	5	6	7	8	9	0
1	a	b	c	d	e	f	g	h	i	j
2	k	l	m	n	o	p	q	r	s	t
3	u	v	w	x	y	z	.	,	-	+

PLAINTEXT: i   n   v   e   s   t   i   g
a   t   i   o   n   s

CIPHER EQUIVALENTS: 19 24 32 15 29 20 19 17
11 20 19 25 24 29

CIPHERTEXT: 19243   21529   20191   71120
19252   42900

Deciphering tips: Using the numerical matrix arrangement, set up the grid as shown above. Separate the characters or numerals into pairs and trace them back to the letters where the row and column designators intersect in the given matrix. The final two zeros serve as nulls.

Multiple Equivalents

The following ciphers contain more than one cipher possibility for each plaintext letter, thereby making cryptanalysis by studies of repetition more difficult.

In the following example the general form of the checkerboard is followed, but each letter has three cipher possibilities. For example, the letter *a* could be encrypted by the numbers 18, 14 or 12, this time reading by column, then row.

	1	2	3	4	5	6	7	8	9
8,4,2	a	b	c	d	e	f	g	h	i
9,5,3	j	k	l	m	n	o	p	q	r
7,6,1	s	t	u	v	w	x	y	z	

Another possibility plays upon the fact that the letters *e t o a n i r s* and *h* appear most often in the English alphabet. This cipher disguises the frequency of their appearance by designating more cipher possibilities to these letters than to the more infrequent plaintext letters. In this example, *e* could be enciphered by six possibilities, 14, 15, 16, 17, 18 and 19, whereas the rarer *q* has only possibility, 29:

	1	2	3	4	5	6	7	8	9
1,2,3,4,5,6	e	t	o	a	n	i	r	s	h
7,8	b	c	d	f	g	j	k	l	m
9	p	q	u	v	w	x	y	z	

Deciphering tips: As in the previous examples, you must possess the same grid as the sender. Compare the ciphertext to the grid, working backwards in order to reveal the message.

Variants

Multiliteral methods also include two or more encryption values called "variants." After filling in the keyword *palindrome*, the cryptographer completes the matrix with the remaining letters in the alphabet (*u* and *v* are combined in a cell). Each column and row has pairs of consonants as variants.

	n	p	q	r	s
	t	v	w	x	z
bh	p	a	l	i	n
cj	d	r	o	m	e
dk	b	c	f	g	h
fl	j	k	q	s	t
gm	uv	w	x	y	z

With this bilateral array, plaintext *c* can be replaced by *dp*, *dv*, *kp* or *kv*. With even a small word like *veil*, there can be multiple concealments:

v	e	i	l
gn	cs	br	bq
gt	cz	bx	bw
mn	js	hr	hq
mt	jz	hx	hw

Variants can also be made within the square itself. As a mnemonic, the word *equivalent* is placed vertically in the first column. A group of ten 10-letter words then fills the matrix:

	0	1	2	3	4	5	6	7	8	9
0	e	n	t	r	u	s	t	i	n	g
1	q	u	a	r	a	n	t	i	n	e
2	u	n	e	x	p	e	n	d	e	d
3	i	m	p	r	o	b	a	b	l	e
4	v	i	c	t	o	r	i	o	u	s
5	a	d	j	u	d	i	c	a	t	e
6	l	a	b	o	r	a	t	o	r	y
7	e	i	g	h	t	e	e	n	t	h
8	n	a	t	u	r	a	l	i	z	e
9	t	h	i	r	t	y	f	i	v	e

For the letter *e* alone, there are 12 dinomes (number pairs) available from the coordinates: 00, 19, 22, 25, 28, 39, 59, 70, 75, 76, 89 and 99.

Deciphering tips: Set up the grid with the digits and words shown above. Using the first number of the pair to locate the row and the second to locate the column, trace across and down until you reach the letter at which they intercept. Continue for all the pairs until the message is fully deciphered.

Multiliteral Ciphers with Keyword

The following example uses the keyword *bankruptcy* in a standard five-by-five matrix. Place the keyword in the matrix and fill the square with the remaining letters. The plaintext is encrypted by giving the row coordinate and then the column coordinate. For example, plaintext *a* is concealed by *b* (row) and *e* (column).

	a	e	i	o	u
b	b	a	n	k	r
c	u	p	t	c	y
d	d	e	f	g	h
f	ij	l	m	o	q

The word *investigations* would be concealed as follows:

i	n	v	e	s	t	i
g	a	t	i	o	n	s
fa	bi	ge	de	ga	ci	fa
do	be	ci	fa	fo	bi	ga

CIPHERTEXT: fabig edega cifad obeci fafob igazz

Deciphering tips: Using the prearranged coordinates on the top and side of the grid, place the keyword in the matrix and fill the square with the rest of the alphabet. Use the first letter in the pair to locate the row, the second to locate the column and their interception point will reveal the hidden letter. The two z's serve as nulls to give the concluding cipher group five letters.

Multiliteral Tables

Another variant is arranged as a table. The plaintext has 25 letters (*i* and *j* are again combined) and the enciphering numerals are four groups of numbers, 01–25, 26–50, 51–75 and 76–00. The starting points of the numerals are arranged by a keyword, in this case *coin*.

The first letter of the keyword, *c*, indicates the position to begin numbering. Starting with 01, the alphabet is numbered consecutively. When *z* is reached, the numbering is looped back to *a* (24) and *b* (25). The next letter of the key is *o*. Locating *o* in the table, numbering is continued in a new column from this point. Thus, *o* becomes 26, *p* is 27 and so forth. When *z* is reached, numbering is continued at the beginning of the alphabet. In this manner the letters of the keyword are worked through to create this table:

A	24	38	68	89	H	06	45	75	96	O	12	26	56	77	U	18	32	62	83
B	25	39	69	90	IJ	07	46	51	97	P	13	27	57	78	V	19	33	63	84
C	01	40	70	91	K	08	47	52	98	Q	14	28	58	79	W	20	34	64	85
D	02	41	71	92	L	09	48	53	99	R	15	29	59	80	X	21	35	65	86
E	03	42	72	93	M	10	49	54	00	S	16	30	60	81	Y	22	36	66	87
F	04	43	73	94	N	11	50	55	76	T	17	31	61	82	Z	23	37	67	88
G	05	44	7	4	95														

Each letter has four variants; communiqués could contain combinations of any of the four. For example, the order "act now" could be concealed by 38 91 61 11 56 34 or 24 40 82 50 77 87, or any other corresponding combination among the table-created numerals.

Deciphering tips: Using the keyword, set out the overall digit patterns as described above. Locate the pairs of numbers from the ciphertext in the table to reveal the plaintext letter.

Russian Nihilist Cipher

Using the basic form of the prisoner's knock cipher, the Nihilists in the 1850s and early 1860s developed a more sophisticated cipher. The Nihilists were a group of Russian anti-Czarists who sought to overthrow the existing order by means of sabotage, terrorism and assassination. However, without the ideology of Marx or the iron discipline of Lenin, they were never as well organized as the Bolsheviks. To protect themselves from the Okhrana, the Czar's secret police, and to pass their messages through the Czar's black chambers, the Nihilists developed their own cipher using the checkerboard and a keyword. The plaintext was numerically enciphered and added to a keyword that repeated through the length of the communication, thus generating the cipher-text.

In the following example, the message "strike czar now" is enciphered with the keyword *unite*.

	1	2	3	4	5
1	a	b	c	d	e
2	f	g	h	i	j
3	k	l	m	n	o
4	p	q	r	s	t
5	uv	w	x	y	z

Both the message and the keyword are numerically enci-phered by locating each letter first horizontally and then ver-tically and recording its coordinates. *Unite* would become 51 34 24 45 15 and the enciphered message is 44 45 43 24 31 15 13 55 11 43 34 35 52.

Adding the repeating keynumbers to the enciphered message numbers, the finished ciphertext would appear as follows:

PLAINTEXT:	s	t	r	i	k	e	c	z	a	r	n	o	w
DIGITS:	44	45	43	24	31	15	13	55	11	43	34	35	52
REPEATING KEY:	51	34	24	45	15	51	34	24	45	15	51	34	24
CIPHERTEXT:	95	79	67	69	46	66	47	79	56	58	85	69	76

This method of adding a numerical key to a checkerboard-generated substitution outlasted the Nihilists. After the Bol-sheviks seized power in the November 1917 revolution, they adopted this method and used it for a number of years to conceal exchanges.

Many variations on this alphabet can be created by the use of different keywords, which could be changed regularly.

Deciphering tips: Numerically encipher the keyword with the checkerboard and place it in a repeating sequence under the ciphertext. Subtract the keyword cipher from each number above it. The numbers remaining will correspond to the checkerboard. By comparing each number to the grid, reveal the plaintext.

STRADDLING CHECKERBOARD

The straddling method is a checkerboard-based cipher used by the Argenti family, cryptologists who developed ciphers for the pope in the 1500s. In modern times, it was used by communist forces during the Spanish Civil War and later conflicts.

The opponents of fascism controlled most of Spain's major cities in the beginning of the struggle and protected their communications, including radio transmissions, with a variety of encryption methods such as the straddling checkerboard.

The communists' version was called straddling because of its unique use of one-and two-number cipher groups which gave the message added protection against a would-be solver. Some letters were enciphered with a single number, while others were enciphered with a number pair. A third party analyst could not be sure of the location of single numbers contrasted with paired digits in a group of intercepted numerals. He or she might match a single number with a member of a numeral pair while analyzing the ciphertext. This "straddles" numbers that should be separate and thereby forms an incorrect pair, ending up with scratch pads full of nonsense.

This method creates one-number versions by deleting a side coordinate from one of the rows. A letter in such a row has only one coordinate (above it) as an equivalent rather than one from the top and one from the side.

In the completed matrix, the coordinates are read from the side and then the top. A keyword initiates the alphabet sequence in the cells of the checkerboard. The cells of the second and third rows are filled with the remaining letters. None of the chosen single numbers (0, 9, 8 . . . 3) begin a digit pair, in order to avoid confusion. The matrix has a period (.) and shift sign (/) to fill out the bottom row. An example follows, using the keyword guernica.

	0	9	8	7	6	5	4	3	2	1
	g	u	e	r	n	i	c	a		
1	b	d	f	h	j	k	l	m	n	o	p
2	q	s	t	v	w	x	y	z	.	/

In order to encipher the message "save Madrid," the cipher maker would begin by locating the coordinates for s, or 29, then a, or 3, and so on through the plaintext, and then place the letters alongside each other to form the digit-covered cipher 293278133197519. Note that a has no side coordinate so it appears as a single cipher digit. The addressee knows that whenever a 1 or 2 appears, it will pair with another digit but will not appear as a single number. Therefore, 29 is obviously a pair while 32 is not, because in this matrix there is no side coordinate 3. A successful interception of such a message by a third party would reveal a series of numbers, but the enemy solution seeker could not be sure which numbers represented singles and which represented pairs.

Deciphering tips: Using your straddling checkerboard, you know that whenever a 1 or 2 appears it will pair with another digit. Therefore, the first two digits will have to be the pair 29. The digit 3 cannot begin a pair and thus only represents the letter a. The pair, 27, is next; v is found at their coordinate point. The digit 8 cannot begin a pair, so it represents the single letter e. This pair-or-single pattern continues through the remainder of the plaintext's recovery.

Polyalphabetic Substitution

Concealments using only one alphabet are relatively easy to solve; by the latter 1400s, cryptographers began to turn to polyalphabetic substitution. Polyalphabetic substitution is a method of creating a cipher through the use of more than one replacement alphabet. This type of concealment method gives cryptographers several levels of letters within which to hide their original words and sentences. As Leon Battista Alberti showed in the 15th century, once letters and numerals had their alignments changed within a given encryption procedure, new alphabets were activated and the method became polyalphabetic.

This type of substitution is achieved by aligning the original letters with various alphabet groups by such means as tableaux (tables), cipher disks and letter-generating grids that create pairings of letters. In such systems, two or more figures, digits or letters equate with the cipher.

DISKS

The first methods of creating polyalphabetic ciphers were disks, which were used to arrange alphabets, numerals or symbols for encrypting and decrypting purposes. Historically, they have been made of materials ranging from paper to metal and have had varied sizes.

As paired dials or rings, they contain letters or digits, sometimes separately or in combinations. Generally the smaller member of the pair is turned on top of the larger one to bring its characters into alignment with those on the outer ring. Occasionally the outer ring has also been movable. By turning the disk, plaintext letters could then be matched with numbers, and ciphertexts could be made in surprisingly varied ways for those days.

By prearrangement, one ring's letters represent the plaintext message, and the other one's figures become the ciphertext. In the simplest form disks can be used to make single monoalphabetic substitutions. However, by changing the alignments during a given message, and thus varying the cipher alphabet, disks can provide polyalphabetic substitutions.

Alberti's Disk

Known as the Father of Western Cryptology, an Italian architect, Leon Battista Alberti took cryptology a significant step forward with the invention of his cipher disk. The Alberti disk was made from two copper plates, one large and one smaller inside ring. On the smaller, movable ring, he randomly placed letters. On the outer one, he placed 20 letters (omitting *H, J, K, U, W* and *Y*) and the numerals 1 to 4. The outer disk represented the plaintext, and the lowercase letters on the inner ring then became the ciphertext.

Alberti's Disk

Each correspondent had a matching disk and could change the alignments of the letters. By mutual agreement, the correspondents would know how to align the circles and at what point in the missive the circles might be changed. Each position shift initiated a new cipher alphabet, creating the possibility of prearranged multiple alphabets and meanings. Because so many alphabet possibilities existed, Alberti is credited with making the first polyalphabetic cipher.

As seen in the illustration, an example indicator *d* is aligned with the numeral 1. Then the letters *A, C* and *E* are equated with *z, n* and *o*. When the cryptographer shifts the movable inner disk slightly ot the right, *A, C* and *E* can be aligned with *a, e* and *b*.

Alberti's disk also incorporated another major contribution to cryptology, enciphered code. With the numbers 1 to 4 on the outer circle, he formulated a system using groups of numbers such as 11, 222, 3333 and so forth in two-to-four-digit combinations. These numerals from 11 to 4444 formed a table of codegroups. For example, 11 could have meant "Buy cloth" or 222 could have signified "Send supplies today."

When these numbers were placed in the missive and concealed, as were the message's letters during shifting, the digits were given an extra layer of concealment and thus became enciphered code. Thus 222 could have been masked by *ohx* at one time and *ivr* at another.

Some 20th-century scholars believe that Alberti used the digits 1 to 4 with an indicator letter to show alphabet changes within the message's contents. By prearrangment, any of these four numerals in the cryptogram could signal a shift to a new cipher alphabet. This was the first example in which a covert key was communicated within the message's own structure.

Deciphering tips: In addition to having a matching disk, you must be informed about an indicator letter such as *d* in the above example. Whenever you find a numeral in the decrypted plaintext, turn the movable disk and align that numeral with the planned indicator letter. This then places the next new group of cryptotext characters in alignment with the next new plaintext letter series. This pattern of turns and indicator/digit realignment continues until all the plaintext is revealed.

Porta's Disk

A second form of disk was created by the Neopolitan Giovanni Porta, who lived between 1535 and 1615. In his version, special symbols on the inner circle served as the ciphertext. The outer disk contained the plaintext letters in alphabetical order, as well as Roman numbers in standard sequence. The designs on the inner disk were brought into alignment with the plaintext originals to create the cipher. This was significant in that it expanded the ways that letters could be concealed and thereby complicated the cryptanalytic process for interceptors.

Deciphering tips: You must be informed about the starting point of the symbols and alphabet matchup. Turn the inner disk to align the successive symbols with their plaintext

Porta's Disk

equivalents. If the symbols are intended to equate with the Roman numerals, this should also be prearranged.

TABLEAUX

Trithemius's Tabula Recta

Known today as tables, tableaux were lists of multiple alphabets used for polyalphabetic encryption. The first such collection is credited to a German abbot named Johannes Trithemius during the early 16th century. In the fifth volume of his *Polygraphiae libri sex* (Six Books of Polygraphy), he proposed a *tabula recta*, the first tableaux in cryptographic writings, which would become the foundation of polyalphabetic substitution.

This table was one of the early geometric patterns used to arrange alphabets, numbers, and symbols to make an encryption, and was a major step forward in that it presented all the cipher alphabets in his system for viewing at one time. Trithemius called his method a "square table" because it consisted of 24 alphabet letters arranged in squares containing 24 rows. A partial table illustration is shown below:

```
a b c d e f g h i k l m n o p q r s t u w x y z

b c d e f g h i k l m n o p q r s t u w x y z a

c d e f g h i k l m n o p q r s t u w x y z a b

d e f g h i k l m n o p q r s t u w x y z a b c

e f g h i k l m n o p q r s t u w x y z a b c d

f g h i k l m n o p q r s t u w x y z a b c d e

g h i k l m n o p q r s t u w x y z a b c d e f

h i k l m n o p q r s t u w x y z a b c d e f g

i k l m n o p q r s t u w x y z a b c d e f g h

. . . . . . . . . . . . . . . . . . . . . . . .

w x y z a b c d e f g h i k l m n o p q r s t u
```

The table is created by transposing the standard alphabet, moving one place to the left with each succeeding row. The letters *i* and *j* were considered equivalent, as were *u* and *v*. This procedure can rightly be called the original *progressive key*, whereby every alphabet is used before any repetition of an alphabet occurs.

The table can be thought of as a matrix of rows and columns. To encipher the missive "she has it," the first letter of the plaintext missive, *s*, is enciphered in row 1, in column *s*. *S* is thus enciphered by itself. The second letter, *h*, is enciphered by the second row alphabet, underneath column *h* in the first row. *H* is therefore enciphered by *i*. The third letter, *e*, is enciphered by the third-row alphabet in column *e*, becoming *g*. This process is continued until the entire message is encrypted, with *t* covered by *b* in the eighth-row alphabet. Thus the message "she has it" is encrypted as follows:

PLAINTEXT:	s	h	e	h	a	s	i	t
CIPHERTEXT:	s	i	g	l	e	z	p	b

For messages of more than 24 letters, the enciphering begins again with the 25th letter enciphered by the first row, the 26th by the second row and so on until the entire plaintext is hidden. Even from this brief example, the advantage of polyalphabetic substitution can be seen: it renders messages more difficult to decrypt by decreasing the level of letter repetition from one communication to the next.

Such polyalphabetic encryption avoids the weakness of using repeating cipher letters to conceal repeating plaintext ones, such as in the message: "she will arrive at noon."

Deciphering tips: Using a matching table find the first ciphertext letter *s* in the first row which still serves as the plaintext. The second ciphertext letter is located in the second row and traced up to the letter on top of its given column. This letter is the plaintext. Every successive cipher letter *i, g, l* and so on is located in the succeeding rows and traced up to the letter on top of its given column. For example, *g* in row 3 is traced up to *e*; and *l* in row 4 up to *h* until *she has it* is recovered.

Vigenère's Table

In the 16th century French cryptographic scholar Blaise de Vigenère developed a somewhat better tableau than that of Trithemius. Vigenère proposed mixed alphabets on the side and top of his table as coordinates. He also advised shuffling the alphabets with keywords and keyphrases, parts of verses and dates.

This advice was not heeded by subsequent practitioners, who failed to mix the alphabets with longer phrases or date varieties, and the Vigenère method came to be accepted in a simpler form. It had standard alphabets along the top and left side of the table, and used a short repeating keyword. His name was mistakenly linked with this weaker version, which was not his original proposal.

	a	b	c	d	e	f	g	h	i	j	k	l	m	n	o	p	q	r	s	t	u	v	w	x	y	z
A	A	B	C	D	E	F	G	H	I	J	K	L	M	N	O	P	Q	R	S	T	U	V	W	X	Y	Z
B	B	C	D	E	F	G	H	I	J	K	L	M	N	O	P	Q	R	S	T	U	V	W	X	Y	Z	A
C	C	D	E	F	G	H	I	J	K	L	M	N	O	P	Q	R	S	T	U	V	W	X	Y	Z	A	B
D	D	E	F	G	H	I	J	K	L	M	N	O	P	Q	R	S	T	U	V	W	X	Y	Z	A	B	C
E	E	F	G	H	I	J	K	L	M	N	O	P	Q	R	S	T	U	V	W	X	Y	Z	A	B	C	D
F	F	G	H	I	J	K	L	M	N	O	P	Q	R	S	T	U	V	W	X	Y	Z	A	B	C	D	E
G	G	H	I	J	K	L	M	N	O	P	Q	R	S	T	U	V	W	X	Y	Z	A	B	C	D	E	F
H	H	I	J	K	L	M	N	O	P	Q	R	S	T	U	V	W	X	Y	Z	A	B	C	D	E	F	G
I	I	J	K	L	M	N	O	P	Q	R	S	T	U	V	W	X	Y	Z	A	B	C	D	E	F	G	H
J	J	K	L	M	N	O	P	Q	R	S	T	U	V	W	X	Y	Z	A	B	C	D	E	F	G	H	I
K	K	L	M	N	O	P	Q	R	S	T	U	V	W	X	Y	Z	A	B	C	D	E	F	G	H	I	J
L	L	M	N	O	P	Q	R	S	T	U	V	W	X	Y	Z	A	B	C	D	E	F	G	H	I	J	K
M	M	N	O	P	Q	R	S	T	U	V	W	X	Y	Z	A	B	C	D	E	F	G	H	I	J	K	L
N	N	O	P	Q	R	S	T	U	V	W	X	Y	Z	A	B	C	D	E	F	G	H	I	J	K	O	P
O	O	P	Q	R	S	T	U	V	W	X	Y	Z	A	B	C	D	E	F	G	H	I	J	K	L	M	N
P	P	Q	R	S	T	U	V	W	X	Y	Z	A	B	C	D	E	F	G	H	I	J	K	L	M	N	O
Q	Q	R	S	T	U	V	W	X	Y	Z	A	B	C	D	E	F	G	H	I	J	K	L	M	N	O	P
R	R	S	T	U	V	W	X	Y	Z	A	B	C	D	E	F	G	H	I	J	K	L	M	N	O	P	Q
S	S	T	U	V	W	X	Y	Z	A	B	C	D	E	F	G	H	I	J	K	L	M	N	O	P	Q	R
T	T	U	V	W	X	Y	Z	A	B	C	D	E	F	G	H	I	J	K	L	M	N	O	P	Q	R	S
U	U	V	W	X	Y	Z	A	B	C	D	E	F	G	H	I	J	K	L	M	N	O	P	Q	R	S	T
V	V	W	X	Y	Z	A	B	C	D	E	F	G	H	I	J	K	L	M	N	O	P	Q	R	S	T	U
W	W	X	Y	Z	A	B	C	D	E	F	G	H	I	J	K	L	M	N	O	P	Q	R	S	T	U	V
X	X	Y	Z	A	B	C	D	E	F	G	H	I	J	K	L	M	N	O	P	Q	R	S	T	U	V	W
Y	Y	Z	A	B	C	D	E	F	G	H	I	J	K	L	M	N	O	P	Q	R	S	T	U	V	W	X
Z	Z	A	B	C	D	E	F	G	H	I	J	K	L	M	N	O	P	Q	R	S	T	U	V	W	X	Y

In this system, with a short keyword such as *sign*, the encipherer conceals the orders "receive the emissary today" like this:

KEY:		S	I	G	N	S	I	G	N	S	I
	G	N	S	I	G	N	S	I	G	N	S	I	G

PLAINTEXT:		r	e	c	e	i	v	e	t	h	e
	e	m	i	s	s	a	r	y	t	o	d	a	y

CIPHERTEXT:		J	M	I	R	A	D	K	G	Z	M
	K	Z	A	A	Y	N	J	G	Z	B	V	I	E

This ciphertext is found by using the table above. Each of the 26 cipher alphabet rows is positioned one letter to the left of the one above it. The plaintext characters are at the top, and the key alphabet is at the side. From the starting position of each, the cryptographer moves down and across to a coordinating point. For key *S* and plaintext *r* the cipher letter is *J*, found at the intersection of row *s* and column *r*. For key *I* and plaintext *e*, the cipher letter is *M*. This procedure continues in the same manner until all the plain letters are hidden.

This method was, surprisingly, considered *le chiffre indéchiffrable* (the indecipherable cipher) for generations. A version of it was found in the Beaufort cipher developed by Royal Navy Admiral Sir Francis Beaufort in 1857.

The cryptographic advantages of polyalphabetics and a progressive key contributed to the widespread acceptance of this method.

Deciphering tips: Using the same alphabet table and beginning with the first keyletter, *S*, move along the horizontal row beside it until you arrive at the first enciphered letter, *j*. Trace this column up to the plaintext letter on top of the column (*S* across to *J*, then up to *r*). Taking the next keyletter, *I*, move across the row until you find *m*, the next cipher letter. Moving up the column, the plaintext letter is *e*. Follow this procedure until the entire message is deciphered.

KEY PROGRESSION

Vigenère's table uses a key, the development of which was one of the fundamental advances in cryptologic history. A key contains instructions for arranging the steps of the general encryption system used, examples of which are cipher tables, codebooks and keywords or phrases. The Renaissance inventors, Giovan Belaso and Girolamo Cardano, along with Blaise de Vigenère, were influential in the key's development.

In 1553 Italian nobleman Giovan Belaso's little book *La cifra del Sig. Giovan Batista Belaso* (The cipher of Giovan Battista Belaso) developed the idea for an easy-to-change key for polyalphabetic substitution ciphers. He called the key a "countersign" and stated that it could be written with different languages as well as with differing numbers of words. Following is an example of Belaso's method. The plaintext words *the proud city* are given the countersign *roman glory*.

```
COUNTERSIGN: R   O   M   A   N   G   L   O   R   Y
R   O

PLAINTEXT:   t   h   e   p   r   o   u   d   c   i
t   y
```

The countersign is written above the plaintext, repeating until it reaches the end of the message. The keyletter thus paired signifies the letter from a chosen alphabet table that will encipher the plaintext letter. In the above example, *t* is enciphered by the table's *R* alphabet, *h* by its *O* alphabet, *e* by its *M* alphabet and so on.

Belaso's small booklet was a significant improvement in written security. It provided surprising variety, since the number of alphabet sequences was expanded considerably from the few elemental systems previously used. Different messages could be given different keys, and broken, forgotten or pilfered keys could be exchanged quickly for others. This advantage contributed to its rapid acceptance, representing another strong block in the edifice of cryptology. Belaso had taken a step toward the modern use of keys, in which several can be alternated during a communication.

Around 1550, Milanese physician and mathematician Giro-lamo Cardano originated the idea of the autokey, a key that was changed systematically with each message to provide greater security than would a key that was repeated. Even though cryptography included keys and multiple alphabets by the 1500s, the autokey was a real advance in form and substance. However, Cardano's example was flawed by a *repeating* key that began again with each new message word. Cardano used the message's first word as the key.

KEY:		F	I	N	D	F	I	N	F	I	N
D	F	I	F	I	F	I	N	D	F	I	N

PLAINTEXT:		F	i	n	d	h	e	r	n	a	m
e	o	r	a	n	a	d	d	r	e	s	s

For example, key *F* applied to plain *F* used cipher letter *a* to cover plain f. But keyletter *I*'s alphabet could also use cipher letter *a* to cover plain letter *i*. Keyletter *N*'s alphabet could use cipher letter *a* to conceal plain *n* or plain *r* as it is aligned in this example when key *N* repeats above *r*. The rapidly repeating keyletters used too many of the same ciphertext replacement alphabet letters for the same plain letters. This caused the recipient great confusion during decryption. This same weakness left the entire mask open to removal should a cipher solver happen to discover the pattern. Therefore, except for the autokey idea, which influenced the work of Blaise de Vigenère, Cardano's system was forgotten.

Born near Paris in 1523, Blaise de Vigenère also worked with keys and was particularly intrigued by the notion of encrypting with plaintext. In his *Traicte des chiffres* (1586), Vigenère introduced his ideas of priming and running keys.

In a running key the plaintext and the key are identical. The key is kept running, or continuous, rather than being repeated as in Cardano's system. An application of the running key may be seen in an example whereby the letters of the plaintext communication provide the letters for the key. A priming key was used as the first keyletter. This letter, or primer, indicated the first plaintext letter from which the addressee could begin to recover the missive's contents. A comparison is shown below:

Cardano

KEY:	the	theth	the	thet	theth
PLAINTEXT:	the	first	was	from	Milan

Vigenère (priming key Z)

KEY:	(Z)H	EWA	S	AFRENCHMA
PLAINTEXT:	he	was	a	Frenchman
CIPHERTEXT:	vp	jbx	u	loctdvgkq

Vigenère's plaintext letters were used as the key letters but were kept running and not used again. Additionally, Vigenère's priming key (Z) was known by the communicants.

Vigenère proposed a second autokey idea which also had a priming key. But this time the ciphertext became the key to encrypt the body of the message. The *Traicte des Chiffres* and these keys were Vigenère's primary contributions to message concealment. Despite their ingenuity, the primer and the autokeys were forgotten until the latter 1800s, when scholars seeking better concealment processes independently discovered the value of Vigenère's ideas.

Deciphering tips: Using the priming key letter *Z*, and Vigenère's table you know that the first ciphertext letter is *g*. Locate *g*'s plaintext equivalent, *h*, among the alphabet equivalents. With *h* then becoming a keyletter, check the equivalents again and find cipher *l* whose plaintext is *e*. Then *e* becomes the keyletter for ciphertext *a*, whose plaintext is *w*. The remainder of the communiqué is revealed using this same pattern.

Marie Antoinette's Cipher

In the early 1780s, Marie Antoinette and her paramour, the Swedish Count Axel Fersen, applied keywords in a method using lists of capital and lowercase letters. Fersen provided a concealment system for a number of these letters; it was a polyalphabetic substitution with keywords. The keywords were located among the capital letters and aligned in columns. According to a prearranged plan, the plaintext and ciphertext letters were placed in pairs horizontally alongside the capitals in an arrangement similar to this:

M	kn	or	sv	wz	ad	eh	il	mp	qt	ux	yb	cf	gj
N	ab	cd	ef	gh	ij	kl	mn	op	qr	st	uv	wx	yz
O	pr	su	vx	ya	bd	eg	hj	km	np	qs	tu	wy	zb

Suppose the key *M* (for Marie) is chosen to hide the word *say*. According to the pairs, *s* = *v*, *a* = *d* and *y* = *b*. *Say* becomes *vdb*. If key *N* is used, *s* is replaced by *t*, *a* by *b* and *y* by *z*. With this key, *say* is *tbz*. Each key would thus provide a different encipherment pattern.

Deciphering tips: Using the cipher and the prearranged keyword(s), find the first keyword letter *M*. Locate the letter pair alongside it that contains *v*, the ciphertext letter. Beside *v* is its plaintext equivalent, *s*. With keyword letter *N*, the cipher letter *t* is found with its paired plaintext letter *s*.

VARIATIONS OF VIGENÈRE'S TABLE

Gronsfeld Cipher

The Gronsfeld cipher was an abbreviated version of the series of alphabets closely associated with Blaise de Vigenère. The method was named after the count of Gronsfeld, an obscure 17th century scholar, who described it to author Gaspar Schott. In 1659 the system was recorded for history in Schott's writings. The cipher entered public awareness in France in 1892 when it was used by a group of anarchists. It was solved by eminent cryptanalyst Etienne Bazeries, and the plotters were arrested.

The Gronsfeld cipher used digits as a keynumber to initiate an encipherment, with the digits indicating a displacement in a standard alphabet from a given plaintext letter to its ciphertext equivalent. For example, the number 8 above an *a* would be an instruction to count eight letters from plaintext *a* to get *i*, its cipher replacement. The keynumber, in this case 5316280794 would be repeated across the entire message. With a chosen keynumber, a multialphabet encipherment can be made (0 involving no change).

KEYNUMBER:	5	3	1	6	2	8	0	7	9	4
PLAINTEXT:	a	n	a	r	c	h	i	s	t	s
CIPHERTEXT:	f	q	b	x	e	p	i	z	c	w

KEYNUMBER:	5	3	1	6	2
PLAINTEXT:	a	r	i	s	e
CIPHERTEXT:	f	u	j	y	g

CIPHERTEXT (FIVE-LETTER GROUPS): fqbxe   pizcw   fujyg

The Gronsfeld method is not dependent on a specific alphabet to accomplish encryption or decryption. Furthermore, readily accessible numbers such as those of a telephone directory can be used as keys. The sender and addressee could use a phone number (at a certain page, column and row) to encipher a given transmission.

Deciphering tips: Using the keynumber, write each number above the ciphertext. This reveals the proper number displacement. Beginning with the first ciphertext character, count back to the point in the alphabet that is equal to the displacement number. For example, for ciphertext *f*, count back five places to *a*. Ciphertext *q* counts back three places to plaintext *n*. The same reverse counting process continues through the remainder of the ciphertext.

Beaufort Cipher

The Beaufort cipher is very similar to Vigenère's table. It was developed by Royal Navy Admiral Sir Francis Beaufort, who becane famous when he developed a numerical indicator of wind speeds.

His cipher, presented to the English public in 1857, took the form of a four-by-five-inch card with a red-and-black printed alphabet square and was advertised as a new method for veiling messages dispatched on postcards and telegrams.

Beaufort created the cipher by arranging the alphabet on four sides of the square, with 27 letters down, 27 across, and *A* at each corner. Its key was derived from a name, place or poetic phrase familiar to the user. The example below shows how to encipher the message:

1. Find the first letter of the message text in the side column (*C*).

2. From this letter, trace horizontally across the table until finding the first letter of the key (*P*).

3. At the top of the column containing *P* will be located the letter *N*, which will become the ciphertext letter.

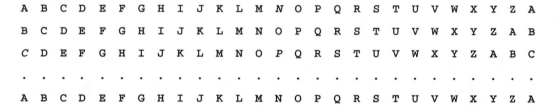

This form is also called "true Beaufort" to distinguish it from a slightly different technique called "variant Beaufort." In the variant form the cipher maker begins with the keyletter, uses it to locate the plaintext, and then, in its column, finds the ciphertext.

With the Beaufort name attached to them, both types of cipher gained respected places in the history of cryptography, although they did not really advance its progress significantly.

Deciphering tips: To recover the plaintext, the receiver follows the same process, this time moving from ciphertext to keyletter and back.

POLYGRAPHIC CIPHERS

Polygraphic concealment encrypts two or more characters at one time. In this system, when a matrix of letters is enciphered, the matrix is considered as a unit. If any part of this unit is altered, the whole encryption is changed.

Porta's Digraphic Cipher

Porta's cipher, which he presented in a table form, is the first known digraphic cipher. His table used a 20-by-20 matrix, with a standard alphabet (minus *J*, *K*, *U*, *W*, *X* and *Z*) running across the top and down the right side of the matrix. The grid could be filled in with any substitute, numbers, letters or symbols—Porta himself used symbols—as long as no one cell is duplicated. In a numeric 26-by-26 matrix, squares would be numbered from 1 to 676, each of which is considered a single unit. In the partial example shown below, numbers fill the rows in a reverse alternating direction:

	A	B	C	D	E	F	G	H	I	J	K	L	M	N	O	P	Q	R	S	T	U	V	W	X	Y	Z
A	1	2	3	4	5	6	7	8	9	10	11	12	13	14	15	16	17	18	19	20	21	22	23	24	25	26
B	52	51	50	49	48	47	46	45	44	43	42	41	41	39	38	37	36	35	34	33	32	31	31	29	28	27
C	53	54	55	56	57	58	59	60	61	62	63	64	65	66	6	68	69	70	71	72	73	74	75	76	77	78
D	104	103	102	101	100	99	98	97	96	95	94	93	92	91	90	89	88	87	86	85	84	83	82	81	80	79
.	.	.	.	.	.	.	.	.	.	.	.	.	.	.	.	.	.	.	.	.	.	.	.	.	.	.
T	495	496	497	498	499	500	501	502	503	504	505	506	507	508	509	510	511	512	513	514	515	516	517	518	519	520

To encrypt the first two characters of the missive "act" find the *A* in the vertical alphabet and the *C* in the horizontal alphabet. Their intersecting point is 3, which becomes the ciphertet which represents both letters. For messages with an odd letter total, a null is used to represent it. The recipient would know that the second letter of the pair is meaningless.

When sending the numbers as a dispatch, the communicants would have to prearrange the points where the single, paired or triple numerals would be separated. One plan would be to have particular triple digit nulls such as 000 or 999 which would signify concluding and beginning points.

Deciphering tips: Using the table, find each ciphertext letter group in turn. Trace its vertical and horizontal coordinates to locate the first and second letter.

Playfair Cipher

A primary example of a polygraphic cipher is the digraphic form Playfair cipher, in which digraphs (pairs of letters) were encrypted in a five-by-five square with their arrangement set by a variable keyword.

Invented by Charles Wheatstone in 1854, this manual polygraphic cryptomethod was the first literal digraphic cipher (Porta's digraphic table used symbols rather than letters.) Developed for telegraph secrecy, the Playfair cipher served British forces in the Boer War and World War I. It was also used by the Islands Coastwatching Service in Australia during World War II. With this system and a keyword transposition, Wheatstone was able to produce a well-shuffled alphabet which was very practical for combat use as it required no charts, lists or mechanical tools, and its special digraphic feature made cryptanalysis difficult.

The system uses letter position relationships in a five-by-five alphabet matrix. A keyword containing no repeating letters is written horizontally from left to right, and sets the pattern of the letters. The unused letters are entered in the remaining cells of the matrix in alphabetical order, with *i* and *j* combined in one cell. In this example Wheatstone's first name, *Charles*, will stand as the keyword.

c	h	a	r	l
e	s	b	d	f
g	ij	k	m	n
o	p	q	t	u
v	w	x	y	z

The message to be enciphered is "the scheme work*s*." To begin enciphering, the plaintext is first divided into two-letter groups. If double letters occur in one of the pairs, an *x* is used to separate them. For example, *will arrive* becomes *wi lx la rx ri ve*. An *x* is also used to complete a pair in words containing odd numbers of letters. The word *reply* would be *re pl yx*. In this case, the phrase ready for encipherment becomes:

th es ch em ew or ks

A given message can be enciphered according to a set pattern every time. The letters in each pair have one of three possible relationships with each other in the matrix; they can be in the same column, in the same row, or neither in the same column nor the same row. The corresponding rules of replacement are as follows:

1. When two letters in a pair are in the same row of the matrix, the one to the right of the message letter is its cipher equivalent. In this example, *m* becomes *n* and *v* is replaced by *w*. The rows are cyclical, with the last letter in a row being replaced by the first letter of the same row. For example, *l* becomes *c*.

2. If two letters are in the same column, the letter below each plaintext letter is its ciphertext. In this case, *h* becomes *s*. The columns are also cyclical; if the letter is at the bottom of a column, it is exchanged for the top letter of the same column. Thus, *y* becomes *r*.

3. If the letter pairs are in neither the same column nor row, each is exchanged with the letter at the intersection of its own row and the other's column. A type of imaginary rectangle exists, with the unseen corners of the rectangle determined by the plaintext pair. For example, the encipherer finds the first message letter in the square. He or she then moves across the row in which it is located until he or she meets the column where the second plaintext letter lies. The letter of the square at the intersection of column and row becomes the first cipher letter. To find the second cipher letter, the encipherer finds the second message letter and moves across the row from it until intersecting the column where the first plaintext letter lies. This aspect of the method is known as "diagonal diagramming." For consistency, the first plaintext letter is always put in the same row as the first letter in the ciphertext pair, so that $xi = wk$ (not kw), $mh = ir$ (not ri) and $zc = vl$ (not lv).

PLAINTEXT: th es ch em ew or ks

CIPHERTEXT: pr sb ha dg sv tc ib

Deciphering tips: Using the keyword, *charles*, place the keyword in a five-by-five matrix and fill in the unused cells with the remaining alphabet in sequence. Decipherment works as follows:

1. If both letters are in the same column, the letter above each ciphertext letter is the plaintext.

2. If both letters are in the same row, the letter to the left of each ciphertext letter is the plaintext.

3. Both columns and rows are cyclical.

4. Letters in neither the same column nor row are simply reversed—if *th* is *pr*, then *pr* becomes plaintext *th*.

Two-Square Cipher

Another type of digraphic cipher is the two-square cipher. While its exact origin is uncertain, it may have been derived from cryptographers' combinations of transposition and substitution arrays.

In this example, the keywords *democrat* and *republican* mix the remaining alphabet letters in two arrays.

d	e	m	o	c
r	a	t	b	f
g	h	i	k	l
n	p	q	s	u
v	w	x	y	z

r	e	p	u	b
l	i	c	a	n
d	f	g	h	k
m	o	q	s	t
v	w	x	y	z

Encipherment is accomplished by finding the first letter of the message in the left matrix(1) and the second plaintext letter in the right matrix (2). This sets the corners of an imaginary rectangle. Their ciphertext equivalents are the characters at the corners of the other two sides of the rectangle, along the opposite diagonal from each other. If the plaintext letters are both in the same row, they are reversed to make their cipher equivalents (for example, *ra* = *ar* and *no* = *on*).

The request "enter pin record" begins with plain *e* across to *n* on a slight diagonal. They are covered by *ba* on the reverse diagonal, in this case *b* above *n* and *a* beneath *e*, to form the envisioned rectangle.

Then *te* (plain) = *im* (cipher), *rp* (next plain pair) = *cd*, *in* = *kt* and so forth. This 14–character message has seven pairs of letters. To be sent as five-character groups, a null such as the infrequent *q* or *x* would be required to make the groups of five.

Deciphering tips: Decipherment is the reverse with the first cipher letter *b* being located in the right matrix (2) and its partner *a* in the left matrix (1). Again, an opposite diagonal is drawn to locate the first letter, *e*, in square 1 and the second letter, *n*, in square 2. This process is continued to uncover the plaintext.

Four-Square Cipher

This is an extension of the two-square digraphic method.

1.

a	b	c	d	e
f	g	h	i	k
l	m	n	o	p
q	r	s	t	u
v	w	x	y	z

3.

r	e	p	u	b
l	i	c	a	n
d	f	g	h	k
m	o	q	s	t
v	w	x	y	z

4.

d	e	m	o	c
r	a	t	b	f
h	h	i	k	l
n	p	q	s	u
v	w	x	y	z

2.

a	b	c	d	e
f	g	h	i	k
l	m	n	o	p
q	r	s	t	u
v	w	x	y	z

In this version, the plaintext letters are located in grids 1 and 2 and their concealments are found in 3 and 4. Again, one must envision imaginary rectangles with opposing diagonals. For example, $dm = co$, $ic = tu$ and $ux = qz$. That is, n from grid 1 and a from grid 2 are replaced by d in grid 3 (in the column with and above a) and by m in grid 4 (in a's row and in n's column, beneath n). Plain i and c are concealed by c (grid 3) and o (grid 4). Here c is in i's row and c's column while o is in c's row and i's column. By the same process, plain u and x are concealed by q and z. Thus the message "enter pin record" would be encrypted as follows:

PLAINTEXT: en te rp in re co rd

CIPHERTEXT: pl to th ck te ui se

Deciphering tips: The process is the same as with the two-square style. For example, with *p* found in grid 3 and *i* in grid 4, plaintext *e* in grid 1 is in *m*'s row and *k*'s column, and plaintext *n* in grid 2 is in *i*'s row and *p*'s column.

Trigraphic Ciphers

Trigraphic encipherment is also possible, with very large tables to enable the $26^3 = 17,576$ trigraphs from *aaa* to *zzz* to be encrypted. Because of the vast number of trigraphs, these methods are impractical for professional and amateur use.

ALGEBRAIC CIPHERS

The first practical method linking algebra and secret writing was developed by mathematician Lester Hill in 1929. A professor at Hunter College in New York City, his article "Cryptography in an Algebraic Alphabet" was published in the June/July issue of the *American Mathematical Monthly.* Hill's system gave numerical values to keys and plaintext letters. It was quite different from other concealment styles, in that there were an equal number of letters and equations in a form called polygraphic. Encryption resulted from solving the equations.

He used the numbers 0 to 25 in a system called *"modulo 26",* which meant that any digit above 25 had to be reduced by deleting multiples of 26 from it. The remainder was equal to the number, modulo 26. For example, 28 is 2 modulo 26, since 28 minus 26 is 2. Likewise, 88 is 10 modulo 26 because there are three multiples of 26 for a total of 78 in 88; then 88 minus 78 is 10. The number 110 is 6 modulo 26. There are four multiples of 26 (equaling 104) in 110, and 110 minus 104 is 6.

Hill made a substitution using a set of four equations with x representing plaintext letters: x_1 for the first plain letter, x_2 for the second, x_3 the third and x_4 the fourth. The y's indicate numerical equivalents of ciphertext letters.

$$y_1 = (8 \times x_1) + (6 \times x_2) + (9 \times x_3) + (5 \times x_4)$$

$$y_2 = (6 \times x_1) + (9 \times x_2) + (5 \times x_3) + (10 \times x_4)$$

$$y_3 = (5 \times x_1) + (8 \times x_2) + (4 \times x_3) + (9 \times x_4)$$

$$y_4 = (10 \times x_1) + (6 \times x_2) + (11 \times x_3) + (4 \times x_4)$$

The first encipherment step involves converting a plaintext message into numbers using a randomly generated alphabet/number chart like the following:

a	b	c	d	e	f	g	h	i	j	k	l	m	n	o	p	q	r	s	t	u	v	w	x	y	z
2	24	11	18	3	16	9	5	17	23	21	8	15	4	1	14	19	7	12	20	6	10	25	13	22	0

For the plaintext message "dusk sign", the number values of the first four message letters are 18(d), 6(u), 12(s), and 21(k). These numbers are placed in the equations for x_1, x_2, x_3 and x_4.

$$y_1 = (8 \times 18) + (6 \times 6) + (9 \times 12) + (5 \times 21)$$

$$y_2 = (6 \times 18) + (9 \times 6) + (5 \times 12) + (10 \times 21)$$

$$y_3 = (5 \times 18) + (8 \times 6) + (4 \times 12) + (9 \times 21)$$

$$y_4 = (10 \times 18) + (6 \times 6) + (11 \times 12) + (4 \times 21)$$

The multiplications and additions within every equation are done in the modulo 26 style, as in the first multiplication for y_1: 8×18 (144) = 14 modulo 26.

$$y_1 = 14 + 10 + 4 + 1 = 29 \ (3 \ modulo \ 26)$$

$$y_2 = 4 + 2 + 8 + 2 = 16 \ (no \ multiple \ of \ 26 \ here)$$

$$y_3 = 12 + 22 + 22 + 7 = 63 \ (11 \ modulo \ 26)$$

$$y_4 = 24 + 10 + 2 + 6 = 42 \ (16 \ modulo \ 26)$$

Then the numbers 3, 16, 11 and 16 are put in literal form and become the cipher letters *e, f, c* and *f.* This process is repeated with the numerical equivalents for the second message word, *sign* (12, 17, 9, 4), plugged into the four equations. Converted into modulo 26, the equations are as follows:

$y_1 = 18 + 24 + 3 + 20 = 65 \,(13 \text{ modulo } 26)$

$y_2 = 20 + 23 + 19 + 14 = 76 \,(24 \text{ modulo } 26)$

$y_3 = 8 + 6 + 10 + 10 = 34 \,(8 \text{ modulo } 26)$

$y_4 = 16 + 24 + 21 + 16 = 77 \,(25 \text{ modulo } 26)$

In the alphabet/number chart, 13 is *x*, 24 is *b*, 8 is *l* and 25 is *w*. For *dusk sign*, then, the complete ciphertext is *efcfxblw*.

Deciphering tips: The receiver of the ciphertext decrypts it using four equations that complement the encrypting ones:

$x_1 = (23 \times y_1) + (20 \times y_2) + (5 \times y_3) + (1 \times y_4)$

$x_2 = (2 \times y_1) + (11 \times y_2) + (18 \times y_3) + (1 \times y_4)$

$x_3 = (2 \times y_1) + (20 \times y_2) + (6 \times y_3) + (25 \times y_4)$

$x_4 = (25 \times y_1) + (2 \times y_2) + (22 \times y_3) + (25 \times y_4)$

Using the alphabet/number chart, insert the numerical equivalents for the first four ciphertext letters, *efcf*, into the four equations:

$x_1 = (23 \times 3) + (20 \times 16) + (5 \times 11) + (1 \times 16)$

$x_1 = 69 + 320 + 55 + 16 = 460 \,(18 \text{ modulo } 26)$

In the alphabet/number chart 18 = *d*.

$x_2 = (2 \times 3) + (11 \times 16) + (18 \times 11) + (1 \times 16)$

$x_2 = 6 + 176 + 198 + 16 = 396 \,(6 \text{ modulo } 26)$

In the chart 6 = *u*.

$x_3 = (2 \times 3) + (20 \times 16) + (6 \times 11) + (25 \times 16)$

$x_3 = 6 + 320 + 66 + 400 = 792 \,(12 \text{ modulo } 26)$

12 is *s*.

$x_4 = (25 \times 3) + (2 \times 16) + (22 \times 11) + (25 \times 16)$

$x_4 = 75 + 32 + 242 + 400 = 749 \,(21 \text{ modulo } 26)$

21 is *k*. This completes the decryption of the first plaintext word, *dusk*. To recover the second plaintext word, its encrypting numerals are placed into the same formulas.

$x_1 = (23 \times 13) + (20 \times 14) + (5 \times 8) + (1 \times 25)$

$x_1 = 299 + 480 + 40 + 25 = 844$ (12 modulo 26)

12 is *s*.

$x_2 = (2 \times 13) + (11 \times 24) + (18 \times 8) + (1 \times 25)$

$x_2 = 26 + 264 + 144 + 25 = 459$ (17 modulo 26)

17 is *i*.

$x_3 = (2 \times 13) + (20 \times 24) + (6 \times 8) + (25 \times 25)$

$x_3 = 26 + 480 + 48 + 625 = 1,179$ (9 modulo 26)

9 is *g*.

$x_4 = (25 \times 13) + (2 \times 24) + (22 \times 8) + (25 \times 25)$

$x_4 = 325 + 48 + 176 + 625 = 1,174$ (4 modulo 26)

4 is *n*. This completes the recovery of the second plaintext word, *sign*.

Hill had another article published in the *American Mathematical Monthly* in 1931, in which he described the use of multiple matrices for encryption. These squares of numbers had their own procedures for addition and for multiplication. The matrix numbers indicated plaintext characters, and several letters could be included in a few matrices. This is possible because, mathematically speaking, a matrix can be considered a single number. Therefore, a few matrices can encrypt more letters with fewer equations overall.

This special type of multialphabet encipherment very securely hid letter pairs, repeating letters or words, and other factors that give solution seekers an angle. Yet, like his algebraic alphabet proposal, the matrices did not prove viable for general applications. Hill's ideas did see one limited use during the Second World War, when the U.S. government applied his mathematics to encipher letter groups of radio call-signs. In general, however, their complexity made solving them too time-consuming even for intended addressees. Consequently, they have been relegated to a niche as admirable but theoretical achievements in the annals of cryptology.

FRACTIONATING CIPHERS

The first fractionating cipher was developed in the 19th century by Pliny Earl Chase, a Harvard graduate and professor of philosophy at Haverford College near Philadelphia. In an article in the periodical *Mathematical Monthly* (March 1859) Chase presented the first known explanation of fractionating ciphers, which were similar to the German ADFGVX and the bifid cipher.

Other alphanumeric methods equated alphabet letters with single numbers, triplets, or combinations of zero and one(the Baudot code, for example). Some of these systems were merely letter-number lists. Others had rows and columns of numbers to create coordinates in a form often called "checkerboard." But these versions treated the digits as whole entities. Chase was the first to alter the top-and-side coordinates approach by separating, or "fractionating," the numbers.

Fractionating ciphers conceal plaintext with two or more letter or number equivalents that are then enciphered, often by transposition. The next stage is a recombination of new multiliterals, which are then replaced by a single encrypted letter. Chase's checkerboard arrangement had 10 columns of the letters *a* through *z* and three numbered rows, and included an ampersand and Greek letters.

	1	2	3	4	5	6	7	8	9	10
1	z	g	u	j	d	a	f	r	&	δ
2	o	c	s	m	q	v	x	b	p	θ
3	h	y	i	t	l	w	k	n	e	λ

To encipher plaintext, Chase would first find the coordinates on the checkerboard for each plaintext letter, beginning with the row number, then the column. Then the two numbers would be placed vertically beneath each letter. For example, to encipher the plaintext word *divide*, the follwing layout would be generated:

D	I	V	I	D	E
1	3	2	3	1	3
5	3	6	3	5	9

The next step was to multiply the lower line by nine and replace it with the product:

5	3	6	3	5	9	
					x 9	
4	8	2	7	2	3	1

This created a new number:

	1	3	2	3	1	3
4	8	2	7	2	3	1

Then Chase returned to his checkerboard's coordinates to change these numbers into letters again, reading them vertically as 4, then 18, 32, and so forth. On the checkerboard, $4 = j$, *m* or *t*; $18 = r$; $32 = y$; $27 = x$; $32 = y$; $13 = u$; and $31 = h$. Thus, the final encryption is *jryxyuh*.

Deciphering tips: Using Chase's special checkerboard, you know that the first letter, *j*, is equated with the numeral 4. Because of the particular nature of this method, you know that 4 will be a single number on the lower left side of the row of numerals. The digits at *r*'s column and row are 8 and 1. Place 8 below and 1 above. The same pattern follows, *y* being 2 and 3, *x* being 7 and 2 and so forth. The number alignments made by the sender are reproduced. You also know that the first row of numerals (132313) form the first row of the plaintext words' "fractionated" digits. The second row is found by dividing 4827231 by 9 to reverse the multiplication step. This results in the number 536359. These numerals are placed beneath the first row's digits to recreate the vertical coordinates 1 and 5, 3 and 3 and so forth. The letters at their row and column coordinates reveal the plaintext word, *divide*.

Aperiodic Fractionating Ciphers

It is possible, using a matrix and a keyword, to initiate an aperiodic fractionation process. A keyword such as *profitable* can generate a mixed alphabet. For a five-by-five square, *v* and *w* are combined in one cell.

	1	2	3	4	5
1	p	r	o	f	i
2	t	a	b	l	e
3	c	d	g	h	j
4	k	m	n	q	s
5	u	vw	x	y	z

As with Chase's fractionating cipher, each letter is enciphered in vertical dinomes (two number sets) beneath the plaintext. The dinomes are found at the horizontal and then the vertical coordinates of the message letters. The message "begin private audit," for example, is enciphered as follows:

PLAINTEXT: b e g i n p r i v a t e a u d i t

DINOMES: 2 2 3 1 4 1 1 1 5 2 2 2 2 5 3 1 2

3 5 3 5 3 1 2 5 2 2 1 5 2 1 2 5 1

The ciphertext is formed by grouping the digits in pairs beginning on the top left and continuing horizontally through the numbers to the lower right. This group of dinomes is then converted to cipher letters by using the matrix.

SECOND DINOME SEQUENCE:

22	31	41	11	52	22	25	31	23
a	c	k	p	v	a	e	c	b

53	53	12	52	21	52	12	51
x	x	r	v	t	v	r	u

For the first pair, locate the first pair, 22, in the matrix using the first digit as the horizontal coordinate and the second as the vertical coordinate. The resulting ciphertext is *ackp-vaecbxxrvtvru.*

Deciphering tips: Using the keyword, set up the matrix with the shuffled alphabet as shown above. Convert the ciphertext back into pairs of horizontal and vertical coordinates. Set up the coordinates horizontally into two equal rows. (At this point, the numerals for *1, 2* and *3* split to make 17 numbers per row.) Your decipherment should now look like the first dinome sequence. Use the first row as the horizontal coordinate and the second row as the vertical coordinate to locate each pair on the matrix.

Periodic Encipherment

Fractionating ciphers can be used for periodic encipherment, so-called because the method is applied to each five-letter group separately. The plaintext is placed in five-letter sets with dinomes below them. The dinomes are generated on a standard five-by-five matrix. The row numeral is on the top and the column number is below it. Using the previously mentioned keyword, *profitable,* the communiqué "secret account located" is encrypted as follows:

PLAINTEXT:	secret	account	located	
SETS OF 5:	secre	tacco	untlo	cated
DINOMES:	42312	2231	54221	32223
	55125	12113	13143	12152

Periodic ciphertext is constructed by horizontally reading pairs of digits within each set of five vertically arrayed dinomes, resulting, with the first set, in the series 42, 31, 25, 51, 25. The process continues with the pairs in the next set: 22, 33, 11, 21, 13; followed by 54, 22, 11, 31, 43; and 32, 22, 31, 21, 52.

Using the same technique as in the aperiodic fractionating ciphers, the new dinomes are converted into letters using the matrix:

DINOMES:	42	31	25	51	25
CIPHERTEXT:	m	c	e	u	e

The resulting ciphertext is *mceue agpto yapcn dactv.*

Deciphering tips: Set up the matrix with the keyword and shuffled alphabet. Convert the ciphertext back into pairs of coordinates. Divide the coordinates into groups of five pairs, and arrange each group in two horizontal rows. Use the first row as the horizontal coordinate and the second row as the vertical coordinate to locate each pair on the matrix.

Tripartite Fractionating Ciphers

Fractionation can also be accomplished with tripartite numerical replacements for the alphabet. The numbers 1, 2 and 3 are put in all of their 27 different possible combinations. To provide numeric replacements for 26 alphabet letters with 27 numeral combinations, the little-used letter *z* is designated as *za* when replaced by 332 and *zb* when replaced by 333. This helps avoid decryption confusion.

a	111	j	211	s	311
b	112	k	212	t	312
c	113	l	213	u	313
d	121	m	221	v	321
e	122	n	222	w	322
f	123	o	223	x	323
g	131	p	231	y	331
h	132	q	232	za	332
i	133	r	233	zb	333

Using, again, the message "secret account located", an encryption is made with trinomes (three-number sets) in columns below the letters.

PLAINTEXT:	s	e	c	r	e
TRINOMES:	3	1	1	2	1
	1	2	1	3	2
	1	2	3	3	2
	t	a	c	c	o
	3	1	1	1	2
	1	1	1	1	2
	2	1	3	3	3
	u	n	t	1	o
	3	2	3	2	2
	1	2	1	1	2
	3	2	2	3	3
	c	a	t	e	d
	1	1	3	1	1
	1	1	1	2	2
	3	1	2	2	1

Encryption is similar to that used with dinomes; the trinomes are read horizontally within each five-letter group—as in the case of the first group, 311, 211, 213, 212 and 332. Returning to the alphabet and number chart, these numbers are replaced by the letters s, j, l, k and za. The rest of the cipher-text is created in the same way.

CIPHERTEXT: sjlkza sdamzb xmjqr   cabpm

Deciphering tips: The addressee will know the special arrangement for za and zb and deciphers accordingly. Using the chart of alphabet/number equivalents, place the numeric equivalents for s, j, l, k and za (311, 211, 213, 212 and 332) in a group of three numbers by five numbers beginning at the top left corner with 31121 and bringing the remaining 1 down to the next row, then 213, 212 and 332 are aligned the same way.

31121	31112	32322	11311
12132	11112	12112	11122
12332	21333	32233	31221

Reading from the top to the bottom of each column beginning on the left, 311 = *s*, 122 = *e*, 113 = *c*, 233 = *r* and 122 = *e*. *T* is recovered from the next number group, 312 = *t*, 111 = *a*, 113 = *c* and so on. The remaining letters of the plaintext are recovered in this way to reveal the plaintext *secret account located*.

Bifid Cipher

Variations on the fractionating cipher, the bifid and the trifid cipher, were created in the late 19th century by Frenchman Felix Delastelle. Delastelle was a bachelor who worked for 40 years as a government tobacco inspector. When he retired, he began writing *Traité Élémentaire de Cryptographie* (Elementary Treatise on Cryptography), in which he categorized and explained the principles of important crytographic methods, most of which were types of ciphers.

Delastelle's study led him to develop a type of fractionating cipher called the "Bifid," using a five-by-five grid. The grid below is filled in with a randomly shuffled alphabet (the letters *i* and *j* are considered interchangeable):

	1	2	3	4	5
1	d	v	m	q	a
2	l	h	y	r	u
3	o	f	p	n	c
4	z	k	w	b	s
5	x	e	t	i	g

To send the plaintext message "order three boxes", write its letters in groups of five. For each message letter, find its dinomial coordinates (e.g., *o* = 31). Place these numbers vertically below each letter.

PLAINTEXT:	order	three	boxes
DINOMES:	32152	52255	43554
	14124	32422	41125

Encipherment is accomplished by combining the coordinated numbers horizontally and in pairs within each group of five dinomes. The first group thus becomes 32, 15, 21, 41, 24—moving across the top row and then the bottom row.

These new dinomes are then reconverted into the letter that is at their coordinates in the grid (32 = *f*, 15 = *a* and so forth) The ciphertext for the original message becomes

`falzr    eutrh    wgbdu`

Deciphering tips: Set up the matrix and shuffled alphabet. Convert the ciphertext back into pairs of coordinates. Divide the coordinates into groups of five pairs, and arrange each group in two horizontal rows. Use the first row as the horizontal coordinate and the second row as the vertical coordinate to locate each pair on the matrix.

ADFGX

The ADFGX, and its more complicated version, the ADFGVX, are two of the best-known field ciphers in the history of cryptography. The ADFG was firt heard on the airwaves during a crucial stages of World War I when the Kaiser's generals were initiating a major German offensive in March 1918.

The ADFG(V)X was created by Colonel Fritz Nebel, a radio staff officer. His cryptographic plan can be described by the following stages. As with some other fractionating forms, a matrix helped encipher plaintext letters or numbers with letter pairs. The pairs were read from the side, then the top (e.g. *XG* enciphers *q*).

	A	D	F	G	X
A	l	r	m	e	ij
D	k	f	v	w	t
F	c	s	a	u	z
G	h	x	g	y	n
X	b	p	o	q	d

Using the sample matrix, the order "advance to the front at the marne with all battalions" was enciphered as follows:

```
FF XX DF FF GX FA AG DX XF DX GA AG
a  d  v  a  n  c  e  t  o  t  h  e
```

```
DD AD XF GX DX FF DX DX GA AG AF FF AD GX AG
f  r  o  n  t  a  t  t  h  e  m  a  r  n  e
```

```
DG AX DX GA FF AA AA XA FF DX DX FF AA AX XF GX FD
w  i  t  h  a  l  l  b  a  t  t  a  l  i  o  n  s
```

The next step involves transposition. The encryption is placed in horizontal rows across a preset number of columns (20 here). The letters are then split and placed one to a column left to right until they have all been placed (in this case 88 letters from 44 pairs). After the rows are completed their columns are numbered randomly. The order of the numerals becomes a transposition key.

14	8	3	15	20	1	7	2	17	9	11	19	13	6	4	16	18	10	12	5
F	F	X	X	D	F	F	F	G	X	F	A	A	G	D	X	X	F	D	X
G	A	A	G	D	D	A	D	X	F	G	X	D	X	F	F	D	X	D	X
G	A	A	G	A	F	F	F	A	D	G	X	A	G	D	G	A	X	D	X
G	A	F	F	A	A	A	A	X	A	F	F	D	X	D	X	F	F	A	A
A	X	X	F	G	X	F	D												

The third enciphering stage involves transcribing the letters from the columns in numerical order, reading vertically downward and combining them in five-letter groups. However, column 4 requires a fifth letter. The cryptographer takes the first letter *X* of the next sequentially numbered column (5) and picks up the first two letters, *G* and *X* from column 6. By the time column 9 is reached, it is comprised of *A* and all four letters of column 10. Column 10 is then built from the letters of column 11 plus one from 12, and so on. This shifting effect added to the number groups makes a very good cover, as can be seen here:

```
fdfax     fdfad     xaafx     dfddx     xxagx
gxfaf     affaa     axxfd     afxxf     fggfd
ddaad     adfgg     gaxgg     ffxfg     xgxax
xdafa     xxfdd     aag
```

Deciphering tips: Decryption is accomplished with knowledge of the key and possession of the original matrix. Count the ciphertext letters and divide by the total key numerals (20 in this case). This will tell you the depth of the columns. In this case, 88 letters divided by 20 gives four with 8 remaining, so the 20 columns are 4 letters deep and the remaining 8 fill in an incomplete row beginning at the base of the far left column.

Transcibe the letters in pairs beginning across the top row, *FF*, *XX*, *DF*, *EF* and so on. By checking the matrix, you will see the letters *a*, *d*, *v*, *a* for the first message word *advance* appearing at the coordinating points of the cipher letters.

ADFGVX

On June 1, 1918 French radio interceptors, Painvin and his fellow cryptanalysts received another shock when they learned that the ADFGX dispatches had been given a sixth letter. Painvin's successses with the ADFGX gave the cryptanalysts important clues as they uncovered the new variation.

Painvin solved enough ADFGVX dispatches to help the Allies stop the last assault of the German's spring offensive, a major attack on June 9 that threatened Paris itself.

TheADFGVX had a six-by-six grid with 26 letters and 10 numerals. were placed in this grid, with the coordinates being *A, D, F, G, V* and *X*.

	A	D	F	G	V	X
A	F	L	1	A	O	2
D	J	D	W	3	G	U
F	C	I	Y	B	4	P
G	R	5	Q	8	V	E
V	6	K	7	Z	M	X
X	S	N	H	Ø	T	9

In the first enciphering step, this configuration is used to sub-stitute letter pairs for each letter of a message, for example "all quiet on this front today." The plaintext letter is found in the matrix, and the letters that are at its coordinates become the pair that replace it, for example, *l* is replaced by *A* and *D*. These pairs always begin with the one in the vertical column at the left, followed by the one in the horizontal row at the top:

a	l	l	q	u	i	e	t	o	n	t	h	i
s	f	r	o	n	t	t	o	d	a	y		

AG	AD	AD	GF	DX	FD	GX	XV	AV	XD	XV	XF	FD
XA	AA	GA	AV	XD	XV	XV	AV	DD	AG	FF		

This results in a bilateral cipher. The second stage involves transposition through superenciphering. The first cipher is placed in a second rectangle, or a transposition matrix, with the encrypted letters written left to right in successive rows according to a keyword, in this case the word *German*, which indicates the number of columns.

TRANSPOSITION MATRIX:

G	E	R	M	A	N
3	2	6	4	1	5
A	G	A	D	A	D
G	F	D	X	F	D
G	X	X	V	A	V
X	D	X	V	X	F
F	D	X	A	A	A
G	A	A	V	X	D
X	V	X	V	A	V
D	D	A	G	F	F

The keyword, and the position of its letters in the alphabet, also indicate the numerical order in which the letters are to be read down each column. For example, if the keyword *German* is arranged in alphabetical order, $A = 1, E = 2, G = 3, M = 4, N = 5, R = 6$.

AFAXA   XAFGF   XDDAV   DAGGX   FGXDD   XVVAV
VGDDV   FADVF   ADXXX   AXA

In the third and final enciphering step, these columns are then taken from the rectangle and written in groups of five letters each until their total number is used to make the ciphertext. Reading vertically down line 1 produces the group *AFAXAXAF*.

Deciphering tips: Using the keyword and the original matrix, place the ciphertext letters in a rectangle according to the same numbered order. Then the letters are taken back through the sequences to the original six-by-six grid. This time, the letter pairs are used to find the original letters (and numbers, if used) at their coordinates.

CIPHER DEVICES AND MACHINES

Although the terms *cipher machine* and *cipher device* have been used interchangably, a cipher device is generally defined as a manual mechanism used to encrypt and decrypt secret messages. Hand-operated devices have their roots as far back as Alberti's disk and include mechanisms such as the disklike Pletts's device and the Wheatstone Cryptograph. These creations came into prominence during the Industrial Revolution when the worlds of communications and security were changing for civilians and the military alike. The term *cipher machine* is more properly reserved for more mechanical, and later electrical, encryption machines which came into use during the 1920s and reached new heights during World War II with the development of Purple, the Purple Analog and Colossus.

Cipher Devices

THOMAS JEFFERSON'S WHEEL CYPHER

Although Thomas Jefferson's cryptographic interests are not well known, he designed a cipher device which was the basis of one adopted by the U.S. military over a hundred years later. Developed during the 1790s, his "wheel cypher" consisted of a cylinder of wood 2 inches in diameter and 6 inches long. Its central shaft held 36 wheels or disks, each ⅙ inches thick with 26 letters in mixed order inscribed on their outer edges. Each section was numbered and placed in numerical order on an iron axis, or spindle, with the assemblage held together by a metal head at one end of the spindle and a screw and nut arrangement at the other.

Using a key to define the sequence of disks, they would be set to align the letters of the desired plaintext sequence. To send the message "Meet us in Virginia," this message would be arranged in line along the iron spindle. After securing its position with the nut and screw, the transmitter looked to any of the other various alignments for a ciphertext. He then chose one of these lines from the cylinder and wrote it in his communication.

When the missive arrived, the addressee used his device to position a matching row of these same letters. This would cause his wheels to be arranged so as to reveal the desired plaintext at another point along the row of disks, and by looking at the other alphabet rows, he would find the intelligible line of plaintext. Additional variety was gained by altering the positions of the wheels on the shaft.

This device was well ahead of its time and in fact superseded a number of other attempts in Europe. Yet neither the United States military nor the fledgling diplomatic corps was to benefit from the wheel cypher because Thomas Jefferson apparently never used it. Occupied with his many activities and presidential responsibilities, he put it aside. Not until 120 or so years later was a similar version of the wheel cypher made available to the U.S. armed forces, and its worth is verified by the fact that the Navy made use of such a mechanism for decades after its introduction.

WADSWORTH'S DEVICE

Born in 1768, Yale graduate Decius Wadsworth was the U.S. Army's first chief of ordnance.

Wadsworth designed a type of disk with an extremely innovative gear process. Wadsworth's particular contribution to cryptology was in making plain and cipher alphabets of different lengths. His instrument was a brass cipher disk mounted in a wooden case that may have been built by his friend Eli Whitney. The disk was composed of an outer ring and an inner ring. The former had 26 letters and numerals from 2 to 8 for a total of 33 characters. These were placed on movable brass tablike pieces that allowed communicants to shuffle the letters and digits. The inner ring consisted of 26 fixed letters.

A brass plate identified the point at which the two alphabets were in perfect alignment. Two openings in the plate revealed two letters, though Wadworth's intentions for plain and cipher distinctions have been lost. Both rings turned and were linked inside the case by two gears, one with 33 teeth, the other with 26.

Presuming that the outer mixable alphanumeric group was intended as the ciphertext, the communicants decided on the tabs'arrangement. They also determined a beginning alignment point, for example, c on the outside and g on the inside. Both parties' devices were then synchronized with the new setting.

The encryptor moved the inner disk using a little knob to bring the first letter of the message into view through the innermost opening in the brass plate. The letter that appeared in the outer aperture was the ciphertext letter. Because this turn and the subsequent moves of the disks were geared in a ratio that reflected the size of the rings, there were varied displacements in the spaces between the letters and numerals applied. This irregularity, which depended on which plaintext letter was used, provided ciphertext variety. This was a better concealment than some standard alphabet shifts which moved predictably and thus were more vulnerable to analysis.

Despite this invention, Wadsworth was not acknowledged for creating cipher alphabets of varied lengths. Instead the credit went to a better-known British scientist, Charles Wheatstone.

WHEATSTONE'S CRYPTOGRAPH

Between 1856 and 1857 the developer of the Playfair cipher, Sir Charles Wheatstone, also created a device he called a "cryptograph." It had alphabets in circular form and "hands" similar to those of a clock that moved by gears. It was officially presented to the public at the Exposition Universelle in Paris in 1867.

This instrument was a small advance over the similar cipher disk, whereby letters, symbols, and digits were arranged in circles on their respective disk faces and aligned in various patterns to produce substitution ciphers. Wheatstone's cryptograph also had two alphabets on its face. The outer, fixed ring held the plaintext with 27 places (26 letters and a blank), arranged in alphabetical order. Aligned beneath the letters *A* through *J* and *N* through *W* were two sets of the numbers 1 through 0. The inner circle had 26 nonsequential alphabet letters. There were no numerals and this ring's movable section was considered the ciphertext.

Its clock-type hands connected by gears in a 27:26 ratio. This meant that for every space the long hand moved clockwise around the outer alphabet, the short hand was shifted one position clockwise around the inner mixed letter sequence. Both hands were moved with the help of small knobs, and the larger was arranged to move over the smaller.

Encipherment was initiated by moving the longer hand and its pear-shaped opening to the blank while a beginning cipher character was aligned with the shorter pointer. The aperture was moved to the first plaintext letter or number. This activated the gears and realigned the short pointer with another letter, which was written as ciphertext. Double letters

Wheatstone's Cryptograph

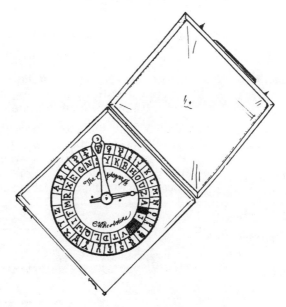

were either varied with an infrequently used *Q* or *Z* or were transmitted as single characters rather than a pair. Decipherment moved in reverse from the encrypted letters found with the smaller hand, to the larger hand that aligned with the plaintext on the outer ring.

While the instrument boasted one of the smoothest operations in manual cryptography (it even had a compact case for easy transport), it never saw any practical use. One reason may have been that its alphabet differential was just one letter. This was by no means a great barrier to a skilled solution seeker knowledgeable about letters pairs and frequency.

SAINT-CYR SLIDE

This Saint-Cyr slide was a device named for the French military academy, where it was used from about 1850 through the early 20th century for instruction in methods of cryptography. It was named by Auguste Kerckhoffs, a Dutch code and cipher scholar.

The slide design was simple and consisted of a long strip of paper or cardboard called a "stator," on which a standard alphabet was printed horizontally. Two openings, one on each side of the alphabet, enabled a second, longer piece of paper or other material to slide under them. The slide, imprinted with two horizontal alphabets, was used to generate polyalphabetic substitution.

Traditionally, the stator letters were the plaintext and the slide letters were the ciphertext. The first plaintext character on the stator was the index letter, *A*, and a slide keyletter was placed beneath it. The other plaintext letters were then located in the stator alphabet and their encryption equiva-

The Saint-Cyr Slide

lents were found on the slide. For example, plaintext *bugle* became ciphertext *jqezg*.

When the groups of letters are in standard A-to-Z order, they resemble the table of letters associated with Blaise de Vigenère. The slide method is also considered a variation on the circular cipher disks such as those developed by Porta and Alberti.

These devices permit numeral substitutions when digits, instead of letters, are printed on the slide. Because the enciphering possibilities are limited, slides did not provide superior security. Still, they proved to be an economical and easy-to-use instrument for both instructional purposes and field use in the years before electromechanical methods superseded manual systems.

CYLINDRICAL CRYPTOGRAPH

This manual substitution device similar to Thomas Jefferson's wheel cypher was created by French army officer and cryptanalyst Étienne Bazeries and presented to French military authorities in 1891. He called it a "cylindrical cryptograph." A system of simultaneous encipherment with multiple alphabets, there were a total of 20 rotatable disks, each with 25 letters on its periphery. The device's key was the arrangement of the disks on its spindle, and the particular order used for a given message was agreed upon by the intended communicators.

The Cylindrical Cryptograph

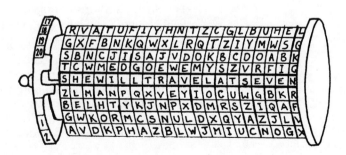

The sender turned the disks so that the first 20 letters of the message's plaintext appeared in a row. For example:

SHEWILLTRAVELATSEVEN

The ciphertext letters were made from the numerous jumbled alignments of 20 letters in any other row, such as these:

ZLMANPQXVEYIOCUWGBKR

If the message contained more than 20 letters, the process was repeated by rotating the disks. The ciphertext was made from any other row of letters. This process continued until the message was complete.

The recipient required an exact replica of the device and the prearranged order of the disk alignment. The letters of the ciphertext were aligned across one row and the recipient simply checked the other rows to see the one and only row that contained the plaintext.

The French military rejected Bazeries's cylinder after the marquis de Viaris, a vengeful rival, broke the cipher Bazeries made with it and dashed his hopes for its acceptance.

PLETTS'S DEVICE

At the beginning of World War I, J. St. Vincent Pletts, of the British Bureau for Cryptanalysis, developed a cryptographic device he called Pletts's device.

An engineer before he joined the Military Intelligence Division of the War Office, Pletts's background enabled him to construct an updated version of the Wheatstone cryptograph. The British authorities hoped to use Pletts's adaptation as a field cipher in the later stages of the war.

Plett's device consisted of two brass disks arranged to function as a stator, or outer disk, and a rotating inner disk. They were held together by an eccentric that was turned around the pivot. White rings on both rotor and stator were divided into sectors, 26 on the rotor and 27 on the stator, on which alphabets could be written, leaving one blank space on the stator. An arm fitted over both the stator and the rotor, with

Pletts's Device

an aperture through which two letters could be aligned. Each revolution of the handle and eccentric moved the rotor one letter position in relation to the stator, thus providing the possibility of polyalphabetic substitution.

Directions for using the device have apparently not survived. It appears likely that a plaintext letter was located on the stator. The arm was then moved to bring this character into the aperture and the aligned ciphertext letter was read from the rotating disk. Using a keyword, a fixed beginning position on the mechanism and a predetermined sequence of movements (such as clockwise), the user could complete an encryption. The intended addressee, using an identical device, would begin at the given starting point and reverse the process to recover the plaintext.

British officials presented Pletts's invention to American experts for testing. William Friedman, with the help of his wife, Elizebeth Friedman, discerned the keywords and broke the encipherment. As a result, the device was rejected for use as a field cipher.

M-138 AND M-138A

The M-138 was a cipher device that was the brainchild of Parker Hitt, a U.S. Army officer and cryptanalyst. Alarmed by the dangerously poor quality of American codes and ciphers,

Captain Hitt developed a hand-operated system of alphabet strips that he introduced to his superiors in 1914.

Hitt based his method on the device created by Etienne Bazeries. Instead of cylinders, though, Hitt unrolled the disks' letters and placed them on 25 slips of paper in a different arrangement. On each piece, he put mixed alphabets that were repeated twice. With numbers added to them, the papers were mounted in a 7-by-3¼-inch holder. These keynumbers governed the strips' position in the holder.

Encipherment was achieved by moving the strips up or down until the message letters of the first line were aligned horizontally. Another row of letters was chosen at random as the ciphertext. This method was repeated to encrypt the entire communication. The recipient of the message reversed the process once the alphabet slips were arranged according to the keynumber coordinated with the sender. The process became known as the "strip system."

A variation of the M-138 was the M-138A, used by the U.S. Army in the 1930s. It also functioned on the principle of mixed, sliding alphabets. The M-138A had 100 sliding strips

The MS-138A

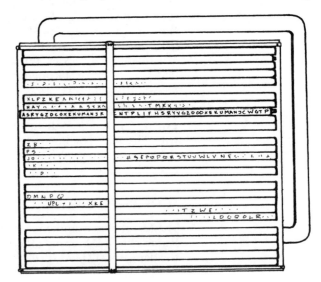

with as many as 30 in service at one time. The slides were placed in horizontal grooves in a hinged board and the grooves were filled with the alphabet slides according to a key.

Encipherment was accomplished by aligning the plaintext letters to read vertically in line with an attached guide rule. The ciphertext column was then selected and the guide rule was moved in and placed beside it for easy copying. The process was continued for any additional parts of the message. Decryption involved using the preset key to arrange the strips and then reversing the procedure to pass from unintelligible letter groups to recognizable words.

The U.S. State Department and Navy were using the M-138A as World War II began. Numerous applications from Parker Hitt's idea made it one of the most frequently applied American concealment systems.

M-94

In early 1917, Hitt's friend at the Army Signal School, Major Joseph Mauborgne, made improvements to the M-138. Mauborgne reintroduced the cylinder form; his disks on a rod gave the alphabets more varied arrangements than did Hitt's original device. This version became known as the M-94 and remained in active service for the U.S. Army until early in World War II.

In 1922 the M-94 was officially accepted for use by the U.S. Army. It was made of aluminum alloy with a 4¼-inch shaft that held 25 disks similar to silver dollars. The assemblage was tightened by a thumb nut at one end of the shaft. Each metal circle had identity markings of letters b through z and the numbers 1 through 25. The position sequence became the key for both the concealment and the solution.

A varied series of alphabet letters were stamped on the rim of the disks. When the disks were rotated, they created multiple alphabets, making the M-94 a polyalphabetic device very similar to Bazeries cylinder. After the cipher maker placed the various disks in a predetermined sequence, he revolved them until the first 25 letters of the communication were arranged horizontally. A choice from any of the other rows provided a line of nonsensical letters that became the ciphertext. This pattern was continued with groups of 25 letters until the entire message was enciphered.

Decryption was accomplished with knowledge of the key arrangements of the disks. Once assembled, the disks were turned until the same line of jumbled ciphertext letters were formed. A glance at the other rows of letters would eventually reveal one that made sense which was the intended message.

The M-94 entered a period of widespread use from its inception through the 1930s. With the name CSP 488, it was provided to the Navy in 1927 for message exchanges with the Army. It was also used by military attachés (1929), naval attachés (1930) and the U.S. Coast Guard (1930).

Advancements in Cipher Machines

By the early 1900s, a number of often impractical gear-and-sprocket inventions had been introduced, which were operated by moving a dial or pushing a button to activate gears that turned themselves. With the advancements first of telegraphy and then of the fledgling radio, cryptology seemed to be falling behind. A large leap forward occurred when American Edward Hebern applied new technological advances to cryptographic needs in 1917 and invented the wired codewheel called the "rotor." This applied the speed of electrical responses to cipher variations to create ever more rapid encryptions. Hebern's rotor, along with its rivals such as those of Boris Hagelin and Arvid Damm, signaled the arrival of cryptology's machine age.

Rotors were generally of two principal types, pinwheel and wired. The former consisted of a platelike disk with a series of projections on its circumference. The projections, or pins, were either fixed or movable in either active or inactive modes.

Pinwheel rotors were used in a version known as pin-and-lug machines, which had numbered or lettered positions on each

Rotors

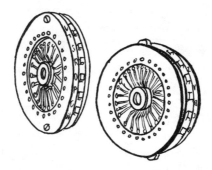

rotor whereby the active/inactive pins were set. When set in motion, a group of rotors advanced one position with each step, extending the period before recycling began. The number of steps, equal to the product of the number of positions in each of the rotors, greatly increased ciphertext complexity and the difficulty of decipherment. One of most widely used pinwheel rotor devices was the American M-209.

M-209

The M-209 was originally called the Hagelin C-36, after its creator, Swedish entrepreneur Boris Hagelin, who had developed it from a series of mechanisms produced at his Stockholm business, Aktiebolaget Cryptoteknik, in the 1930s. When the Nazis invaded nearby Norway in 1940, Hagelin and his wife left Stockholm with blueprints and two dismantled cipher machines and made a perilous train journey through Berlin to Genoa, Italy, where they boarded a ship for the United States.

The U.S. Army became interested in the Hagelin device, and after passing a number of tests, the C-36 was given the designation Converter M-209 and was activated for medium-level cryptographic concealment in military units. The M-209 gave U.S. forces better transmission protection than they had had with the majority of World War I era cryptosystems.

The M-209

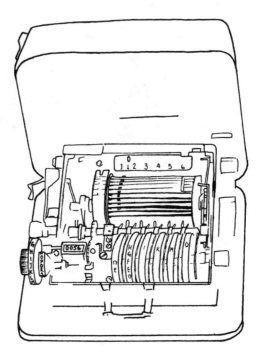

The machine was a gearlike mechanism that used varied teeth to move a cipher alphabet through multiple positions to achieve polyalphabetic substitutions. The machine's works appeared to be a maze of metal to the casual observer, but there were really just a few primary aspects of the mechanism's function:

1. A cylinder (the "cage"), containing 27 horizontal sliding bars. These bars revolved on the cylinder and each was a sextuplet of 1s and 0s. The 1s were projections, also known as lugs, with two per bar. The lugs could be fixed in two of eight possible positions on the bar (two inoperative and six operative). The bars could be put in positions (1 through 6) that corresponded with six key wheels. When the bars were moved to the left and engaged, they formed the cogs of a gear with a varied number of teeth.

2. Guide arms (vertical rods controlled by the keywheels). These could be nonoperational or moved into one of six active positions. In an active mode a guide arm contacted lugs as long as the lugs were also in an active position. If either guide arms or lugs on a bar were in an inactive position, no contact would occur. The guide arm was positioned to meet an active lug turned downward by the cage apparatus. The angle of the arm pushed the lug to the left, whereby a tooth was added to the gear. The number of teeth thus engaged set the displacement number (shift in the alphabet sequence) for each encryption.

3. Six-letter keywords, arranged as an external setting to create a starting point for another major part of the device called the keywheels. The keywords, made from the letters on the wheels, were varied for each communication and sent along with the transmission to the intended receiver. The beginning position of the wheels was an integral part of a series of encipherment stages.

4. Six keywheels, each of which directed a guide arm and had a varied number of letters on its rim. These numbers had no common divisor (such as 17, 21, 25). Projections (pins) on the periphery could be active or inactive, as they indicated a letter on one of the half-dozen keywheels. The total number of pins was 131, with each wheel having 17, 19, 21, 23, 25, and 26.

The keywheel pins could be inactive (0) or active (1). On each wheel was a pin that could affect the encipherment of a letter. Called the current pin, it was one of six such projections, collectively called the current sextuplet. With the encryption of a letter, each wheel moved forward one place, and the next character was enciphered with a new group of inactive and active pins. The first keywheel returned to its starting point after completing 26 such actions, as did the others with their differing numbers of letters and pins. Character by character, the message was concealed, with each plaintext letter being hidden in an ongoing alphabet displacement.

5. Printing mechanism. A knob on the left-hand side moved an indicator disk having 26 plaintext characters. The turning

of this knob also moved a typewheel that printed the output letters on paper tape. The knob also was connected (through a gear in the typewheel and an intermediate gear) to the end of the slide-bars that became the teeth of the variable gear. A plaintext character was thereby set for encryption.

Decipherment was very similar to encipherment. The lugs' prearranged key locations and the preset key arrangements for the pins had to be positioned to match those of the sender's mechanism. When the encryptor moved the half-dozen keywheels, the letters on the rims were his indicator. He then arranged these letters at a planned place in the message to give the recipient the necessary beginning position to decipher the message.

The cipher alphabet was the same as the plaintext, only reversed. With a key displacement of 0, $a = z$, $m = n$, $q = j$ and so forth.

Because the two letter sequences were reversed, the concealment process was reciprocal. As can be seen in the previous example, a was encrypted as z and z as a. During concealment, spaces were placed to section the ciphertext into five-character groups. In deciphering, compensation for the spacing was made by not printing the letter plaintext z. The resulting words were complete except for the z, which appeared very infrequently.

The M-209 had weaknesses similar to those of rotor-based machines. Keywheels might be set close enough to cause message letter overlaps, permitting analysts to compare letters and parts of words. The discovery of similar or matching ciphertexts could lead to plaintext revelations.

The key settings were also vulnerable because of the periodic nature of their pins. For example, a mistaken pin setting on the 21-letter keywheel would result in an incorrect letter surfacing every 21 letters and would also cause the guide arm to initiate other lugs in a chain reaction of mistakes. Knowledge of this weakness, combined with facility with trial-and-error displacement sequence arrangements, could allow a solution solver to find the chink in their armor.

Though deemed to have provided only moderate security, the M-209 proved itself many times under battle conditions around the globe. The M-209 remained sturdy while issuing many millions of enciphered characters. More complex mechanisms and even electrical devices have no better service record.

Electrically Wired Rotors

The other major type of rotor was the electrically wired wheel. Described in a patent of Swedish inventor Arvid G. Damm in 1919, the wired rotor was brought into practical use by American Edward Hebern, who had rotors in mind as early as 1917 and received his U.S. patent in 1924. The modern rotor was developed from the Hebern "electric code machine."

Wired rotor types and their uses multiplied through the 1920s. Dutch inventor Hugo Koch, German engineer Arthur Scherbius and Willi Korn, of a company called Chiffrier-maschinen Aktiengesellschaft, all contributed design advances to the Enigma machines that were later used by the armed forces of the Third Reich.

The Enigma (with improved variations) became the first commercially viable cryptographic machine and was purchased by different countries. Poland's General Staff used a version of it before World War II. The Japanese Enigma, labeled the Green machine by cryptosystem solvers in the United States, was anything but secure. The United States produced an advanced wired rotor machine, known as M-134-C/SIGABA by the U.S. Army and ECM by the U.S. Navy, which apparently was never broken by the Axis powers.

Rotor-based machines remained in prominent use by cryptologists until computers came to dominate the field with their tremendous possibilities for both generating and analyzing data.

EDWARD HEBERN'S ELECTRIC CODE MACHINE

The wired body of the rotor was made of rubber or another insulated, nonconducting material. Later, Bakelite was used in the disks, which ranged in size from two to four inches in diameter, with a standard thickness of half an inch. On the disk's flat surface were 26 equally spaced electrical contacts. These studs were often made of brass and were randomly interconnected by a web of wires to the same number of contacts on the disk's opposite face. This formed a network by which an electric current could be conducted from one facing surface to the other and the rotor thus became a commutator.

The contacts were equated with the letters of the alphabet. At the input stage they were plaintext and on the output step they became ciphertext. The different wire links formed a

An early Hebern Electric Coding Machine

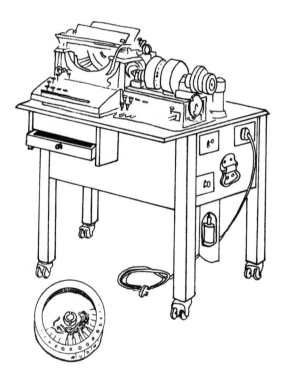

monoalphabetic substitution alphabet subject to electro-mechanical variations.

The encipherment process was initiated by positioning the rotor between two insulated plates. These held 26 studs in a circle on each of their circumferences that corresponded to those on the disk. The contacts on the input plate were attached to a typewriter key (plaintext letter), and the output plate was linked to an indicator like an illuminated glass bulb containing a letter. By touching typewriter key l, the cipher maker caused an electric current to move from its power source (batteries or wall outlet) into the input plate's l contact. The electricity then passed to the rotor at the l stud, and through the wire maze to the rotor's output contact for a potential ciphertext letter, such as z. This contact connected to z on the output plate. The particular indicator mechanism was activated so that when a key was touched, current passed through the plate, rotor and the opposite plate and a different letter appeared on a ciphertext-indicating device. A bulb lit up under z, and a clerk transcribed it as ciphertext z. In some machines the ciphertext was printed on a paper tape.

Increasing the complexity introduced by the wire maze was the fact that the rotor moved forward and also backward in some machines. The current begun at l, and reappearing at z, could emerge at q because a different contact-wire-contact sequence had occurred with a forward or backward step. Engineers also expanded the number of rotors and grouped them in what was called a "basket." The turning of one rotor caused the next to turn, with the multiple contact points and wires creating multiple encipherments or polyalphabetic substitutions. This provided yet greater secrecy by masking the frequency of letters used.

Hebern also initiated the "interval method" of rotor wiring, in which specific contacts were chosen rather than being established randomly. This resulted in a polyalphabetic distribution that kept enemy analysts from making an easy frequency study of common English letters.

INPUT CONTACT	A B C D E F G H . . . Z
OUTPUT CONTACT	K O G N J U P W . . . I

The difference or shift between input contact *A* and output contact *K* was called a displacement value. It measured the shift of the electrical current as it moved through the wire web of the rotor. The displacement in this example from *A* to *K* is 10; from *H* to *W* it is 15.

In addition, Hebern's machine contained four other rotors. They were similarly wired to have one displacement value omitted from the possible total of 26. Each also had one value repeated twice due to the fact that with an even number of studs it was not possible to have every displacement represented exactly once.

ELECTRONIC CIPHER MACHINE (ECM) AND ECM MARK II

In 1925 Navy officials began a search for a machine that incorporated the Hebern wired code wheel, or rotor, but that would be better suited for inclement weather, battle and other conditions. Experimentation led to a new cryptograph in the period 1932–34.

A very early version of ECM reportedly had five rotors (wired disks). These rotors were each controlled by a pinwheel having 25 pins. Their projections were set in either an active or an inactive mode, thus moving or not moving the rotors. An additional plugboard was attached to pass control from one rotor to another.

At this same time the Army's Signal Intelligence Service (SIS) was also studying encrypting mechanisms. In June 1935, SIS leader William Friedman and cryptologist Frank Rowlett developed the complex idea of cascading groups of rotors, also known as the "stepping maze." In laypersons' terms this could be compared to the cascading flow of a waterfall. The movements or turns of one rotor, with each new character enciphered, affected some rotors wired to act in similar ways. Other rotors were preset to move by intervals or stages in more irregular patterns after the first one in the group had shifted forward. Thus, like falling water, the effects of the early electrical flow and regular/irregular movements built up a series of variations of substituted letters in the resulting ciphertext.

The Electronic Cipher Machine Mark II was developed by the Navy around 1937–1938, according to documents released by Project Opendoor in April 1996. During October or November 1935, William Friedman of the SIS had given the details of the "cascading" or "stepping maze" process to Navy cryptologist, Lieutenant Joseph Wenger. Unbeknownst to Friedman and Rowlett, Wenger and Commander Laurence Safford, another cryptology expert, began discussing the possibility of the Navy making a multirotor mechanism. By the late 1930s, with corporate advice from the Teletype Corporation, the ECM Mark II became a reality. In February 1940, the Mark II's details were disclosed to the Army.

The Mark II had 15 rotors placed in three banks—five cipher–making rotors (back row), five control rotors (middle row), and five index rotors (front row). Initiated at the keyboard, typed plaintext characters became electrical currents.

The ECM Mark II

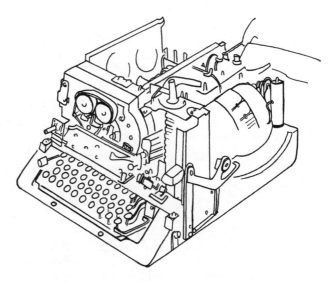

They entered the control (middle) bank and then the index (front) bank via a series of contact points. These two rotor groups governed the cipher-making rotors' movements at the back. The five cipher and five central rotors were larger and had 26 electric contacts on each side with each contact representing an alphabet letter. The five index rotors were smaller and had only 10 contacts on each side.

The center three (2, 3 and 4) of the five control cipher disks stepped in a metered pattern. Control rotor 3 was known as the fast rotor and moved forward once for each character typed. Number 4 was called medium and moved once when 3 had completed its full rotation. Control 2 was designated the slow rotor which moved once only after number 4 finished a full rotation (26 turns for the 26 contact points). The other control rotors, 1 and 5, did not step. The front five index rotors were placed only in the index bank apertures in a pre-planned order once a day (e.g., 4, 1, 5, 2, 3).

Based on the declassified documents, the Mark II's had many built-in complexities. Only four contacts on the first control rotor could be energized to transfer pulses from the keyboard. Connections between the last wired wheel of the control bank and the first index rotor had nine groups of connections, with between one and six wires in each group. Some versions did not use one of the index rotor's contacts. The ten

Interior of the ECM Mark II, with the rotor cage, the control rotor (left), the index rotor (right), and the shaft that holds the cipher rotors in place

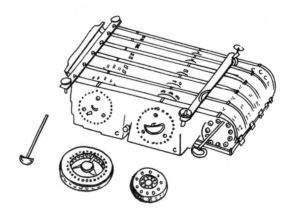

output contacts of the last index rotor were connected in pairs to five magnets that were energized to step the cipher rotors, and from one to four cipher wheels were stepped for each character typed on the keyboard.

According to ECM expert Richard Pekelney and the ECM's instruction manual, entitled *SIGQZF,* and its revisions numbered 2 and 3, the ECM Mark II had a detailed series of key lists and rotor arrangements for deciphering processes.

The encryptor and decryptor had to know that the index rotors were reassembled (rotors' order altered) once a day during the majority of World War II. The recipient also received a daily secret key list which contained the order of alignment (rotation) of the index rotors. These alignments were changed according to potential changes in specific security levels of the given dispatches, such as secret, confidential and restricted. A top-security communiqué required the altering of index rotors.

The control and cipher rotors were also reassembled on a daily basis throughout the Second World War. Their order was also governed by the secret key list, but both types of rotors were always changed with each message received. The recipient site required all the information to decrypt a given message.

The ECM Mark II was retired by the Navy in 1959. It was finally too slow to serve the needs of modern transmission. No currently available, valid record demonstrates that it had been broken by a third party.

SIGABA

The U.S. Army adopted the ECM Mark II with minor changes and named it the Converter M-136C or SIGABA (SIG for Signal Corps and ABA as one of a series of concealment designations.)

The SIGABA became a primary dispatch security system for the Army during World War II. For portability, it even had a special safe, designated CH-76, for longer journeys and a specific wooden box for shorter moves. A version called the

The SIGABA

M-134A (SIGMYK) was made secure with a type of one-time tape like that developed for teletypes by Gilbert Vernam. The tape system of combinations of marks and spaces (e.g., contact, no contact) helped regulate its rotors.

To enhance Allied communications during the war, adapters were built to make the SIGABA/ECM compatible with British devices such as the Typex. One such adapter, a U.S. Navy device called the CSP 1600, produced a hybrid device with five rotors designated the CSP 1700.

In the later stages of the conflict, an advanced SIGABA/ECM had 15 rotors, including 5 that gave multiple position changes to the other 10 which made the complex encryptions.

The SIGABA had applications after World War II and remained a U.S. possession, while other machines were developed for alliances like NATO (e.g., the KL-7 Cryptograph, one of the last rotor-based machines produced). The SIGABA was probably retired in the mid-to-latter 1950s.

ENIGMA

In the early 1920, German engineer Arthur Scherbius developed a series of devices based on a machine the size of a cash

The Enigma

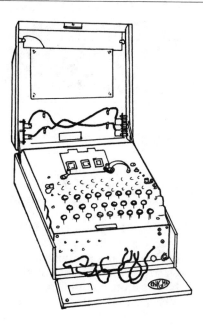

register to a more manageable, portable model. He called this machine Enigma. It enciphered messages by applying the principle of the electrically wired rotor. As Scherbius improved his machines' switch systems and typewriter keys, he added variations to the rotor, including different wire contact arrangements and gears to change the progression sequences of the rotors.

The Enigma had a number of appealing qualities, and a corporation was formed to manufacture and promote it. But a war-weary world and the financial instability of German's Weimar Republic limited the chances of financial success for a new firm based on a secrecy device. The invention failed to be profitable and another group of business partners assumed control.

Both political and economic conditions favored such businesses after 1933. When the Nazi hierarchy and Germany's military leaders concluded that the Enigma provided satisfac-

Closeup of the Enigma plugboard
with cables removed

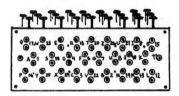

tory communications security, production of the machines increased.

The Enigma machine itself was an electromechanical device that was improved through successive versions designated *A* through *D*. The rotor was made of a wheel of thick rubber or a similar material. Its circumference was lined with peglike contacts that made electrical connections by wire in a linkage that was not even but varied. When the rotor was put between plates connected to typewriter keys and given an electrical stimulus, new multiple alphabets began to be possible. As more complex rotor machines were developed in the 1930s, additional rotors were added to create an immense number of electrically generated possibilities. Model *A* began with four rotors driven by four geared wheels. Different numbers of teeth in the gears allowed variations in the rotors' positions for producing polyalphabetic encipherments. The presence or absence of the teeth channeled the electrical pulses into different paths to the output stage of the machine.

Both the *C* and *D* versions were cryptographic advancements over *A*. They also had four rotors with 26 contact points each. The fourth rotor was a "half," also called a "reflecting," rotor that had contact points on only one of its faces. These

Initial rotation position of a
three-rotor Enigma

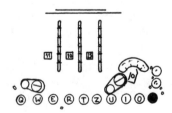

Enigma rotors

points sent ("reflected") the incoming pulses back through the rotors whence they came.

This process meant that the cipher alphabets produced were reciprocal. If plaintext q enciphered into n, then plaintext n would become ciphertext q. None of the letters represented themselves, however, because the reflected pulses did not follow their original passage through the wires. Models C and D both had these aspects, with some minor variations. C had a reflecting rotor that had only two positions in the machine; the keyboard was set in the regular alphabet's A-to-Z sequence. The D model had a reflecting rotor that could be changed but did not move during the encryption process. By arranging the rotors in the same order and original position for encipherment, one could simply type the cipher letters to recover the plaintext.

Mechanisms such as an added plugboard, ratchets, notches and pawls, along with variable turning speeds of the rotors, were all part of the Enigma. Decipherment was accomplished by knowledge of the rotors' order (fast, medium or slow) and their starting positions during encipherment. Typing the ciphertext letters enabled recovery of the plaintext, which was indicated by glow lamps under the given letters.

TYPEX

Typex, or Type X, was the British adaptation of the German Enigma. British authorities had bought and analyzed commercial Enigmas in the 1920s. After about a decade of tests and improvements, they considered their own machine ready for service.

The Typex had five typical rotors and a sixth, specialized one that resembled the half rotor, or reflecting rotor, of the Enigma. The sixth disk reversed the electric current it received, sending it back through the other rotors along a different pathway through their wire mazes.

This reflector enabled the machines to turn out cipher alphabets that were reciprocal. For example, if *c* was enciphered by *n*, then *n* was enciphered by *c*. No letter represented itself, because the direct rotor pattern never crossed the reflected one.

The Typex also had ratchet wheels that resembled those on advanced models of the Enigma. The ratchets had notches that affected the movement of the rotors. At the point where the notch on the most rapid rotor came to a certain place in its revolution, a pawl pushed the medium-speed rotor a step forward. The same operation repeated for the next-fastest rotor. This increased the number of rotor letters available for multiple-alphabet encipherment.

On one version of the Typex, the first two rotors were stators. Once preset, these did not move again during the encryption

The Typex

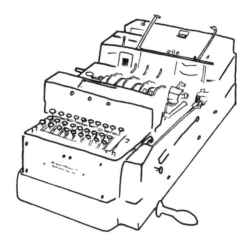

process. In practice, the function of these was similar to that of plugboards on German machines, adding another level of complexity to the cipher. The three other moving wired disks had ratchets with several notches (from five to nine per wheel). At the other end of the sequence was the reflector.

Some Enigma versions had so-called alphabet rings, imprinted with either the standard alphabet or the numbers 1 through 26, to assist in setting the rotors. The Typex, however, did not have an alphabet circle similar to that of the Enigma. On the Typex, the primary encryption cylinder body was generally a slug that could be removed from the rotor housing to enable differently notched rims to be used (5, 7, 9 and even-notched). Some slugs were arranged with 52 contacts on each side, or two rings of 26, or one set of a pair that only performed a back-up function.

Some Typex versions had rotors at the entry point of the electric current that moved automatically, creating still greater diversity in polyalphabetic encryptions. Some of their rims had standard alphabets, but others had varied alphabet sequences. Experts believe that the starting positions of the rotors were encrypted and were used as a key for beginning the process.

The keyboards of the Typex machines often had both numbers and letters. The infrequently used letters x, z and v served for word divisions, figure shifts and letter shifts respectively. The complete ciphertext was printed on paper tape.

The addressee was informed about which Typex version to use as well as the settings of the rotors and slugs. When ready, the sender's ciphertext was typed on the keyboard. If all was properly set up, the original plaintext issued forth, printed on the receiving machine's tape.

The Typex held up against German code and cipher breakers, guarding Allied communications against Germany's operation of interception and analysis.

COLOSSUS

A computer predecessor, Colossus was the result of the arduous efforts to break the encryption systems of the Third Reich. Sometimes reported as a machine used to attack the ciphertexts produced by the German's Enigma, the Colossus is now considered the primary mechanism that was more often directed at helping solve the Nazis' advanced enciphered teletype transmissions. According to historian Friedrich Bauer's detailed study of these systems, the German's SZ40 and SZ42 *Schlüsselzusatz* teletype, built by the Lorenz Company, were the prime targets. Breaking the Siemens-built teletype encryptor the T52 *Geheimschreiber*, was a less important objective.

Huge by today's standards, the Colossus Mark I appeared in December 1943 and was a marvel of its time. It was the brainchild of Alan Turing, a noted mathematician in England's Bletchley Park staff, who also developed the Turing machine, another landmark in the development of the computer. Described as being the size of three large wardrobes, it had numerous specially developed vacuum tubes, ranging from

Colossus

1,500 at its inception to 2,000 in the Mark II model (June 1944). The Mark I's photoelectric sensing capacity was an astounding 5,000 characters per second (figures about the Mark II differ). This meant that it could count and tabulate data that was punched in tape and which was fed in for analysis. This process was augmented by an electronic switching capacity and plugboard programming for some mathematical operations.

The machine's high development cost was quickly justified by the information it provided Allied commanders.

RED

U.S. cryptanalysts gave the name Red to a Japanese cipher machine built around 1934–35. Also called *angō kikai taipu A*, the apparatus depended in part on the assumption that the Japanese language was a formidable obstacle to most Westerners. It consisted of a typewriter for message input, a telephone exchange plugboard, a rotor, a 47-pin wheel and a typewriter for ciphertext output. (Some analysts believe that the Red, or at least one version of it, had two rotors.)

The one clearly identifiable wired code wheel was made in the half rotor design. Its exit points or contacts aligned with the standard fixed end plate for electric current transfer. The half rotor's input contacts were governed by slip rings, 26 of which were positioned on its shaft. The shaft, in turn, contained 26 input points that were contacted by the rings to transmit an electrical current. A step of the rotor and a turn of the shaft caused the rings to slip around the shaft as they maintained position with their input wires.

The stepping of the rotor was dependent on the 47-pin wheel, which had 47 teeth, the arrangement of which determined the rotor's movement and produced the encryption. The current entered the plugboard, first between the keyboard and the rotor input and a second time between the rotor exit and the output typewriter. Daily resettings of the plugs added to the polyalphabetic nature of the machine and further complicated decryption efforts.

In 1936, cryptanalysts of the American Signal Intelligence
Service, particularly Frank Rowlett and Solomon Kullback,
broke the Red shield when they found that this machine
encrypted Romaji, a Japanese type of Latin alphabet containing 26 letters. They were even able to predict what keys
would be used to encrypt messages nine days before they
were used. This first U.S. solution of an electromechanical
device was an important accomplishment that paved the way
for future decryptions of Japanese machine ciphers by U.S.
analysts during World War II.

ALPHABETICAL TYPEWRITER '97
AND THE PURPLE ANALOG

Late in 1938 decryptions of Red system messages announced
the planned inception of a new machine cipher. A decipherment on February 18, 1939, announced that the new mechanism would be activated for Tokyo and its embassy
transmissions two days later.

Known as the *97-shiki O-bun-Injiki* to Japanese cryptographers, and "Purple" to United States cryptanalysts, the Alphabetical Typewriter '97 was an advance in Japan's cipher-
making efforts. The device was numbered 97 for the year it
was invented, 1937, which on Japan's ancient calendar was
the 97th year of the 25th century.

In cryptological history, the '97 is important in two respects.
First, it was a clearly different device from the American
SIGABA and the Nazis' Enigma, both of which depended on
the rotor to create their cipher variety. And second, the core
of the Alphabetical Typewriter was a battery of six-level, 25-
point telephone exchange switches. When combined with a
plugboard, an intricate wiring pattern resulted. The many
plug-in variations of the board helped with key changes and
allowed for millions of encipherment combinations. A significant feature of this plugboard arrangement was that it was
double-ended so that the output was the inverse of the input.

To operate the '97, the chosen key was set (a three-letter code
was used to indicate numerals); a message was entered on the
keyboard of the first electrical typewriter; the communication

Stepping switch for the
Alphabetical Typewriter '97

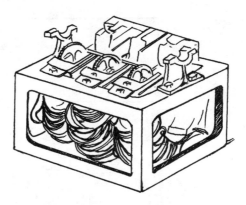

was sent through the maze of stepping switches enhanced by the keyed plugboard; and the enciphered message was printed on a second electrical typewriter.

The '97 was believed to be unbreakable. In fact, so confident were the Japanese that they installed it for transmissions between Tokyo and their embassies in Washington, London and Berlin.

William Friedman led the Army SIS team that included Robert Ferner, Samuel Snyder and Genevieve Grotjan, the last of whom is credited with finding important patterns or intervals that revealed the '97's secrets.

For a time, this group joined Laurence Safford and the Navy's OP-20-G in a frantic attempt to decrypt the '97. By autumn 1940 the SIS team, including Japanese-language specialist Frank Rowlett, had developed a blueprint for what they hoped was duplication of the Japanese system (by this time the OP-20-G members had been withdrawn to work on Japanese naval codes). By September, the first Purple Analog had been built by technician Leo Rosen. Four later versions were built with Navy help.

A maze of wires and clattering relays housed inside a black wooden box, the analog was an astounding achievement of cryptanalysis and engineering and it did indeed succeed in decrypting Japanese diplomatic exchanges.

The Purple Analog

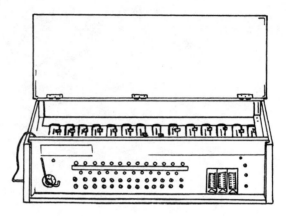

The extent of SIS's success was realized at the end of the war, when it was discovered that five Purple Analogs were actually less likely to garble encryptions than were the Japanese originals. Furthermore, a captured Japanese, 97/"B" machine was found to differ from the original U.S. Analog by only two wire connections.

Unbreakable Ciphers

Unbreakable ciphers include one-time methods which are considered unconditionally secure. There are now no known cryptanalytic processes which can solve them. Although Blaise de Vigenère was credited for years with having created *le chiffre indéchiffrable* (the indecipherable cipher), his system was not a perfect form of message protection. It was not until the 20th century that truly impenetrable cryptomethods were devised.

BAUDOT CODE/VERNAM CIPHER

The foundation for unbreakable ciphers was laid by Gilbert Vernam, an AT&T engineer who was studying security problems with the teletypewriter in 1917. He developed a process with paper tapes and sequences of marks (holes) and spaces (no marks). Encipherment of standard teletype characters was accomplished with a key of holes and spaces that was added to the marks and spaces of the plaintext message. The sum of this electromechanical process was the ciphertext.

Vernam's method was based on the Baudot code for standard teletype communications, developed in 1875 by French telegrapher Émile Baudot. This code replaced the letters of the alphabet with electrical pulses called "units." Every character was given five units that signified either a pulse of electric current (marks) or its absence (spaces) during a given period. This made 32 combinations of spaces and marks, 26 representing the standard alphabet and 6 used for operational processes such as number and letter shifts and carriage return. The transmission was accomplished by commutators that rotated and produced the sequence of electrical pulses denoting the depressed key.

By 1917 the teletypewriter had evolved into a second mode called "indirect." In this version the initiation of typing and activating the electrical processes caused the signals to be changed into marks (holes) that were punched in a paper tape. The absence of pulses left spaces in the tape strip, thereby creating a pattern of spaces and marks that put the Baudot code on paper. The tape was read by a device with metallic fingers that slipped through the strips at the openings, made contact with another part of the mechanism and sent the pulse. The spaces (unperforated segments) kept the contact points apart, so that no pulse was sent. The circuit remained interrupted until the next opening permitted another contact. Here is a table of the Baudot code with alphabet letters and teletype process symbols. The number 1 represents a mark and 0 a space.

BAUDOT CODE:	A	11000	B	10011	C	01110	D	10010
	E	10000	F	10110	G	01011	H	00101
	I	01100	J	11010	K	11110	L	01001
	M	00111	N	00110	O	00011	P	01101
	Q	11101	R	01010	S	10100	T	00001
	U	11100	V	01111	W	11001	X	10111
	Y	10101	Z	10001				

Well aware of this process, Gilbert devised a cipher using the very same items. He took a tape of randomly generated key characters and added its pulses to the pulses of the plaintext letters. The total of the two became the ciphertext that concealed the teletyped communication. Vernam also developed a convenient method for reversing this addition, so that the intended addressee could recover the original plaintext.

With the mark-and-space process as the foundation, he had four variations:

PLAINTEXT	KEY	CIPHERTEXT
space +	space =	space
space +	mark =	mark
mark +	space =	mark
mark +	mark =	space

When the plaintext and key pulses were both the same (two marks or two spaces), the ciphertext was always a space. Two different combinations always meant a mark as the cipher. The Baudot letter *I* was designated as *space mark mark space space* (in digital terms, 01100). The key letter *O* was *space space space mark mark*, or 00011. The cipher sum became 01111, or *V* in the Baudot alphabet.

PLAINTEXT: 01100(I)

KEY: + 00011(O)

CIPHERTEXT: 01111(V)

At the recipient site, the ciphertext pulses were added with the prearranged key pulses for the particular message sequence in order to find the plaintext. The previously mentioned rule of space and mark equivalents (*mark + mark = space*) still applied:

CIPHERTEXT: 01111(V)

KEY: + 00011(O)

PLAINTEXT: 01100(I)

The electrical adding of the pulses was accomplished with a device that included magnets and relays. Key tape and plaintext tape-reading mechanisms generated pulses which were fed into Vernam's enciphering device. Incoming pulses with a combination of 1 and 0 closed a circuit and made a mark; like pulses resulted in a space. The encrypted spaces and marks were then sent as a regular teletype transmission. The Vernam mechanism at the transmission's destination was arranged so as to restore the plaintext pulses, which were then put into a teletype receiver that in turn printed the plaintext.

By introducing "on-line encipherment," Vernam brought automation to a process formerly dependent on clerks with varying degrees of skills. Nevertheless, the method was vulnerable. The Baudot code was publicly available and its alphabet variations, though polyalphabetic, were limited. An eminent cryptanalyst, Joseph Mauborgne, recognized these weaknesses and, building on Vernam's work, developed a truly secure one-time system.

MAUBORGNE'S ONE-TIME SYSTEM

U. S. Army Major (later General) Joseph Mauborgne applied his experience in military security and cryptography to Vernam's AT&T system and in 1918 proposed a nonrepeating, unintelligible key to encrypt the plaintext. This advance came to be known as the "one-time system."

Based on studies of the Baudot code, the special element of Mauborgne's system was this randomly chosen key, which Mauborgne defined as a series of numbers, electric pulses and spaces or letters. This endless sequence of cipher characters was generated with a series of pulses that were electrically added to the Vernam tape's mark-and-space patterns. As a matter of primary importance, Mauborgne advised that the particular group of characters chosen to conceal the given communiqué be used only once.

With the Vernam/AT&T equipment, this procedure became the one-time tape. The random key characters added to the original message pulses created a lack of predictability that confounded the standard procedures of outside analysts. The possible sequences would be endless. A key would be recorded for the use of the sender and recipient, but once used, it was never repeated. A list of keys for different days existed. Careful records were kept at both ends of the message transmission to avoid mistakes. Barring accidental repetition, which could give third parties solution angles by comparing letters, or security breaches, the method is still considered unbreakable.

Suppose that the key was a series of numbers like 44207786358173 etc. Enciphering would begin by converting the letters of the plaintext message into numbers. The simple example below uses the word *Pacific* as the plaintext with the cipher alphabet equated with alphabetical positions: $a = 01$, $b = 02$, $c = 03$. Based on this cipher alphabet, $p = 16$, $a = 01$, $c = 03$, $i = 09$, $f = 06$, $i = 09$, $c = 03$. Pacific becomes 16010309060903. This number is placed beneath the random key 44207786358173. These two numbers are added using noncarrying addition (for example $108 + 102 = 200$):

```
  44207786358173
+ 16010309060903
  50217085318076
```

This sum is the cryptogram. After the receiver gets it, she or he writes this number above the prearranged key (44207786358173) and subtracts the latter from the former:

```
  50217085318076
- 44207786358173
  16010309060903
```

In this step, the subtraction is done without borrowing from the column to the left (for example, $70 - 77 = 03$).

Mauborgne had achieved the cryptographer's dream of an unbreakable cipher. Its random uncertainties prevented any cryptanalyst of the day from gaining any clues which might lead to decryption. Despite this remarkable achievement, Mauborgne's discovery did not achieve widespread military use, as the armed forces often needed to send a number of communications in rapid succession. Because keys were sometimes inadvertently reused and since fresh key groups were always needed, it was not practical for the military except for very limited high-level dispatches.

ONE-TIME PADS

Cryptography experts in different countries were conducting research similar to Mauborgne's, and between 1918 and the early 1920s other independently developed one-time techniques appeared. In Germany, the German Foreign Office formalized a single-use system that became known as the "one-time pad." Its name was derived from the two sheets of paper typed with a sequence of random numbers that became the key. A series of these pages were placed in two identical groups, or pads, one for the sender and one for the receiver. The numbers on the pads' pages were intended to be used just once and then discarded.

When an encryptor used a pad for encipherment, he included a prearranged means of identification called an "indicator group." The indicator changed with each transmission and identified the new pad sheet to be used. The numbers on each page provided a random key that was added to a second group of digits formed from the plaintext words of the message.

The words of the communication could be changed into numerals by a "monome-dinome table," a checkerboard or similar matrix that converted letters to numbers. The monome-dinome table aligned letters with single and paired numbers by using a keyword to initiate and mix the alphabet. This style also used configurations of numerals at the top and side of the matrix to make the single and paired numbers possible (similar to the straddling checkerboard). To encipher a message, a checkerboard could be set up with the frequently used English letters *e, t, a, o, n, r, i, s* mixing the alphabet and top and side digits. A full stop (.) and diagonal (/) for letter/number shifts often appear in such checkerboards as necessary functional devices.

0	9	8	7	6	5	4	3	2	1
e	r	t	o	n	i	s	a		
1	b	c	d	f	g	h	j	k	l	m
2	p	q	u	v	w	x	y	z	.	/

In this example, the frequently used letters in the top row are replaced by the numerals above them. The letters in the other two rows are replaced by coordinates derived from reading the corresponding numbers from the side and then the top (for example, $d = 18$ and $v = 27$). The plaintext letters are encrypted from this checkerboard.

If, for example, an international security agent (codenamed AO) located a wanted weapons smuggler (given the codenumber 20) he could send an encrypted communication with both digits in the number 20 being tripled to distinguish it from cipher pair 20.

PLAINTEXT: f o u n d / 2 0 / i
n p r a g u e . a o .

CONVERSION: 17 7 28 6 18 21 222 000 21 5 6 20 9 3
16 28 0 22 3 7 22

To these digits the agent would then add a series of nonrepeating numbers from the onetime pad using a noncarrying form of addition called "modulo 10", or the Fibonacci system, in which the 10's digits are not carried over. The sum of $9 + 2$ is written only as 1, not as 11. This increases the rapidity of encipherment and decipherment, helps prevent carrying errors and permits encryption from left to right. An example is shown below:

CONVERSION:	1	7	7	2	8	6	1	8	2	1
	2	2	2	0	0	0	2	1	5	6	2	0	9
	3	1	6	2	8	0	2	2	3	7	2	2

PAD DIGITS:	4	4	1	4	2	3	8	0	5	2
	7	6	7	5	0	8	1	2	5	9	5	2	6
	1	8	1	6	2	8	7	2	3	1	8	9

CIPHERTEXT:	5	1	8	6	0	9	9	8	7	3
9	8	9	5	0	8	3	3	0	5	7	2	5
4	9	7	8	0	8	9	4	6	8	0	1

TRANSMITTAL: 51860   99873   98950   83305   72549
78089   46801

Even if agent AO's transmission was intercepted by the arms dealers, the odds are that the dangerous smuggler known as 20 would be captured and convicted long before the pad secured communiqué could be revealed—if it ever was.

A second type of monome-dinome table can be arranged as follows. The keyword *consul* is written horizontally with the remaining alphabet characters placed beneath it in rows. Each of the numbers 0 through 7 is aligned with one of the often-used letters *a, e, o, n, i, r, s,* and *t* according to its location in the columns, reading down and left to right. Number pairs are assigned to the other characters in combinations starting with 8 or 9 and ranging in order down the columns from 80 to 99. These are placed in the table that also has a diagonal for digit-character shifts in the plaintext and a period for stops.

c	o	n	s	u	l
80	1	4	6	92	96
a	b	d	e	f	g
0	84	86	7	93	97
h	i	j	k	m	p
81	2	87	89	94	98
q	r	t	v	w	x
82	3	5	90	95	99
y	z	.	/		
83	85	88	91		

The next stage is to begin substitution of the message letters with the numbers from this table. For transmission clarity, all numerals in the plaintext are repeated in triplicate.

PLAINTEXT:

c o u r i e r / 10 / h a s it

80 1 92 3 2 7 3 91 111 000 91 81 0 6 2 5

These numbers are then added to the key's nonrepeating digits and noncarrying addition is applied.

PLAINTEXT:	8	0	1	9	2	3	2	7	3	9	1	1	1	1	0	0	0	9	1	8	1	0	6	2	5
KEY: +	1	7	0	3	4	8	5	3	9	5	4	2	6	0	2	3	9	1	4	6	0	8	5	1	7
CIPHERTEXT:	9	7	1	2	6	1	7	0	2	4	5	3	7	1	2	3	9	0	5	4	1	8	1	3	2

Deciphering tips: To decipher the one-time pad, the recipient has to possess the same monome-dinome table or matrix and one-time pad keynumbers. He or she subtracts the key digits, without borrowing numbers, from the ciphertext numbers to recover the numerals representing the plaintext:

971261702

−170348539

801923273

He or she then finds the numbers in the table or matrix and locates the plaintext letters equated with them, discovering that 80 = c, 1 = o, 92 = u, 3 = r, 2 = i, 7 = e, 3 = r and so forth to decrypt the message.

During World War II, U.S. agents of the Office of Strategic Services (OSS) and their allies in Britain's Special Operations Executive (SOE) used one-time methods that combined alphanumeric grids and numerical key sheets. The grid form was often made of easily-disposed-of silk and contained 26 alphabet letters with columns of sequential numerals and cyclical alphabets aligned beneath them. The key sheets of five-digit groups were printed on paper or a very flammable synthetic material. By 1944 technicians had developed pages made of film that could be read with a hand-held magnifying device.

By the early 1960s, pads had become sheets the size of postage stamps or scrolls the size of large pencil erasers. Some were printed on paper that was photographed and sent as microfilm for extra concealment. Pads have also been made of foil-like material and have had pages of extremely combustible cellulose nitrate for rapid destruction in case of emergency.

When properly prepared and distributed, one-time pads are considered unbreakable ciphers. The odds against solving them are astronomically high and the time required for decipherment, even if successful, would be so long that any information gained would be obsolete. The absence of any predictable aspect in the key undermines standard analytical tools such as same-key encryption matchups, frequency counts or searches for familiar words and typical opening and closing phrases. Even an attempt to check every possible key would fail, because this trial-and-error method would produce an impractical number of possibilities, even for today's computer technology. Concerns about relevant "current time" data are always present too, as secrets can quickly become stale and irrelevant.

Despite their resistance to solution, one-time pads are not universally applied because they have some weaknesses. By their very nature, pads need huge quantities of continuous random digits. This can be accomplished with computers, but since the communicators cannot memorize the stream of numerals, the pads must be printed and distributed, rendering them vulnerable to discovery. Thus one-time systems have still not proven widely applicable as military field ciphers. Businesses also generally prize efficiency and economy over such high-level security and for these reasons, one-time pads have tended to be used only in the specialized worlds of the diplomat and the spy.

VIC CIPHER

The VIC cipher derived its name from the codename of Soviet spy and defector Reino Hayhanen. After his defection in 1957, Hayhanen's cooperation led to the solution of a puzzling encryption that had been in the possession of the FBI for almost four years. The mystery had begun in 1953, when a Brooklyn newspaper boy found an unusual nickel among his change. It had fallen, split in two and revealed a piece of microfilm $5/16$ inches square, covered in tissue paper. Turned over to the New York police and then the FBI, the microfilm was enlarged, revealing 21 rows of numbers in a series of five-numeral groups with a total of 1,035 digits.

The VIC cryptosystem's key terms and numbers are not purely random, as are those of a properly used one-time pad; nevertheless, several experts consider it to be a virtually unbreakable cipher. It proved so in actual use.

In technical terms, a cipher like the VIC is a straddling bipartite monoalphabetic substitution superenciphered with modified double transposition. As Hayhanen explained it, the method was built upon four mnemonic keys that varied for each Soviet agent. Hayhanen's keys consisted of the first seven letters of the Russian word for snowfall (*snegopad*); part of a folk song ("The Lone Accordion"); the date of victory over Japan in World War II (3/9/1945 in Continental date style); and the number 13 (Hayhanen's personal digit, changed to 20 in 1956).

These four keys controlled the development of the alphabet for the substitution process as well as for two transposition tables. The keyword *snegopad* affected the arrangement of the letters and symbols in a type of straddling checkerboard. The other three mnemonic devices helped develop a series of numbers that governed the numerical keys for the checkerboard and transposition tables.

Because the mnemonic keys do not require key lists (a weakness of one-time pads) and because of its complex encryption procedures, a cipher like the VIC remains a very formidable challenge to analysts.

Computer Cryptography

Computers are crucial to any modern nation's security. The use of machines for collecting and analyzing intelligence was first initiated by military personnel in the mid-1930s, when U.S. Navy and Army signals intelligence (SIGINT) groups began using tabulating machines. These devices used a punch-card format and were considered extremely advanced in their day. The costly inventions were limited, however, because each was built to attack a specific cryptosystem. If that particular code or cipher was changed or canceled, the device was relatively useless.

With the outbreak of World War II, the SIGINT teams worked with Kodak, IBM and Bell Laboratories, among others, to develop machines with greater speed and increased flexibility. Other Allied nations were involved in similar pursuits, but after the defeat of the Axis powers, the research in this area was taken over by academia. In 1946 the University of Pennsylvania and its Moore School of Electrical Engineering were responsible for an important advance with their "Eniac," an electronic numerical integrator computer with 18,000 electron tubes housed in a 30-by-50-foot room with a total storage capacity of just 20 numbers. The seminal Eniac was followed by the Naval Security Group's Atlas (1950), the Army Security Agency's Abner (1952), and then the National Security Agency's Harvest (1962) and Stretch (1962).

Today's smaller and smarter progeny of these behemoths are the primary means by which diverse intelligence data, from spies' transmissions to satellite surveillance, is compiled. Nations protect their communications with computer-generated encryptions to achieve the highest level of communications security.

The computer is also vital to modern decryption techniques. Its speed and capacity to process volumes of material enable cryptanalysts to consider the millions of letter, syllable and number combinations and other frequency possibilities necessary to uncover clues in an encryption shield.

DATA ENCRYPTION STANDARD

In 1977, appreciating the increasing security needs for secure government records such as taxes and social security, the U.S. National Bureau of Standards chose the Data Encryption Standard as the official process for securing files in federal departments.

Developed by IBM, the DES is a single-key encryption process with the same key used for encrypting and decrypting purposes. The DES is an algorithm (a set of procedures) that secures information stored in binary code. Its increased application in integrated electronic computer chips has made the DES a primary part of commercial information security.

The DES is a product block cipher. Its key begins as a sequence of eight decimal numbers chosen from 0 to 127. For the computer language of bits (zeros and ones) the decimal numbers are turned into their seven-bit binary equivalents, including an extra eighth bit added at the end for verification purposes, not concealment. When activated, the key is eight decimal numerals multiplied by seven bits, or 56 bits in length. The message also undergoes conversion to binary digits and is represented in ASCII form.

At the time of its inception, the 56-bit key was itself a very substantial defense. A third-party solution seeker using a computer to try all the combinations of 56 ones and zeros, even at a million tries a second, was facing 1,142 years of attempts. Though this time frame has been narrowed over the years by computer advances, a number of experts still consider the DES a strong transmission guardian.

The block cipher undergoes 16 iterations, or rounds, of message letter rearrangements, transpositions and substitutions. The encryption processes include: blocking, compaction, expansion, permutation and substitution. With the DES a 64-bit block of plaintext can be encrypted as a 64-bit block of ciphertext at any one time. The key that governs the series of processes also is made of 64 bits, but only 56 are at the user's disposal. Eight of them are used for a parity check in which binary-coded data is totaled as being odd or even, odd 0s or even 1s, or vice versa, as a way to find transmissions errors.

Segments of the encryption are bonded at points during the procedure by an added process known as "exclusive OR." This operation is signified by a circled plus sign. It is a function of a logic gate in a computer's circuitry. In this operation, when adding like bits (two ones or two zeros), the result is zero. If adding unlike bits (zero and one), the total is one:

```
  1  0  0  1
+ 0  1  0  1
  1  1  0  0
```

The OR process is a primary part of the rearrangement of the keys known as subkeys. The subkeys in turn affect sections of the communication's bits. The OR operation bonds the subkeys and message bits and gives extra variation to the transmission.

Along with this special bonding, a series of procedures occurs with the message block and the key. The former has its bits rearranged in a permutation stage, similar to letter transposition. The new arrangement is divided into a right-side and a left-side block and the right-side block is changed by a formula before it joins the key. The key's bits (decimals changed to binary numerals) are transposed and split into two blocks; these blocks then undergo another permutation. Specific bits move from the blocks and are then realigned as a single sequence in a compaction stage. A segment of a flow chart is shown in below.

This sequence of steps is a part of a single round. There are 15 additional iterations in the complete process for concealing a message block including nonlinear substitution.

The DES algorithm achieves 70 quadrillion encryption possibilities for a given communication. The planned receiver can decipher it by using the chosen key and reversing the order of the stages beginning with the 16th round.

However, this process is dependent upon key secrecy, and the protection of keys can be a security problem. Other processes developed in the late 1970s and 1980s now rival the DES as security measures.

TWO-KEY CRYPTOGRAPHY

This new cryptographic process became publicly known in 1976 when Stanford University's Martin Hellman and research students Whitfield Diffie and Ralph Merkle created what they called the two-key, or public key cryptosystem. National Security Agency officials claim now that they had discovered two-key 10 years earlier, but secrecy restrictions prevented them from making it known at the time.

The two-key system was developed as a response to the problems inherent in the distribution of keys. In a single-key system, the message transmitter and addressee both keep the key a secret. The distribution of such keys is a significant security problem, especially when more than two people need the same key. The public-key cryptosystem avoids key distribution difficulties by having encryption keys published in a directory available to the public. Security is maintained by a second decryption key. Using the national directory, one person can send secret data to someone else's computer. Barring theft, only the recipient would have the decryption key. Because it is computationally impractical to recover the decryption key by possessing the encryption key, the encryption key does not need to be concealed. It should be protected from being covertly exchanged or modified, however, since a fraudulent encryption could be used, in conjunction with a decryption made for it, to steal data.

The Stanford researchers built their concept on a series of mathematical "secret entrances." Also known as trapdoors, they are planned openings in computer systems which allow legitimate access in order to bypass security procedures. This is normally done for tests of new equipment and software or for repairs. However, they have also been used by hackers to gain illegal entry.

The Stanford scholars proposed a trapdoor version known as a knapsack to exemplify their cipher algorithm. As a mathematical principle, the trapdoor only "swings one way," that is, it is difficult to solve when trying to work back from the end result to the beginning steps. Solving one is frequently compared with trying to guess the total number of cylinders in a knapsack. The cylinders are of the same width but varying lengths and are placed end to end in a cylindrical knapsack of the same diameter. The power of the algorithm lay in the difficulty for the analyst who tried to discover the exact sequence of the components as they were arranged to make the sum.

Hellman and Merkle developed complementary key pairs. A message encrypted with one member of the pair was decrypted by the other member. Encryption involved assigning a digit to each message letter and then adding these digits together. The decryption key separated the original numerals from their sum, and the plaintext was recovered. For the third-party analyst, unable to discern the secret decrypting key, the only alternative was believed to be a brute-force, trial-and-error computer attempt to find every possible key, necessitating costly and time-consuming electronic guesswork.

As the public key concept became well known, it led to new suggestions and methodologies. As with the DES, debates arose over whether universities and companies should be allowed to develop such systems, and whether organizations such as the NSA should be entitled to have knowledge of their mathematics and to approve their applications.

RSA CRYPTOALGORITHM

The Hellman/Diffie/Merkle system was lauded as one of the very few unbreakable ciphers. This seemed to be the case until 1982, when Adi Shamir of the Massachusetts Institute of Technology solved the trapdoor knapsack. Not long thereafter, Shamir's former MIT associate Leonard Adleman used Adi's formula to solve the knapsack on a home computer.

Shamir, Adleman, and associate Ronald Rivist of MIT had a competitive two-key process that became known as the RSA cryptoalgorithm. Its public and private keys were also based on a trapdoor, but had combinations of prime numbers (numbers that can be evenly divided only by themselves and the number 1, such as 3, 5 or 7).

When a prime integer is multiplied by another, the product is one that cannot be produced by any other pair of prime numbers. The inventors of the RSA system recommended very large integers to make the private and public keys. Thus an RSA-based computer program would randomly select a pair of prime numbers around 100 digits long, then multiply them together, with the product becoming the basis for encrypting the two keys. As with the complementary pair system of Hellman and Merkle, an analyst trying to remove such a concealment was faced with the immense difficulty of factoring these large integers in a "brute force" attempt.

Throughout most of the 1980s, the RSA cryptoalgorithm was described as being computationally cryptosecure; the term *unbreakable* is more difficult to apply. In fact, most experts do not define the RSA as completely unbreakable due to advances in both factoring theory and computer-enhanced decryptions which can supply answers in a more practical time frame than was previously possible.

AUTHENTICATION

With two-key, or public key cryptography, one of the problems revolves around establishing the authenticity of senders or buyers, long-distance agreement signatories and so forth. In this regard, the single secret key (or symmetric key) method had its advantages. The sender, who possessed the secret key, was expected to be the recipient's friend, business partner, or war ally.

As multiple access to communications through the Internet is now possible, authentication is a growing issue. Since the public part of the key is available in a key directory, senders and receivers must be sure that their correspondents are indeed who they claim to be. To respond to such concerns, present-day security procedures include digital signatures, certificates and certificate authorities.

A digital signature is today's electronic version of the hand-written signature. Both are supposed to connect the signer to the document, purchase deal or communiqué. Digital signatures are intended to guarantee the authenticity and integrity of a transmission as well as the veracity of the sender.

A digital signature has an encrypting process known as a hash function. Hash functions link the length of a message's contents to a briefer, fixed length number known as a hash value. The hash value becomes a compact representation of the communication. In mathematical terms, a hash function is the result of an algorithm that works in the same manner as a check digit functions in other electronic transmissions. For example, suppose that to indicate an internal company product shipment, the number is 4392072 is exchanged. The hash algorithm could be arranged to multiply each numeral in turn (excluding the 0). In this example, the first digit of any multiplied product above 9 is dropped.

1. Multiply 4×3 12
2. Drop the first numeral (1) leaving   2
3. Multiply $2 \times$ the next numeral in the group (9)   18
4. Drop the 1   8
5. Multiply 8×2 16
6. Drop the 1   6
7. Exclude the   0
8. Multiply 6×7 42
9. Drop the 4, the result is   2

The check digit or hash value is *2*, the last numeral in the message. A temporary change or sending error in the original transmission would alter the hash result and alert the recipient of a garble or an outsider's alteration.

A hash function is applied to secure the contents when the transmitter (Anna) computes the hash value of the message (a sequence of numbers). The hash is then encrypted with the recipient's (Zena's) public key or a previously picked symmetric key (privately shared between them only). The combined hash and one or the other encryption key is added to the message. To prove her identity, Anna signs the hash with her private key. The complete contents of the communication

Hash function

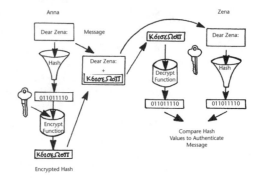

should also be encrypted to ensure secrecy (a number of other algorithms serve this purpose). Anna then sends her message to Zena.

On the receiving end, Zena applies Anna's public key to extract and verify Anna's authenticating signature. Zena then decrypts the communiqué using her own private key or the prearranged symmetric key which enables Zena to extract the hash function and compute its value. If the result exactly matches Anna's hash value Zena knows that the message came from Anna, that there were no transmission errors and that the message has not undergone successful tampering. If this integrity check fails, then the transmission has been changed or Anna is being impersonated.

Part of the security of this procedure relies on trust in public keys. With a public key Certificate Authority (CA), the recipient can be further assured of the transmitter's authenticity.

A certified public key, also known as a public key certificate, is the electronic counterpart of a passport, driver's license, marriage license or bank signature card. A public key certificate verifies the identity of the holder and usually contains the owner's name, public key and expiration date, among other pertinent information.

They are created by a certificate authority, whose verification processes are trusted. CAs develop certificates for groups and computer networks and have a directory that lists the various certified public keys. When a sender wishes to send a certificate, the CA makes a digital signature of the sender's public key including the sender's name and other pertinent data.

The CA secures its processes by digitally signing the certificate through computation of a hash value of the message. The hash value is then signed with the CAs own private key. Along with the certificate, the CA provides a copy of its public key.

When Zena receives this data, she can confirm the validity of the signed certificate and the authenticity of Anna's public key by using the CA's public key to verify the CA's signature, computing the certificate's hash value and comparing it to Anna's hash value. If the hash values do not match, an error or deception has occurred.

In short, the authentication steps for Anna's and Zena's message are as follows:

1. Anna makes a hash value for her communication.

2. Anna digitally signs the hash with her private key.

3. Anna obtains Zena's certificate from a directory or from a Certificate Authority. Zena's certificate includes her digitally signed public key and the CA's accompanying public key.

4. Anna verifies Zena's certificate by applying the CA's public key to check the CA's signature. This includes extracting the CA's signature and hash from the certificate, validating it with the CA's public key, computing the certificate's hash value and comparing it with the extracted hash. If the hash values match, Anna can trust Zena's public key.

5. Anna encrypts her message with Zena's now verified public key.

6. Anna adds her digitally signed hash value (from 1 and 2) to her message and sends it.

7. Zena receives it and decrypts it with her private,key.

Certificate Authority

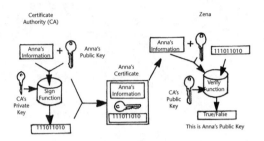

8. Zena extracts Anna's digitally signed hash value.

9. Zena validates Anna's signature using Anna's certified public key, obtained from Anna's CA and verified by the CA's public key.

10. Zena computes the hash value of the message and compares it to the hash value she extracted from Anna's communiqué. A match means no errors and no deception.

In addition to digital signatures, certificates and certificate authorities, there are other authentication procedures that complement encryption. They include passwords, photographs and biometric qualifiers such as fingerprints, retinal scans and voiceprints.

KEY GENERATION

As hackers have sometimes attempted to circumvent authentication with forged digital signatures or by bribing CA employees to alter documents, authentication also includes varied forms of key generation.

One procedureinvolves machine generating specific random keys for each communication. All direct parties know these session keys but they cannot be predicted ahead of time and therefore cannot be controlled or altered in any currently foreseeable way. The random session key is added to the sender's and recipient's private keys to further enhance their security and make attempts to subvert the process more difficult.

A similar arrangement uses machine generated session keys known by only one party. A rather recent development is one in which the exchangers' interactions generate a key. In this procedure, the key is one for transaction only. It cannot be biased by foreknowledge or actions as the transactions are occurring since that would be an obvious attempt to defraud.

The production of large amounts of sequences of numerals to create keys for encryption purposes has required the development of new technologies to keep pace. Although secrecy-conscious groups like the National Security Agency often do not make their discoveries known, some publicly available literature can provide additional information.

For example, cryptologist Gustavus Simmons has described contemporary cryptomachine technology in the form of the Fibonacci generator. It was named for the Fibonacci sequences in number theory.

With the Fibonacci series of integers such as 21, 13, 8, 5, 3, 2, 1, each succeeding leftmost numeral is the sum of the two digits on its right (for example, $3 = 2 + 1$, $13 = 8 + 5$). In a loose comparison, a Fibonacci sequence can be considered one in which each numeral is the sum of a series of earlier numbers in fixed locations in that series. In the sequence 68,

The *n*-stage Fibonacci generator

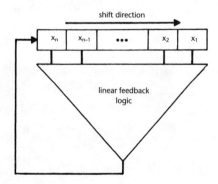

37, 20, 11, 6, 3, 2, 1, for example, 2 + 1 = 3 and 1 + 2 + 3 = 6. This series then shifts one place: 1 + 2 + 3 + 6 are *not* 11. The sequence has undergone a set shift so that 2 + 3 + 6 = 11. Another shift transpires with the next digits whereby the sum of 3, 6, and 11 is 20.

In what is described as an "*n*-stage" Fibonacci generator, the numbers forming the contents of the number-holding register are shifted right one position with each step. The numeral (bit) at the extreme right of the shifting group is moved out and lost to the sequence. The next new left-hand bit that is generated is determined by the sum of the numbers at prescribed points in the number register's contents. This shifting and adding process provides a well-shuffled series of digits, exemplified by the Fibonacci scramble of the numbers 1 to 31 in the diagram.

This Fibonacci process has been proven to be vulnerable to cryptanalysis because it is based upon linear feedback logic. This means that the successive states or orders of the numbers are in a predictable sequence and can be determined by certain formulas. Furthermore, the Fibonacci generators are not a secret and their shifting functions are known.

Cryptographers have countered with nonlinear feedback logic machines that have a greater capacity for random number generation. The nonlinear form uses a key that jumbles the numeral combinations and diffuses the information in the

Numbers shuffled with the Fibonacci generator

message text concealed by the numerals. The hope of the cryptographers is that these transmissions will prove essentially unsolvable, in that they cannot be revealed by an analyst trying to work backward from the output ciphertext to the numbers' original states. The immense difficulty in trying to break such ciphers is called "computational infeasibility"— the cryptographer's almost perfect security ideal.

QUANTUM CRYPTOGRAPHY

As further research continues into new cryptomethods, a new possibility has opened in the field of quantum cryptography. Classical cryptography employs basic mathematical theorems to protect keys from being read by potential eavesdroppers. RSA, one of the most popular public key encrypting systems relies on the difficulty of factoring large integers to ensure the safety of its keys. The assymetry between verification and determination of possible factoring is what makes messages so easy to encode but very difficult to decode.

Quantum cryptography is based on the possibility of discovering an algorithm that could determine whether a number was prime just as easily as it could verify that two factors correctly multiply to give that very number. If this was discovered, the multiplication of large primes as a mathematical foundation for encryption would prove insufficient to conceal communications.

Fortunately for the government and many secure servers across the world, this algorithm has not yet been discovered. One of the fastest classical computer algorithms for factorization is known as the "number field sieve factorization method." This method can factor a number with L digits in time that grows exponentially with $L^{1/3}$. Currently, the largest number that has been successfully factored as part of an RSA challenge is a 130 digit number factored in April 1996 using a distributed network of computers which shared information across the World Wide Web. Although this was a remarkable feat, adding more digits to the key will continue to make it exponentially harder to unlock. By contrast, just a few years ago a quantum algorithm was developed which could, theoretically, factor numbers in time that scales as L^2. This means that a quantum computer performing only 100 calculations per second could factor a 100-digit number in approximately two minutes. If one could build a quantum computer, then it might be possible to crack codes and ciphers transmitted classically across the Internet and phone lines in a very short period of time.

Fundamentals of Quantum Computation

A classical computer like a typical PC or Macintosh stores information in bits. This binary representation simplifies many of the internal computations that every computer must perform in order to compute. For example, in all digital computers, the number 15 would be stored as 1111, while the number 10 would be represented as 1010.

Although any number N can be represented using approximately $\log^2(N)$ digits, it takes N memory locations each of length $\log^2(N)$ bits to store all possible numbers between 0 and N. For example, in order to store all the numbers between 0 and 15 requires 15 memory locations, each 4 bits long. In contrast, a quantum computer can represent all 16 numbers between 0 and 15 inclusive using only a single memory location of 4 quantum bits.

A quantum bit, or qu-bit (pronounced cubit), is based on the fundamental notion of superposition of quantum states. The idea of superposition has attained many popularized forms, the most famous of which is contained in the paradox of Schrodinger's cat. In this paradox, a cat is placed in a box containing a relatively stable radioactive source. The cat will die if the source emits any radiation. In real terms, the cat is either dead or alive, regardless of whether or not we look inside the box. According to quantum mechanics, the cat is both dead and alive; it exists in the superposition of dead and alive states. Until we open the box and look inside (i.e. perform a measurement), we must think of the cat as existing in some sort of metaphysical limbo. The only physically correct statement is that there is a certain probability that the cat is dead and a certain probability that the cat is alive. This is known as the probability amplitude.

As far-fetched as this idea seems, it is partially responsible for a whole range of observed physical effects, the most basic of which is interference of photons in light rays which pass through small slits in an otherwise perfectly reflective wall. The paradox of Schrodinger's cat has often been used in order to understand the idea of a qu-bit. The cat is the qu-bit. If the cat is dead it is designated 0, if the cat is alive it is represented by a 1. Since the cat is neither dead nor alive, the cat is both in state 0 and in state 1. In this sense two numbers (0 and 1) have been represented using only one qu-bit (in this case, one cat). A classical bit would require 2 bits, one off (dead) and one on (alive) to simultaneously represent the two numbers, 0 and 1.

A quantum computer manipulates qu-bits in such a way that the probability of measuring the correct outcome increases during the course of the computation. The information must be stored in some quantum state which is relatively isolated from the environment. The most promising approach is to use the nuclear spin—which in quantum mechanics is analogous to the spinning of a top—that can store information for times on the order of seconds. Nuclear spin is used as small fluctuations of the environment are less prone to affect the quantum state and therefore the quantum coherence of the system.

Quantum coherence is essentially a measure of the degree of "conversation" between the computer and the environment. An atom's nuclear spin can occupy one of many states. If the quantum states interact with the environment, then information about the states such as the spin of the nucleus and the positions of atoms leaks out, causing the quantum states to lose their purity and therefore lose their ability to interfere. Interference patterns contain relative information about the configuration of quantum bits. Since interference properties of quantum states form the basis of most theories of quantum computing, the ability to limit decoherence is the study of many scientific papers.

A group of researchers at MIT and the University of California at Santa Barbara have already built a quantum computer using basic principles of nuclear magnetic resonance (NMR). These NMR-based quantum computers use redundant spins where one can actually visualize the answer to simple computations arising from the density of certain spin configurations.

Quantum Cryptography

Quantum cryptography can be thought of in two ways. Quantum computers could potentially be used to break classical ciphers which rely on the difficulty of factoring large numbers. Alternatively, methods of quantum mechanics could be employed to send quantum-encrypted messages which are theoretically impossible to break even using a quantum computer.

In 1994 Peter Shor, a researcher at Bell Labs, proposed a quantum algorithm for factoring large numbers in polynomial time. Polynomial time algorithms would make the breaking of ciphers based on 100- or even 1,000-digit keys a fairly simple operation. The algorithm does not factor the number directly but rather relies on certain mathematical properties of greatest common divisors to separate possible from impossible solutions. By beginning the algorithm with the qu-bits set in a superposition of all possible states the algorithm manages to find the "order" of a given number in a short period of time. Since factorization canbe related to finding the order of a given number, one can then extract the factors using relatively simple methods. As of yet, no one has actually managed to factor a reasonably simple number for a classical computer (e.g., 29 or 45) using Shor's algorithm on a quantum computer.

The other branch of quantum cryptography involves the actual transmission of information via states which are secured from eavesdropping by the laws of physics. Though this might sound far-fetched, every few weeks a new journal article appears which discusses the possibility of implementing a different method of quantum cryptography. One of the most recent works by Dominic Mayers and Andrew Yao, computer scientists at Princeton University is entitled "Unconditional Security in Quantum Cryptography." Although the titles might appear audacious, the results might not be too far off in the future.

The basic protocol behind Mayers and Yao's paper, and many other cryptographic systems, is based on earlier proposals by S. Wiesner in 1970 and C. H. Bennet and G. Brassard in 1984. The quantum cryptosystem can be understood as follows. Consider a sender, Alice, and a receiver, Bob. Alice may transmit photons in one of four possible polarizations—0, 45, 90, or 135 degrees. Polarization indicates that the electromagnetic field of the photon is oriented in the given direction. According to quantum mechanics, a detector may distinguish between polarized photons of 0 or 90 degrees, or between those of 45 or 135 degrees. Since these states are incommutable, it is impossible to distinguish between both types simultaneously.

In order to agree on a key, Alice transmits a series of photons whose polarization is chosen at random. For each photon, Bob chooses a type of measurement (0 vs. 90 or 45 vs. 135) at random and then records the result of the measurement. Bob then announces the order of measurement and Alice informs Bob which measurements were of the correct type. For example, if Alice sends a 0-oriented photon and Bob measures along diagonals then Alice informs him that he was incorrect. For all measurements of the correct type, both Alice and Bob write down a series of 1s and 0s depending on whether the signal sent was 0 and 90, or 45 and 135. This string of numbers becomes the key.

According to quantum mechanics it is impossible to observe a state without modifying it and altering the polarization. Therefore any eavesdropper will modify the transmitted photons and Alice and Bob will no longer have the same key. By revealing a random subset of the key in public Alice and Bob can verify that in fact there was no eavesdropping. Once they are satisfied, they can proceed to transmit information, encrypted by the key. Although this is just one of many quantum cryptosystems it is perhaps the easiest with which to understand the basic mechanism of quantum cryptography.

CRYPTANALYSIS

Cryptanalysis refers to the process of solving a code or cipher. Coined by the brilliant cryptologist William Friedman in 1920, this term is derived from the Latin *crypta*, the Greek *kryptē* < *kryptós*, "secret, hidden," and *ana*, "up, throughout" and *lysys*, "a loosing." Terms such as *decode*, *decipher* and *decrypt* are sometimes used interchangeably, but they are not synonyms. As the names imply, decode refers to breaking a code and decipher means to solve a cipher. Decryption, a more general term, can refer to either. Cryptanalysis refers to the interception and solution of a message by a third party.

In ancient times cryptanalysts were thought of as practitioners of black arts. It took centuries for their work to be understood and appreciated. Eventually, however, with organized government-supported efforts such as the black chambers, cryptanalysis became a respected profession.

Organized cryptanalysis has been a hidden part of diplomatic relations for several hundred years. In the past, each black chamber had its nimble-fingered specialists who took impressions of wax seals, and made molds and forgeries. Others softened the seals and opened the letters with heated wires or knives. Once the contents were copied, the letters were carefully resealed and sent on to their destinations.

Cryptanalysis continues to be a part of security and diplomacy today, as groups of skilled analysts intercept and decipher missives, using the information garnered to advance the interests of their government. Instead of the extraordinary pen-and-paper efforts of the past, however, satellites, specialized sensing devices, computers and other technology all now play a crucial role in the cryptanalyst's art.

CRYPTANALYTIC TECHNIQUES

Cryptanalysis requires skill and experience with letter frequency counts, contact charts, statistics, characteristics of different languages, as well as in military, diplomatic and commercial fields. Following is an outline of the basic methods of cryptanalysis:

1. Knowledge of the situation for which the particular masking system has been created. For instance, if the assignment were to solve diplomatic exchanges, searches should be focused on terms frequently used in diplomatic missives. In addition, knowledge of the plaintext language is essential, as all established languages have unique characteristics of grammar, syntax, sentence structure and spellings which can yield valuable clues.

2. Detailed study and arrangement of the material to highlight patterns or repetitions.

3. Application of number and letter frequency charts.

4. Linking patterns to the proper cryptographic method. Cryptanalysts must use all their deductive and inductive reasoning to determine which cryptographic style has been used.

There are many methods of cryptanalysis, and the cryptanalyst can apply his or her own methods to tackle the problem. The first step, however, is usually an analysis of ciphertext frequency.

The recording of letter repetition came to be called "frequency counts." The term *frequency* refers to the number of times a letter or digit appears in the encryption. This is a primary clue for a cryptanalyst, because every legitimate language has discernible patterns for vowels, consonants and syllable pairs. An analyst generally begins an assault on a cryptomethod by counting the letters, numbers and any other symbols used and by noting with which other letters or figures they appear. Some letters can be expected to appear more often than others, although if the sender's shield is strong, no discernible pattern may be apparent at first.

Standard tables of letter usages are consulted for different languages. The contrived word *etaonrish* is a favorite of cryptanalysts working with the English language. This word is made up of letters most often used in English. The precise ordering of these letters varies with different tabulations, but the letter *e* tops almost every list and is usually followed by *t, a, o, n, r, i, s* and *h*. Infrequently used letters such as *q, x* and *z* are also important since they may be covered by enciphered letters that occur only a few times.

The check for frequency is set up with a full alphabet and marks known as *tallies* are placed above each letter every time it appears. Patterns begin to be obvious as the digits are totaled.

X	G	L	W	C	Q	U	B	K	I	V	A	P
R	D	N	Y	E	J	T	M	S	Z	O	F	H
9	9	8	8	8	7	6	5	4	3	2	2	1
30	28	27	27	26	24	23	20	20	18	16	15	11

Codes and ciphers cannot be solved by simply comparing frequencies to standard lists of letter usage. Still, once letters have been found to show some frequency tendencies, these patterns are studied.

When an analyst sees tendencies in the ciphertext's characters that seem to match examples in the sender's language, he or she tries testing the characters by matching them up in pairs with other letters they contact. Contact charts take advantage of the fact that most languages have letters that typically operate in pairs and other letters that rarely meet. The characters or numerals that conceal them may therefore show some of the same tendencies. When groups of letters in a cryptogram begin show certain relationships, they are arranged in a contact chart.

In the chart below the letters being counted appears at the left and the same letters are placed in rows to the right. The number of contacts between the letter on the left and each letter on the right is noted and the total number of contacts is listed on the far left. Usually a mark is placed above a letter to show that it has come before the subject letter, and a mark is placed beneath to show that it has followed the subject letter.

20	Q	Q	C̲	P̲	K̲	G
18	C	Q̲	C	P	K	G
15	P	Q̲	C	P	K	G
12	K	Q̲	C	P	K	G
9	G	Q	C	P	K	G

In this chart *C* has preceded *Q* three times creating the digraph *CQ*, and has followed *Q* twice making the digraph *QC*. This pattern also appears for *Q*; it is shown preceding *C* twice and following *C* three times. Because *Q* and *G* do not contact, there are no contact marks in the *G* and *Q* comparison.

In a properly constructed contact chart the letters follow the characteristics of the language being used, which gives an analyst clues to the plaintext. The most common English letter *e* has contact with many other letters. The next three high-frequency vowels are *a*, *i* and *o*, which not frequently found together. In English, *ai*, *oi* and *ao* are rather infrequent, for example, while *io* is used more often. When a digraph in the chart appears to have an infrequent association, it may prove to be a rare pair in the plaintext.

Consonants also have predictable tendencies. Some consonants stand next to each other in the standard alphabet but are not standard pairs in words (e.g., *r* and *s*). The letter *r* has many contacts with vowels, while *s* tends to pair with consonants. The digraph *he* is quite common while its reverse, *eh*, is rare. The letter *n* is often preceded by a vowel, so a ciphertext letter in the chart often preceded by the same enciphered letter might signify a plaintext *n* preceded by a vowel. Pairs of consonants like *th* may also be revealed by a frequent digraph pair.

Single-letter frequency is, in part, a function of digraph frequency and a second frequency pattern can therefore be drawn for digraphs (see Appendix). These groups appear frequently because of English word-formation patterns. Some of the most common pairs are *en, re, nt, th* and *an*. Frequent tri-graphs (three-letter combinations) are *ent, ion* and *tio*.

After finding such contacts, the analyst tries these findings in the ciphertext to see whether identifiable patterns emerge. Frequency lists are particularly helpful when there is a pattern of repetitive letters. For example, *zz* or 11 may very well be *ee* or *ss*. A suspected transposition can also be attacked by plac-ing the numbers or letters in various geometric designs. Letter counts can also help as the number of characters in a transpo-sition ciphertext is often the same as the number in the plaintext.

These trial-and-error tests are known as anagramming. Unlike the hobbyist who deciphers anagrams from known letters (such as "now"–"won," "dear"–"read"), the cryptanalyst begins with the unknown, considering such factors as the language in which the message is written, its length, the pat-terns and numbers of the groups of letters or digits used and any clues to the subject matter.

Once the analyst makes a breakthrough with an opening phrase, date or letter pair, more involved substitution begins. Known as parallel reconstruction, this rebuilding of the key alphabet can be accomplished by placing ciphertext letters beneath those of the standard alphabet. Once the analyst dis-cerns some of the plaintext/ciphertext equivalents, he or she begins comparing the possibilities.

PLAINTEXT: a b c d e f g h i j k l m n
o p q r s t u v w x y z

CIPHERTEXT: • 1 • n • • • v w • z • • •
• • x • • • c • • • • •

Sometimes the analyst is fortunate enough to find clues in such primary arrangements. In this example, the apparent matchups of *b* and *l*, *d* and *n* indicates that ciphertext *m* may be plaintext *c*. Since ciphertext *x* is suspected to be plaintext *q*, ciphertext *y* becomes a possible match to plaintext *j*.

Such similarities of position may lead to further solutions. Suppose that a different alphabet arrangement apparently indicates that ciphertext *o* is plaintext *h*. If the analyst finds a *p* in the encrypted message, he or she could try *i* as its plaintext with the assumption that if *p* follows *o*, then *i* follows *h*.

Pairs of identical letters, frequently used characters, common pairs such as *th*, and vowels that appear with particular consonants can all be useful in the anagram-solving process. The final proof, however, is in the discovery of sensible words.

In addition to the techniques mentioned above, a cryptanalyst often uses tables with words listed by length, rhyming sequences, popular phrases and even jargon or slang. Using these basic techniques, a cryptanalyst is well on the way to breaking a cipher's defenses.

HINTS FOR BREAKING POLYALPHABETIC CIPHERS

A substitution can be best analyzed by frequency when it is monoalphabetic. Substitutions often involve multiple alphabets, however, so these must be reduced to simpler forms. As well as frequency tables, word lists, and anagramming, other analytical aids have been developed for specific types of ciphers.

In 1863 Prussian army officer Friedrich Kasiski's text *Die Geheimschriften und die Dechiffrir-kunst* (Secret writing and the art of deciphering) described a way to break polyalphabetic ciphers with repeating keywords. Kasiski realized that when a repeating keyword encrypts repeating letters in the plaintext, this causes repetitions in the ciphertext. For example:

KEY:	s	k	y	s	k	y	s	k	y	s	k	y	s	k
	y	s	k	y	s	k	y	s	k	y	s	k	y	

PLAINTEXT:	t	o	b	e	o	r	n	o	t	t	o	b	e	t
	h	a	t	i	s	t	h	e	q	u	e	s	t	i	o	n

CIPHERTEXT:	q	c	v	n	a	e	x	a	p	q	c	v	n	y
	1	k	z	r	g	y	1	h	v	c	j	x	p	d	a	u

When the repeating key letters *s*, *k* and *y* contact the repetitious letters in the message *t*, *o*, *b*, *e t* and *h*, the ciphertext sequence *q*, *c*, *v*, *n*, *y* and *l* is repeated. This tells a cryptanalyst that a keyword has repeated one or more times. This discovery led Kasiski to notice patterns of repetitions and spaces between letter clusters.

The sequence from the first letter of the series *q, c, v, n, y, l* until the next time it repeats is called an *interval*. This particular group has a nine-letter interval. Kasiski found that this length of the interval was correlated with the number of times the keyword repeated. In this example, the nine-letter interval indicates a three-letter keyword used three times. By the same token, a 16-letter interval could signify a four-letter keyword used four times. Such clues enable the analyst to locate other points in the encryption where the specific sequence appears again.

The discovery of the number of characters in the keyword is an important breakthrough because it indicates the number of alphabets in the encipherment. For example, if the keyword has three letters, three alphabets have been used. The analyst then groups all of the letters encrypted by the first keyletter, all the characters enciphered by the second keyletter and so on. These groups of characters, each enciphered by only one letter, can then be treated as a monoalphabetic substitution.

Cryptanalysts must be careful of leaping to conclusions, however, as finding the pattern of repetitions and determining the actual number of letters in the keyword are not the same thing. For example, an interval of 20 letters could signify a keyword four letters long and repeated five times, or it could be a 10-letter keyword repeated twice. To distinguish between these possibilities, Kasiski developed a procedure which has since become known as a Kasiski examination.

Using Kasiski's process, the analyst works out the numerical distances separating the letter clusters in the cryptogram. To do this, he or she calculates the distances separating the repetitions from each other by listing the positions of the repetitions and finding their factors (any two or more quantities which form a product when multiplied). An example is shown below:

REPETITION	FIRST	SECOND	INTERVAL	FACTORS
rdp	3	165	162	1 × 3 × 6 × 9
dnlur	6	186	180	2 × 2 × 5 × 9
plxw	24	402	378	3 × 3 × 7 × 9

The repeating letter group *dnlur* occurs six letters into the cryptogram and again 186 characters later. The interval is 180 (186 minus 6). The factors of 180 are 2, 5 and 9 (2 × 2 = 4 × 5 = 20 × 9 = 180).

Other repeating letters and their intervals are also studied. While a small number like 2 might be a frequent interval, its brevity means that it is not a good candidate for a key length. Since very short keys have proven to be unlikely possibilities, analysts look for longer keys. As well as genuine key-induced repetitions, solvers also have to consider accidental repetitions that give false leads and complicate the search for factors.

Suppose the best potential factor is 9. This means that the number of letters in the key is nine. The analyst then writes out the first nine letters of the message. Beneath them he or she places the characters assumed to be encrypted with the same keyletter. Trial-and-error searches are made for plaintext equivalents in these columns and a frequency count is developed from this step.

As the frequencies are tallied, patterns develop. It is important at this point to discover a monoalphabetic substitution, which will prove that the key length is valid. Unexpected results, such as an unusual frequency profile, can indicate a mistake. For example, the regular English profile has expected areas of low frequency as segments such as *pqr* and *uvw-xyz*. A marked deviation from this outline could be a warning of a problem.

If the frequency count shows a profile that conforms to normal distribution, the analyst can then begin the process of anagramming to reconstruct the plaintext. If no recognizable system basis emerges, the analyst must return to linguistic clues, contact charts and more detailed frequency guesswork.

Another important contribution to cryptanalysis was made by Dutch language scholar Auguste Kerckhoffs in his 64-page *La Cryptographie Militaire* (Military cryptography) published in 1883.

In addition to discussing encrypting methods like the Saint -Cyr slide, mechanical encrypting mechanisms, the telegraph and military signaling needs, Kerckhoffs was the first writer to distinguish between a general cryptographic method (e.g. a codelist or a cipher disk) and its key. This meant that as long as the key was strong and was kept secret, the general method could be learned without endangering a particular dispatch.

Kerckhoffs also introduced two important tools for analyzing polyalphabetic ciphers: Superimposition and symmetry of position. Superimposition is a primary way to solve polyalphabetic substitutions. Using a number of communications known to be hidden by the same key, the cryptanalyst arranges one cryptogram on top of the other. This sets up a series of columns in which encrypted letters and their potential keyletters can be sought.

Kerckhoffs demonstrated that since all the dispatches were encrypted by the same key, each of the letters in the first row were encrypted by the same keyletter. This was true for each column and each key element in turn. He set up a chart that is briefly exemplified as follows:

COLUMN:

CRYPTOGRAM:	1	2	3	4	5	6	7 . . .
1	b	g	j	d	x	c	r
2	b	g	n	d	a	z	u
3	b	c	m	i	x	z	q
4	h	g	n	y	e	r	v
5	b	l	c	d	h	p	r
6	b	g	o	q	n	h	r
.							
.							
.							

With this one-to-one relationship established, each column can be considered a monoalphabetic substitution. It can then be probed with such analytical tools as letter frequency counts, typical letter pairs and contact charts.

Kerckhoffs developed a second major decryption tool, symmetry of position, which helps determine the locations of ciphertext letters. Kerckhoffs set up a table with a plaintext alphabet and placed the cipher messages beneath them. He then looked for ciphertext characters' locations and relationships with each other. He found that there were similar letter placements as well as set patterns of intervals or spaces between them.

This can be demonstrated with the words *world language* (letters used only once), inserted into a cyclical alphabet in a table. The letter *b* is five places from the letter *a* throughout the alphabet sequences, as is the letter *h* from *p*. Eight cells are always found between *l* and *c*.

PLAIN ALPHABET:

a b c d e f g h i j k l m n o p q r s t u v w x y z

CIPHER ALPHABET:

w o r l d a n g u e b c f h i j k m p q s t v x y z
o r l d a n g u e b c f h i j k m p q s t v x y z w
r l d a n g u e b c f h i j k m p q s t v x y z w o
l d a n g u e b c f h i j k m p q s t v x y z w o r
d a n g u e b c f h i j k m p q s t v x y z w o r l
a n g u e b c f h i j k m p q s t v x y z w o r l d

After careful tabulations Kerckhoffs argued that letter-to-letter relationships could be made between two characters. If one of these letters could be identified in a second cipher alphabet encrypted with the same key, the known interval could help locate the position of the second letter in that cipher alphabet. This tool for predicting the positions of ciphertext characters also helps to locate potential plaintext equivalents using the idea of the interval.

Assume that a transmission made from this table has been solved and it is known that cipher characters *r* and *n* conceal plaintext *c* and *g*. The cipher letters have an interval of four. When the analyst discerns *r* in a different cipher alphabet, he or she can also find the location of the second letter in this alphabet because of the known number of places between them. If *r* is found to be plaintext *h*, an interval of four aligns cipher *n* with plain *l*. New cipher equivalents can then be located throughout the new alphabet and create other breaks in its shield.

William Friedman's monographs have also contributed much to the science of cryptanalysis. In monograph number 15 titled, *A Method of Reconstructing the Primary Alphabet from a Single One of the Series of Secondary Alphabets* (1917), Friedman presented a clever way to analyze polyalphabetic encryptions.

The primary alphabet was exemplified as a mixed alphabet, such as the Vigenère tableau, which was used to make a polyalphabetic encipherment. The secondary alphabet was found in organized stages by code and cipher solving. A primary alphabet built upon a keyword such as *absolute* could be arranged as shown:

PLAIN ALPHABET: a b s o l u t e c d f g h i j k m n p q r v w x y z

CIPHER ALPHABET: e c d f g h i j k m n p q r v w x y z a b s o l u t

The analyst would not know the order generated by the keyword unless he gained it by other intelligence. The recovery process is therefore set up alphabetically, with the lower row of letters becoming the secondary alphabet and the keyword hidden.

PLAIN ALPHABET:	a	b	c	d	e	f	g	h	i	j	k	l	m
	n	o	p	q	r	s	t	u	v	w	x	y	z

CIPHERTEXT:		e	c	k	m	j	n	p	q	r	v	w	g	x
	y	f	z	a	b	d	i	h	s	o	l	u	t	

The original alphabet could be drawn out of the maze of letters by creating links of letters and spreading them out in arranged patterns. The solver makes the connections by starting with the letter beneath plaintext *a*, which is *e*. The analyst then finds *e* in the plaintext alphabet and uses the letter below it, *j*, as the second part of the link. After completing the first chain, the analyst writes the letters out in trial intervals with increasingly wide spaces until plaintext parts of the keyword can be discerned.

	1	2	3	4	5	6	7	8	9	10	11	12	13	14	15	16	17	18	19	20	21	22	23	24	25	26
1	e	j	v	s	d	m	x	l	g	p	z	t	i	r	b	c	k	w	o	f	n	y	u	h	q	a
2		e		j		v		s		d		m		x		l		g		p		z		t		i
3			e			j			v			s			d			m			x			l		

Discovery of a primary alphabet is important to finding the solutions of other messages believed to be based upon it. It can help reveal other keywords of different lengths, and can provide clues to the keying methods in different communications from the same source.

Cryptanalysts still apply Friedman's discoveries in the form of modern algorithms used to program computers.

As high-security ciphers become more complex, decryption becomes correspondingly difficult and cryptanalysts rely on clues provided by espionage and miscues to aid their research. Many of these cryptanalytic techniques are now performed with the aid of computers, which help review millions of possible letter combinations. For most of history, however, analysts relied on their own mathematical and intuitive abilities to break ciphers. Even after computers have provided many clues, the lessons learned in the past are still essential to modern cryptanalysts.

5 7 1 6 3
CODES

A code, from the French and Latin *codex*, meaning "writing tablet" or "tree trunk," is a cryptographic system by which a word, number, letter, symbol or phrase is substituted for plaintext words, letters and phrases. While codes and ciphers are often confused, their primary difference is that ciphers transpose or substitute individual letters or letter pairs, whereas codes substitute entire words, phrases or number sets. A chart comparing codewords to a reverse transposition cipher is shown below:

PLAINTEXT	CODE	REVERSE TRANSPOSITION
Washington, D.C.	200	cdnotgnihsaw
London	castle	nodnol
Madrid	###	dirdam
Paris	gourmet	sirap
New York	trial	kroywen
Bombay	@c2	yabmob

Secret codes developed from individual name replacements to become small lists and then larger collections of substitutes. Formal concealment codes used by papal emissaries and diplomats of Mediterranean city-states have been found in early Vatican archives dating from the 1300s. As the security needs of national representatives grew, larger lists of coded names accompanied by cipher alphabets were combined in nomenclators. These collections were the primary form of concealment from the 15th to the mid-19th centuries.

At first nomenclators were kept on large folded pieces of parchment; later, covers were added to protect them and hide them from prying eyes, creating codebooks. Codebooks are essentially code dictionaries, in which codewords and the plaintext are equated. Unlike ciphers that often carry their key enciphered within them, codebooks are necessary for encoding and decoding because complete codelists are usually too long for easy memorization.

As codebooks grew to accommodate increasingly technical needs of the military and their agents they became less practical for secret activity. The possession of a codebook could draw suspicion to an undercover agent, and the loss or theft of a codebook would compromise the entire system. In addition, codes were not always practical on the battlefield as commanders did not always have time to consult lengthy general tomes nor the right personnel to form codes in combat conditions.

To counter these problems, cryptographers created other book-linked methods. These included using specialized codebooks for different sectors, such as navy or army, to facilitate encoding and decoding. They also created codes using the words found in novels or dictionaries, so that agents could keep them without attracting undue attention. For military commanders, it became necessary to provide specialized texts for battle preparations and actions. These took the form of headquarters' codebooks and front line field codes.

The development of Morse code and the electric telegraph (1844), and the completion of the transatlantic cable (1866) greatly increased the applications of codes for commerce. To reduce telegram and cable costs, businessmen consulted thick tomes listing brief codewords, five-character groups of letters or numerals that replaced lengthier word or sentence groups.

Codes can be made up of numeric or alphabetic substitutions. The 26-letter alphabet, however, provides more permutations for a five-character group than a numeric system. It is possible to create a total of over 100,000 permutations of pronounceable five-letter "words," all differing by at least two letters, using the normal alphabet with six vowels. The two-letter differential between each code word is important to compensate for mistakes during transmission or encoding.

Recent codes have also been constructed so that accidental transposition during transmission does not change the meaning of the plaintext word. For example, the commercial code in which *BOCKI* means "Do not arbitrate, if necessary submit to what is asked but endeavor to get some advantage," would not have codewords *OBCKI*, *BCOKI*, *BOKCI* or *BOCIK* contained within the code.

Codes are classified under two general groups: one-part and two-part codes. The former are also known as alphabetical or numerical codes, while the latter are known as randomized codes. As the names imply, one-part codes are set up in numerical or alphabetical order. Both the plaintext and the codewords are in alphabetical order so the same codebook serves both encoding and decoding purposes. Two-part codes have plaintext in alphabetical and numerical order, but their coded equivalents are random. Chance decides the matching factors for two-part codes. Separate codelists are required for encoding and decoding.

Like ciphermakers, codemakers often create additional layers of protection to make their codes more difficult to break. This process is known as superencipherment. During superencipherment a codeword or codenumber is enciphered by transposition (especially for number codes) or substitution (for words and numbers). In a numeric code, numbers known as additives can be added to a codenumber. Additives were introduced in the 1800s, when increased secrecy was needed for cables and telegrams. For example, suppose that a codenumber, also called the plain code or placode, was 1162 7824 5286. The additive 2040 was then added to this codenumber to create an enciphered code, or encicode:

PLAINCODE:	1162
ADDITIVE:	+2040
ENCICODE:	3202

7824
+2040
9864

5286
+2040
7326

After all the codenumbers in the message had been super-enciphered, the resulting encicode was ready for transmission. While the additive was not special in itself, as long as its secrecy was maintained it was easy for the communicators to use and did increase the difficulty for potential interceptors.

Alphabetic codes can also be subject to superencipherment through substitution. Some businesses have words that are difficult to conceal with other terms and the use of one phrase unique to a business may provide clues to the rest of the message. Under these circumstances, enciphering methods are added for more security.

In the chart, both a plaintext word and an original number are shown being concealed by an encicode. When the codeword signatory is replaced by substituted letters, the original word blueprint has been encicoded.

The codenumber for the digits 4916 is 2370. The numerals' positions can be transposed, then substituted with other numbers, resulting in the encicode 5068.

PLAINTEXT:	blueprint
CODEWORD:	signatory
TRANSPOSITION:	(no transposition of letters)

SUBSTITUTION:	dxqkmjvfz
ENCICODE:	dxqkmjvfz
ORIGINAL NUMBER:	4916
CODENUMBER:	2370
TRANSPOSITION:	7203
SUBSTITUTION:	5068
ENCICODE:	5068

Decoding tips: Using the agreed-upon substitution alphabet or transposition method, find the original codeword by reversing the steps. Consult the codebook to find the corresponding plaintext.

Another means of introducing variations into codes is through a system known as the "syllabary." As its name suggests, the syllabary is based upon on a table of syllables and is designed to avoid repetition of codewords which may compromise the entire code system. For example, if the plaintext word *ammunition* is frequently used, it can be broken into different syllables and a codeword assigned to each. In one instance, the word could be broken into *am mun i tion*, the next time as *amm unit ion*, another time *as a mmu nit ion*. As the codewords for each syllable would be different, this tactic lowers the number of repetitions during transmissions and thereby decreases the possibility of enemy decryption.

While codebooks are often depicted as the object of military espionage, they are also used in industrial spying. In today's global competition for technology, businesses have had to conceal their national and international correspondence, particularly in Western nations, where there is freer access to information than elsewhere. Such protection efforts must be constantly updated due to the efforts of cryptanalysts in science, industry, diplomacy and espionage.

The advantages of codes compared to ciphers primarily lie in the ease of encoding and decoding. Strong ciphers are far more time-consuming to encipher and decipher, and a few enciphered mistakes can cause the entire missive to be garbled. If a cipher is intercepted and decrypted by an enemy, the entire system is compromised, while the breaking of one encoded missive does not necessarily reveal the entire codebook.

On the other hand, a codebook is generally impractical for espionage, as it is difficult to carry around without attracting attention. A codebook is also susceptible to loss or theft and it is a more arduous task to create, print, bind and distribute a new codebook than it is to create a new cipher.

As with ciphers, the security of a code lies in frequent changes. Repetition of codewords can disclose the foundation of a code as surely as it can a cipher. While one-part codes offer very little security, a two-part code can be an effective concealment. The larger the code vocabulary and the more frequently it is changed, the more secure the missive.

Nomenclators

From about 1400 to 1850, the nomenclator, from the Latin *nomen*, "name" and *calator*, "caller," was the primary means of masking communications. Beginning as a list of names with their equivalent codewords used by the scribes of popes and kings to protect personal and diplomatic correspondence, the nomenclator developed into a collection of syllables, words and names similar to a code, with a separate cipher alphabet.

The Renaissance trade rivalries of the Mediterranean city-states increased the need for organized masking systems. With a growing number of names and phrases to encode, symbol cryptography was incorporated into the nomenclators. Also used were meaningless symbols called *nihil importantes*, or nulls, which were placed in the nomenclators to confuse would-be solution seekers.

A portion of a 15th century nomenclator similar to one from Siena, Italy, with 23 letters equated with symbols and names concealed by codewords, symbols, nulls and double letters is shown below:

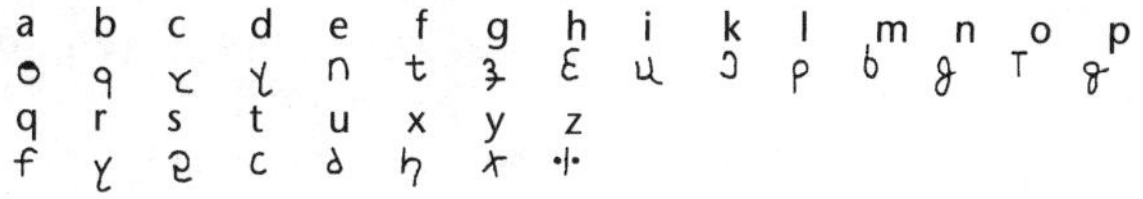
a b c d e f g h i k l m n o p
q r s t u x y z

Antonello da Furli	Forte
Cardinales	Florenus
Comes Urbini	Lux
Dux Calabrie	Ventus
Dux Venetiarum	Celum
Florentini	Terra
Gentes Regie	Stelle
Gentes Venetorum	Arena
Marchio Montisferrati	Arbor
Napoli	Nobile
Pitigliano	Duram
Sorano	Fortuna
Dominus Alexander Storza	
Ildibrandinus	

Duplices: bb cc dd ff ll mm nn pp rr ss tt

Nihil. Importantes:

Nomenclators were used by royal houses throughout Europe to conceal diplomatic and personal dispatches. However, they continued to list both their plaintext words and the nomenclator in alphabetical order until the 17th century, when the superb French cryptologist Antoine Rossignol recognized this security weakness and developed two-part nomenclators. The two-part version contained two sections: one listed plaintext elements in alphabetical order with shuffled code elements. The second part reversed this order, putting the code lists in order and mixing the plaintext elements. Although Rossignol's improvement was not always applied, nomenclators continued to grow, having as many as 3,000 elements by the 1700s. Nomenclators remained in use until around 1850, when the invention of the telegraph and its partner, the Morse code, superseded them. Rather than revising their nomenclators or creating new ones, governments phased out the long-used masking systems.

HOW TO MAKE A NOMENCLATOR

A nomenclator is a very simple code to create. Write a list of the words you most frequently use in your correspondence. These may be names, terms or addresses. Create codewords or symbols for each and record them in a list which equates each word to its codetext. This is the foundation for your nomenclator. The next step is to create a cipher alphabet which you will use for all the words that are not included in your codetext. (Refer back to Monoalphabetic Ciphers in the Ciphers section for ideas.) To encode a missive, isolate all the words which have code equivalents and replace them with their codewords. The remaining words should be enciphered with your cipher alphabet.

Decoding tips: Both sender and recipient should have a copy of the nomenclator. When you receive the encoded missive, look up the codewords on your nomenclator list and replace them with their plaintext equivalents. To decipher the remaining text, compare each cipher symbol or letter to the cipher alphabet to discover the plaintext.

Benjamin Tallmadge's Nomenclator

Benjamin Tallmadge was a Continental army officer and congressman during the American Revolution. He led the New York–based Culper Ring of spies, which used various methods of cryptography. In order to protect their missives, Tallmadge developed a one-part nomenclator of around 760 components. He chose the *New Spelling Dictionary* by John Entick, and applied its most frequently appearing words. Patterned after the nomenclators used during the Renaissance, the system was ready by July 1779. After writing selected words in a column, Tallmadge gave each word a number, while the names of specific locations and people were placed in a separate part of the code. The nomenclator was arranged in alphanumerical order and also included a mixed alphabet to encrypt terms and digits not listed. The system was somewhat vulnerable because its components were listed in sequence. Still, it provided a reasonable level of security.

Below are sample words with their numerical equivalents:

37	attone		143	defense
38	attack		144	deceive
39	alarm		145	delay
40	action		146	difficult
306	industry		550	ruler
307	infamous		551	rapid
308	influence		552	reader
309	infantry		553	rebel
711	George Washington		726	James Rivington
723	Robert Townsend		727	New York
724	Austin Roe		728	Long Island

Following is Tallmadge's mixed alphabet cipher:

```
a  b  c  d  e  f  g  h  i  j  k  l  m  n  o  p  q  r
s  t  u  v  w  x  y  z

E  F  G  H  I  J  A  B  C  D  O  M  N  P  Q  R  K  L
U  V  W  X  Y  Z  S  T
```

With this method, the message "rapid infantry attack" was indicated by 551.309.38. Periods signaled the completion of a full number so that digits would not be misconstrued. To form the past tense, a flourish (~) was placed over the number; 1$\widetilde{44}$, for example, meant "deceived."

A message about a new recruit, Arn, not in the nomenclator's words, were made with the existing words and the cipher alphabet. The message "delay rebel Arn" was encoded as 145.553.elp.

XYZ Nomenclator

The infamous XYZ Affair of 1797 involved secret demands by French officials for monetary commitments from the young United States. Details of these discussions sent by U.S. delegates in France to Secretary of State Timothy Pickering were encoded with a number-letter nomenclator that included the following coded words and fragments:

26	ped	493	ce
27	pen	494	ced
28	people	495	cieve
29	per	496	cent
449	no	1176	sy
450	nob	1177	sive
451	nom	1178	six
452	non	1179	sixteen
456	N. Carolina	1191	sort
457	not	1192	south
458	noth	1193	a
459	notify	1194	ab

Outraged by these proposals of bribery, U.S. commissioners sent this encoded answer:

449.449.457.1193.1178.27.493.

The decoded message spells out one of the most famous replies in American history:

no	no	not	a	six	pen	ce
449.	449.	457.	1193.	1178.	27.	493.

James Madison's Nomenclators

James Madison used several different nomenclators during his long service to the United States. In 1781, U.S. Secretary of Foreign Affairs Robert Livingston had created a nomenclator that contained an alphabetized group of words and syllables on one side and a list of the numbers 1 through 1,700 on the other. Thomas Jefferson and James Madison built their own private security screen using the Livingston system. It was practical because its flexibility permitted the letters or numbers to be paired with the plaintext in any pattern chosen by the sender and the addressee. Some examples of the Jefferson-Madison method included:

nal	119	o	527	P	941	qua	103
name	717	oa	746	pa	290	quest	1386
nant	42	oach	559	pan	381	question	799
nar	60	oad	217	paper	207	quin	1202

While this may not appear very complicated, the following missive would be incomprehensible to an interceptor who did not hold a copy of the matching lists:

`Don't 799 anyone. Look for the 717 on the 207.`

The recipient would know not to question (799) anyone, but to find a name (717) on the paper (207).

As a member of the Virginia delegation to the Continental Congress, James Madison used a nomenclator of about 846 items to send private messages to Governor Benjamin Harrison of Virginia. This system consisted of a list of digits, alphabet letters, syllables and place-names such as vienna (not capitalized in this usage), as in the following excerpt:

27	public	101	rhode island
28	found	112	carolina
29	perhaps	118	america
30	ing	144	britain
42	only	437 ß	section
43	farther	764 ∂	paragraph
44	at	782 –	hyphen
45	ci, cy	846 "	quotation

During his term as secretary of state (1801–09), Madison also used a 1,700-element nomenclator to correspond with Livingston, who had been chosen in 1801 as minister plenipotentiary to France. More than 40 communiqués were encoded in the style shown below:

118	could	551	place
119	council	552	plai
120	count	553	plan
121	country	554	play
302	monarch	689	with
303	monday	690	within
304	money	691	without
305	month	692	wn

Madison also communicated with his representatives Livingston and James Monroe, using a new code that came to be called the "Monroe Cypher." Although it was called a cipher, this system had general nomenclator characteristics, and its 1,600 elements were still arranged alphabetically:

118	consider	551	those
119	constitu	552	thought
120	consul	553	thousand
121	cont	554	thr
302	fa	689	arp
303	fab	690	art
304	fac	691	ary
305	fact	692	as

Book Codes

In addition to creating codelists from scratch, codemakers can also be use a preexisting text to encode the messages. Book codes probably developed as codemakers sought new sources of grouped words while compiling lists of words and equivalent terms for use in nomenclators.

In a book code, the desired word is found in the text and given a number, according to its location in the text. This number becomes the code. For example, the codetext 100.1.7 would indicate the seventh word found on page 100, line 1. Both message sender and recipient must have identical copies of the chosen book in order to avoid confusion.

The books were chosen according to individual tastes and needs. Language was an obvious factor since mutual under-standing between correspondents was a necessity. Content and level of complexity were almost as important, as it was essential that the plaintext words that the cryptographer was encoding could be easily found in the text. Spanish spies would not logically be using a chemistry text from Stock-holm, unless they were stealing chemical industry secrets.

Types of books also affected applications. Novels provided many words and imaginative titles and characters' names, but finding the specific words sought while scanning scores, and perhaps hundreds of pages, was not an easy task. Still, novels for creating book codes were frequently included within a spy's repertoire.

Encyclopedias, with their numbered, alphabetized volumes and standard column-arranged style, gave a better order to code arrangements. The volumes and alphabet groups were also encoded and placed in transmissions to indicate which book was to be used. For example, vol.1, A-B, could be con-cealed with a 9 for 1 and a 1 and a 2 for the A-B alphabet group.

Arguably the most popular books for these purposes over the years have been dictionaries. Their exact listings of words, definitions, word origins and pronunciations made them popular among codemakers. Their alphabetical organization and their extensive vocabulary made it easy for codemakers to find desired words. Their two-columned arrangement facil-itated decoding by page, column and alphabetized word.

In general, three numbers were used to signify a particular word: the first for the page number, the second for the line, and the third for the word. Terms not found in the text could be given a separate alphabet, or equated with other words or codenumbers. Variations on this method involved counting paragraphs, using *L* or *R* to indicate the right or left column, and making an enciphered code by adding preset numbers to the three standards, page, line, and word.

A distinct advantage of a book code is that it eliminates the need for a suspicious-looking codebook which could threaten an agent's cover or be lost or stolen, thereby compromising the entire system. However, along with problems in locating some words and in concealing oddly spelled names, dictionaries, encyclopedias and novels each have one obvious vulnerability. Unlike protected limited-edition codebooks, these texts are publicly sold. If discovered as the foundation for a code, that encryption form is completely compromised.

HOW TO MAKE A BOOK CODE

Choose the text which will form the foundation for your book code. A dictionary is easiest, but a novel, encyclopedia or other texts will also serve the purpose. The only restriction is that the vocabulary you will need to form your messages can be easily found within the text. Before you send a message, make sure that both you and the recipient agree on the order of the indicators: page, line, word; or page, column, line, word. Locate the first word of your missive in the text and record the page, line and word number, for example 47.12.2. Continue until the entire message has been encoded.

Decoding tip: Make sure that you are using the same book and same edition as the sender. If your edition is different, the total number of pages and number of words on a page may vary. As you already know the order of indicators, use each number to locate the page, line and word or page, column, line and word. The indicated word is the plaintext. Continue until you have decoded the message.

Benedict Arnold's Book Code

Perhaps the best-known case of treason in American history is the infamous actions of Benedict Arnold and his British collaborator Major John André during the American Revolution.

Arnold corresponded with André using at least three texts for book codes. At first they used a code coordinated with volume 1 of the fifth Oxford edition of Blackstone's *Commentaries*. Using a book method standard, the first number indicated the page, the second the line and the third number located the word.

Difficulties arose as those words that could not be found in the book had to be spelled out in their entirety, and the conspirators began to encounter problems when important words used in military parlance could not be found whole. Matters became further complicated when they were forced to mix increasing amounts of spelled and numbered phrases. Much time was also lost counting the lines and letters, and Arnold discarded this system after learning that his recipients were having difficulty with the long, involved cryptograms.

The plotters decided to shift their coding to the lexicon of Nathan Bailey's *Universal Etymological English Dictionary*. Its alphabetically listed terms were easier to locate that those in the *Commentaries*, and the Bailey book enabled Arnold and André to formulate a large portion of their conspiracy.

The pair then shifted their code base to a small, unidentified dictionary using a three-part number foundation of page, column and word. They also began superenciphering their new codenumbers, adding seven to each of the three parts of the numbers. For example, if the number needed was 11.2.1, the superenciphered encryption became 18.9.8 (11 + 7, 2 + 7, 1 + 7). An obvious weakness with this same single digit application was that the middle number would always be either 8 or 9 (7 added to either column 1 or 2). An astute cryptanalyst could have used this repetition to break through the additive mask on the other numerals as well.

Nicholas Trist's Book Code

A Virginian born in 1800, Trist was President James Polk's special intelligence-gathering agent to Mexico during the United States' war with Mexico from 1846 to 1848. Although it began smoothly, Trist's mission faltered after a misunderstanding with General Winfield Scott, who had conquered

Veracruz. Trist's cover was later severely compromised when reports of his mission were printed in national newspapers.

According to the Library of Congress's Trist Papers, Trist designed a code based on a former instructor's text. Its prefatory address initiated a letter-and-number code wherein almost all the letters in one alphabet were given three numerals or numeral pairs for equivalents, such as:

a	19, 23, 27	g	16, 21
b	61, 99, 215	s	4, 26, 40
c	36, 105	t	1, 5, 29
d	7, 64	u	6, 22, 38
e	3, 14, 25, 30	v	200, 448
f	10, 28, 110	w	52, 102, 167

Trist also developed a book code similar to earlier dictionary methods, although his code indicated letters rather than words. In his version, any digit after the third signified other letters in the same line. Trist's text was later determined to be a small text titled *True Principles of the Spanish Language*, written by Joseph Borras, the U.S. consul to Barcelona. Its prefatory address, which began, "The study of foreign languages . . . ," was the same as that used by Trist to make his letter-number codes.

While Trist was sending coded dispatches to Secretary of State James Buchanan, he also conducted negotiations with the Mexican government on his own initiative. Even after being recalled by Buchanan, he continued to negotiate on behalf of the United States. His perseverance and U.S. military successes resulted in the Treaty of Guadalupe Hidalgo in February 1848.

Though the war was a success for the United States, Trist was disgraced for disobeying orders and lost his State Department employment.

Presidential Election of 1876

A book code also figured in the 1876 U.S. presidential election controversy. Representatives of Samuel Tilden, the losing Democratic candidate, were found to have used the *Household English Dictionary* (London, 1876) and *Webster's Pocket Dictionary* among other methods, to mask telegrams about electoral fraud and bribery.

According to Ralph Weber's extensive studies, the *Household English Dictionary* was used by Democrats to mask their messages regarding Oregon's Electoral College Voters.

A dispatch called the "Gabble" telegram became famous when published in the *Detroit Tribune* and the *New York Times* in February 1877. The message signed by GABBLE was sent from Portland, Oregon to Samuel Tilden's address in New York City on December 1, 1876. It was worded as follows:

```
Head scantiness cramp emerge peroration hot-
house survivor browse of piamater doltish hot-
house exactness of survivor highest cunning
doltish afar galvanic survivor by accordingly
neglectful merciless of senator incongruent
coalesce.
```

Testimony before the U.S. Senate's investigative committee in February 1877 led to revelations about the dictionary code's system. The decoding process included:

1. Finding the message word on the codebook's particular page and column.

2. Counting the number of words from the top of the column down to the encoded word (e.g., seven).

3. Counting forward eight columns (four pages at two columns a page).

4. In the eighth column, counting down the page using the same number of words (seven). The seventh term was the plaintext equivalent of the codeword.

The decoded dispatch read:

I shall decide every point in the case of post-
office elector in favor of the highest democrat
elector, and grant the certificate accordingly on
morning of sixth instant. Confidential

    Governor.

The governor was Democratic Party loyalist E. F. Grover of Oregon. While his career was not directly threatened by the exposure of the telegrams, other hearings and lengthy newspaper coverage eventually ruined Samuel Tilden's second run for presidency.

Codebooks

Codebooks are the primary means of encoding and decoding. They function like foreign-language dictionaries, in which the codeword and the plaintext are equated. In most cases codebooks have an encoding and decoding section, although for security purposes the sending and receiving parties will sometimes only have the section applicable to their needs. Codebooks can be broken in two general categories: one-part and two-part codes.

One-part codes are codes in which the plaintext words or numbers and their concealment equivalents are both listed alphabetically (for letters) or sequentially (for numbers). As illustrated in the segment below, the plaintext is listed in alphabetical order, using a root word and listing its variations in order. The codetext is also listed in alphabetical or numerical order. A base code is used to designate the root word and variations equate to the variations in the plaintext.

As both codetext and plaintext are in sequential order, only one codebook is required for both encoding and decoding. This same feature, however, makes one-part codes susceptible to analysis and decryption. Analysis of repetition can disclose a root word, which then also reveals root variations. From this small chink in the armor an entire code can potentially be undone.

Despite this flaw, one-part codes have been used by both businesses and governments to conceal communiqués from interceptors, as well as to condense them to save on transmission costs. Below is a segment of a one-part business code:

```
cebai    Send
cebco    Send as follows
cebeh    Send as usual
cebgl    Send cheapest route
cebif    Send newest route
cebkm    Send quickest route
cebmu    Send rail route
cebok    Send road route
cebqx    Send water route
```

Two-part codes have code equivalents that are out of sequence in relation to the alphabetically organized plaintext words. More complicated than a one-part code, a two-part code requires a second list or a codebook that contains the codewords compiled in numerical or alphabetical order for decoding. Friendly communicators have both versions. The following is an example of a segment of a two-part code with codenumbers used for U.S. Army communications:

ENCODING		DECODING	
advance	2400	3812	attack
attack	3812	3815	refuel
battalion	3826	3820	front line
battle	3647	3826	battalion
begin	3849	3830	east
besiege	1390	3837	weather
binoculars	2629	3842	company
bivouac	7812	3849	begin
bridge	9151	3853	supply route
brigade	4170	3857	river

To encode a dispatch, the transmitter simply looks up the required words in the codebook to locate their codeword equivalents. As the codes are not in sequential order, the decoder requires a codebook that lists codegroups in numeric and alphabetic order. He or she would locate the codenumber or codeword in this separate list to decode the message.

The general of an army in battle may give hundreds of orders that in turn initiate thousands of radio messages and replies along the chain of command. An example is shown below:

`3826  3849  3812`

According to the decoding section above, these numbers contain the command:

`battalion` (`3826`)

`begin` (`3849`)

`attack` (`3812`)

For added security, the senders could superencipher the code by adding a prearranged series of digits to the codenumbers with noncarrying addition: for example, $3849 + 5136 = 8975$. Having been informed about the use of these additives, the intended receiver would subtract them with nonborrowing subtraction. The remainder is the codenumber originally equated with the plaintext word.

HOW TO MAKE A ONE-PART CODE

Codes are time-consuming to create as they often encompass a large vocabulary. To make a one-part code, write all the words you are likely to need in alphabetical order. Ensure that all phrase variations of the same root are placed in the same section for easy reference. In a column next to the plaintext vocabulary, create codewords for each. The codewords must also be in alphabetical order, or in sequential order if it is a numerical code. Following the example of the one-part business code, keep the same root—the first digraph or trigraph—for variations of the same plaintext words. You have now created a type of dictionary, in which both the plaintext and the codewords are in sequential order.

To encode a message, look up each word in your codebook to find its code equivalent. In some cases, there may be one codeword for an entire phrase. In the business code example, the instruction to "send by quickest route" can be replaced by the single codeword *cebkm*. Continue until the message is completely encoded.

Decoding tips: To decode the message, ensure that you are working from the same codebook as the sender, and that both the plaintext and the code equivalents are in sequential order. Look up each codeword in your codebook to find the plaintext equivalent. Continue until the message is decoded.

HOW TO MAKE A TWO-PART CODE

Two-part codes are fairly complicated to create, and a large code can take months to compile. A small two-part code can be made by writing in an alphabetical list all vocabulary relevant to your messages. In a column next to the plaintext, assign random numeric or alphabetic codes. Make sure that no combinations are repeated, and that your codenumbers or codewords are not in alphabetic or sequential order. To encode a message, simply look up each word and write down its codeword or codenumber equivalent.

You will also require an accompanying decoding book. To make this, compile all your codewords or codenumbers in sequential order and write their plaintext equivalents alongside them. Both correspondents will require an encoding and decoding book.

Decoding tips: Using your decoding book, locate the codenumber or codeword in the list to reveal its plaintext. You must also be informed if the codenumber has been superenciphered by any means. If additives have been applied, subtract the additive from the superenciphered number before locating the codenumber in the text.

EARLY STATE DEPARTMENT CODES

State Department Code of 1867

After the Civil War, the U.S. State Department went through a series of changes in its encryption policies in response to internal personnel differences and external technological innovations.

William H. Seward, Secretary of State from 1861 to 1869, is best remembered for his "folly" in encouraging the U.S. government to purchase the Alaskan "wasteland" in 1867 from Russia. In November 1866, however, he was also criticized for sending a very costly cable message to the U.S. ambassador to France, John Bigelow, which addressed U.S. concerns about French involvement in Mexican independence. The dispatch was sent via the transatlantic cable—a new and stunning achievement in the history of communications. Construction of the cable had cost millions of dollars and the early fees to use this advanced form of telegraphy were extremely high. Because of their sending complexity, code or cipher communiqués cost double the standard landline rates of Western Union.

Seward had used the Monroe Cypher, the U.S. diplomatic code that had been in use since 1803 to encode his 3,722-word message. After the reprimand that followed, Seward was determined to create a more modern and frugal concealment and commissioned the 1867 State Department code. Not all State officials agreed with this decision, and with good reason, as events transpired.

The code consisted of 148 pages with 23 of the most commonly used words in diplomatic exchanges each equated with a single alphabet letter. Six hundred and twenty-four of the next most commonly used words were designated with two alphabet letters. A cipher list was made for names or terms not in the codegroups, and the letter *w* was omitted, except in cipher form, because some European telegraphers did not have it in their national languages. Cipherletter *a* was distinguishable from codeletter *a* because cipher *a* would only be used to make words not contained within the codelist. Following are some examples of the 1867 State Department Code:

CODELETTER:	PLAINTEXT:
a	the
b	it
c	have
d	part., passive, or imperfect, indicative
e	and
f	of
g	ing.
h	see-sea
i	is
x	if
y	which
z	not
aa	me
ab	be
ac	my
ad	at
ae	old
af	now
ag	here
ah	so
ai	as
aj	all
ax	are
ay	with
az	such

PLAINTEXT LETTER:	CIPHER LETTER:
a	c
b	f
c	h
d	j
e	l
f	a
g	e
h	i
i	m
j	o
k	d
l	b
m	g
n	y
o	r
p	u
q	n
r	x
s	v
t	w
u	k
v	p
w	q
x	t
y	z
z	s

According to historian Ralph Weber, decoding proved to be more complex than Seward had intended. In his model, three-letter groups had been designated for lesser-used terms in diplomatic cablegrams. However, the three-letter groups were repeated in a number of sequential alphabets, and many an embassy worker spent long hours searching for the correct plaintext equivalents. The cable telegraphs also had problems creating spacing between the one-, two- and three-element codes and often ran letters together. For example, if a code-word was mistakenly omitted, the flanking codeletters *a*, meaning "the" and *f*, meaning "of" could be transmitted together as the two-letter code *af*, the code for "now." A simple cipher or transmittal mistake could inadvertently become an order for rapid action.

Although the 1867 code eventually proved to be defective instead of thrifty, bureaucratic inertia kept this flawed method active until 1876.

Chief Signal Officer's Code (Circa 1871)

In the early 1870s Colonel Albert J. Myer's Signal Corps office prepared a one-part code, apparently on their own initiative, for the State Department. The 88-page text included codes for hours of the day, dates, months and, for the first time in code history, years. The Chief Signal Officer's code was also the first State Department code to equate one codeword with a long phrase or a complete sentence, although this had been common practice in commercial codes for some time. Among the numerous codewords, women's names represented hours of the day and men's names signified days of the month. Cities and countries were replaced by digits. This method did not become an official replacement for the awkward 1867 code, although it proved to be far more comprehensive and easy to use.

One of the advances of the Chief Signal Officer's code was that it minimized repetition by offering alternate codewords for frequently used words. The plaintext words or numerals were written in script in the center of the page and hand-printed lists of the choices of cipher replacements were written in columns on both sides of the plaintext. As the list below illustrates, the time 10:30 could be replaced by either *Anna* or *Ida*.

CODEWORD	PLAINTEXT	CODEWORD
Anna	10:30	Ida
Agnes	12	Jane
Alice	8	Jenny
Amelia	9:30	Kate
Emily	10	Maggie
Emma	2:30	Nancy
Ellen	5	Nora
Hanna	8:30	Sally
Hilda	7:30	Sarepta
Henrietta	6	Sophia
Indian	six	Norfolk
India	ninety	Newark
Italy	forty	Norway
Ireland	five	Nashville
Inverness	one hundred	Nassau
Illinois	seventeen	Naples
Charles	13th	Mason
Calvin	23rd	More
Clark	4th	Grimes
Cameron	9th	Green
Cole	27th	Grant
Achieve	Act of Congress	Achieve
Active	Your action is approved	Active
Acute	Your action is not approved	Acute
Adage	Answer by telegraph in cipher	Adage
Adapt	Addressed a communication to	Adapt
Add	Authorize	Add
Adder	You are authorized to	Adder

The Cipher of the Department of State (1876)

In the centennial year of the founding of the United States, a codebook that combined real economy and better secrecy was finally produced. Although really a code, the State Department was still using the word cipher in its codebooks' names, so this codebook was titled *The Cipher of the Department of State*.

The 1876 "cipher" was developed by John H. Haswell, a native of Albany, New York, who had risen through the ranks at the State Department. He eventually became chief of the Bureau of Indexes and Archives in 1873. For a number of these years, Haswell had been bothered by State's problems with insecure and costly telegrams and cables, and began creating a replacement code on his own initiative.

As he began formulating *The Cipher of the Department of State*, Haswell studied methods such as Albert Myer's and Anson Stager's Civil War ciphers and business codes such as Robert Slater's *Telegraphic Code* (1870), incorporating some of the best aspects of each into his own creation. His 1,200-page one-part codebook was eventually nicknamed Red from the color of its cover.

Following the typical one-part code outline, Red contained plaintext words, phrases and short sentences in alphabetical order. While this setup made the code vulnerable to analysis, Red was considerably more practical than previous State Department codes.

Haswell's code contained groups of meaningless terms, known as arbitrary words, to express words and sentences. Like the Slater business code, it used five-digit codenumbers as well as codewords as concealments. Haswell also added an aperiodic changing key that used the names of animals as indicators to change a code. For example, if the word *eagle* was positioned as the first word in a coded dispatch, it instructed the recipient to add the number 20 to each group of digits. The numbers 15025 would then be read as 15045, and the recipient would locate codenumber 15045 in the codebook to find the appropriate plaintext.

This aperiodic key could also be changed as the months passed. For example, if January was represented by *Jaguar*, February by *Fox* and March by *Mink*, the arrival of a new month changed each communication's arrangement. This was the first time in State Department history that a cipher or code allowed for automatic code variation.

Aware that his orderly patterns made his alphabetical listings vulnerable, Haswell added the *Holocryptic Code, An Appendix to the Cypher of the Department of State*. The appendix listed 50 "rules," or methods of superencipherment that could be applied to encoded messages. Each rule was named for an animal in a list from "Ape" through "Zebra." The transmitter indicated the rule controlling a given missive by prefixing the codeword or codenumber for that animal to the message. For example, if the rule designator was "Falcon" and its code-name was *Deer*, the presence of the codeword *Deer* at the beginning of a cable told the transmittee which rule controlled the message.

The 50 rules fell under three general classifications: Route, Miscellaneous and Addition. In route forms, the recipient placed the plaintext words up or down different columns and read them from left to right across the columns. The Miscellaneous group involved a type of codeword shuffle. For example, the Tiger rule required starting the communiqué with the last codeword or codenumber and concluding it with the first.

The Addition classification warned the recipient that an additive had been used. The 12 numeral additions ranged from 33 to 322, each with an indicator term that identified the additive to the recipient. For example, the codeword *Horse* could indicate that the additive 203 had been added to each codenumber. The recipient would then subtract 203 from each number to discover the original codenumber.

To give additional protection to a message, encoders often used an additive and then located the new encicode number in the codebook, sending this codeword in the message. For example, if the indicator "Hawk" (codename *Hacon*) was used, signifying an additive of 100, the arrangement would look like this:

PLAINTEXT	FIRST CODENUMBER	ADDITIVE (PLUS 100)	CODEWORD
The	10050	10150	absolute
Joint	36823	36923	literal
Committee	20960	21060	cost
from the	31049	31149	marble
Senate	49221	49321	safely
and	12792	12892	bargain
House of Reps.	47260	47360	reconcile
of the	52622	52722	fasten
United States	54426	54526	topical
called upon	18992	19092	crescent
the President	44369	44469	planned
and	12792	12892	bargain
informed	35328	35428	know
him	33400	33500	imminent
of the	52622	52722	fasten
organization	42039	42139	outside
of the	52622	52722	fasten
Forty-fourth	30723	30823	glad
Congress	21895	21995	decide
and their	52638	52738	subtract
readiness	45995	46095	profit
to hear	33217	33317	important
from him.	33401	33501	improve

With the codeword indicator first, the message would then read:

HACON ABSOLUTE LITERAL COST MARBLE SAFELY
BARGAIN RECONCILE FASTEN TOPICAL CRESCENT
PLANNED BARGAIN KNOW IMMINENT FASTEN OUTSIDE
FASTEN GLAD DECIDE SUBTRACT PROFIT IMPORTANT
IMPROVE

To decode this missive, the recipient would identify the indicator, in this case *Hacon*, and its additive (100). Locating the codenumber equivalents for each word in the encoded message, he or she would then subtract 100 from each digit group to recover the original codenumber. He or she would look up each of these reverted codenumbers in the codebook to find their plaintext meanings.

The last section of the codebook contained a spelling section that functioned as a cipher substitution. This was used to conceal words that were not covered in the codebook's main vocabulary list. The spelling/cipher substitution section had three divisions. Each division contained 10 mixed alphabets that were identified by the digits 0 through 9. A standard alphabet was written above each cipher alphabet. The key was a five-digit number, usually the first codenumber in the message or the codenumber equated with the very first codeword. The first time the keynumber was applied, it indicated division number 1 of the mixed alphabet and its first mixed alphabet. The second use of the keynumber initiated division 2 and its first mixed alphabet.

To transmit the name of Hamilton Fish, secretary of state from 1869 to 1877, using the keynumber 78125, the sender would arrange the encryption as follows, repeating the keynumber to cover the plaintext:

7	8	1	2	5	7	8	1	2	5	7	8
H	A	M	I	L	T	O	N	F	I	S	H

Using cipher alphabet 1 and the keynumber 78125, the sender would locate the plaintext letter *H* in line 7. Suppose that its cipher equivalent was *D*. In line 8, under *A* in the regular alphabet is cipher letter *X*. In line 1, beneath *M* was cipher letter *C* and so forth until 7 reappeared. The sender then turned to division 2, which contained a new cipher alphabet. For the final two letters (*S* and *H*), division 3 was used when 7 and 8 repeated. The final encryption looked like this:

7	8	1	2	5	7	8	1	2	5	7	8
H	A	M	I	L	T	O	N	F	I	S	H
D	X	C	Y	Q	G	V	H	L	R	J	U

To decipher the message, the addressee placed the key numbers beneath the enciphered letters and reversed the process. Above *D* in line 7 was plaintext letter *H*, above *X* was *A*, and above *C* was *M*. With the repetition of the keynumber 7, the recipient referred to division 2's mixed alphabets and continued until he or she had deciphered the message.

The 1876 codebook continued to be active until November 1899 when John Haswell completed *The Cipher of the Department of State*. Haswell died that same month, but his efforts to improve the State Department's concealments had a lasting effect there.

The Cipher of the Department of State (1899)

The *Cipher of the Department of State* was nicknamed Blue after the color of its cover. It closely resembled the 1876 State Department system, although the Blue version did have some noteworthy differences. Blue was a more extensive code, having some 1,500 pages while the Red book only contained 1,200. Empty spaces for the addition of extra plaintext terms were made available at the start of each group of codes. Haswell also tried to include the most essential plaintext terms and phrases for U.S. embassy and consulate correspondence.

Haswell developed a small 16-page book to accompany the 1899 tome, called the *Holocryptic Code, An Appendix to the Cipher of the Department of State*, which contained 25 more rules than the 1876 appendix. These new rules provided numerous ways to mask communiqués with changed routes, additives or subtracted numbers, and substitutions of codenumbers for other codenumbers. Animals' names were again used as rule indicators. An example of the arrangement of a page from Blue follows:

CODEWORD (P)	CODENUMBER (609)	PLAINTEXT
Promotes	00	Russia
Promoting	01	Agreement between Russia and
Promotion	02	Agreement with Russia
Promotions	03	Ambassador from Russia
Promotive	04	Ambassador of Russia
Prompt	05	Ambassador to Russia
Prompted	06	And Russia
Prompter	07	Army of Russia
Prompters	08	Authorities of Russia
Promptest	09	Authority of Russia
Prompting	10	By Russia
Promptings	11	Cabinet of Russia
Promptly	12	Charge d'affaires of Russia

Green Cipher

After Red was outdated and Blue had been stolen in St. Petersburg in 1905, both codes were supplanted in 1910 by the Green Cipher, the first codebook published 1910 by the Department of State since the theft of Blue. A 1,418-page codebook, the Green reflected a growing concern for transmission concealment, with some composition and design changes. The pages of the book were divided into five columns, as shown below:

DAB 101	DAB	101	DAB	A
Aa	da	00	ad	A
Aalen	fa	01	af	—treatise
Aach	ga	02	ag	—treaty
Aadord	ka	03	ak	—treaty port

For example, the codeword for the city of Aalen was Dabfa. The plaintext Aalen was given this codeterm made from the top (*Dab*) and column letters (*fa*). The plaintext words "a treaty" had the codeword *Dabag*.

A process for dating telegrams was also revised:

MONTH		DAY TENS		UNITS		HOUR		
B	January	0	0	B	1	B	1	
C	February	A	1	C	2	C	2	
D	March	E	2	D	3	D	3	
F	April	U	3	F	4	F	4	i-Morning and noon
G	May			G	5	G	5	
K	June			K	6	K	6	
L	July			L	7	L	7	
M	August			M	8	M	8	y-Afternoon and midnight
P	September			P	9	P	9	
R	October			R	0	R	10	
S	November					S	11	
V	December					V	12	

The date and time were encoded with the alphabet letter representing the month, one letter each from the tens and units columns according to the day (for example, the fifth is *O* and *G*), and one letter each from the hour and morning or afternoon designation, such as *C* and *y* for 2 P.M. Using this process, May 6, 1 P.M. becomes the date-time codeword *GOKBy*. When used to conceal the date and time of a communication, the date-time codeword was added to the first message word to make a 10-letter codeword.

For additional protection, the instructions restricted transmitting the plaintext of any missive encoded by the Green cipher. Such communications had to be paraphrased and a copy sent to the State Department.

After May 1919 the Green had a slight title change, when Frank L. Polk, acting secretary of state, discarded the term *cipher* in reference to codebooks. With a new title, *Department of State Code A-1*, for a new code list, the Green became the Green Code. It remained in use in combination with other codes through the 1930s, when World War II increased the need for security and necessitated several changes in diplomatic and military cryptosystems.

CODES IN WORLD WAR I

Zimmermann Telegram

The World War I–era document known as the Zimmerman Telegram proposed a military alliance with Mexico against the United States. On January 16, 1917, German Foreign Minister Arthur Zimmermann sent a telegram from Berlin of about 1,000 codenumbers in groups of digits. It was wrapped in a two-part code called 0075 that was used by German diplomats.

The telegram was intercepted by Room 40, who deduced these facts from further intercepted exchanges commenting on the telegram as well as from other German transmissions using 0075. The leader of Room 40, William Hall, and his associates devised a plan to fill in the blanks of the dispatch. They knew that the exchange had been sent twice: once through American diplomatic cable lines and again through the "Swedish Roundabout" to the German ambassador in Mexico City. They guessed that the cable to Mexico City was probably not encoded in Code 0075 because the latter embassy had not used that cryptosystem in its previous communications. Based on this assumption, Hall tried to obtain the telegram that arrived in Mexico in order to study an encryption that they hoped was less secure than Code 0075.

A British agent codenamed T was said to have obtained a copy of this telegram from the Western Union office in Mexico City. Historians have later suggested that this account was a ruse to cover Room 40's cryptanalytic success. However it was obtained, the text was found to be masked by Code 13040, with which the Room 40 analysts were already familiar. Code 13040 had 75,000 codenumbers and approximately 25,000 plaintext items in a hybrid form of a one-part and a two-part code.

The solution of typical words and proper names in 13040 gave Hall and his staff more information to verify their work on 0075 and eventually enabled them to decrypt the German's Mexico transmission.

The encoding segment had hundreds of codenumbers arranged alongside sequential words, but the numbers were in purposely disarranged segments. The decoding section had words in mixed arrangements similar to these and a standard numerical order:

ENCODING		DECODING	
3827	ambassador	1289	signatory
3910	ambassadorship	1301	treaty
4156	consul	1320	treaty port
5161	consular	1361	ceremony
5304	consular agent	1423	courier
6177	consulate	1440	document
10728	consul general	1517	emissary
10949	consulship	1556	legation
8293	diplomacy	1602	minister
8605	diplomat	1644	mission
7012	diplomatic corps	1680	negotiation
7134	embassy	1715	plenipotentiary

The mixed blocks of numbers made analysis of this code more difficult than that of a simple one-part code. Still, the position of some of the words in the list helped reveal others that were bracketed by the known terms in 0075.

American Trench Code

With the American entry into World War I, security needs required new codes which could be easily used in battle. Some codebooks that developed at this time included the *American Trench Code* and the *Front-Line Code*, the first American codes to be created "in the field." Both were one-part codes with single letter superencipherment, but the former consisted of 1,600 elements compared with the latter's 500.

Heading the project was Howard Barnes, whose qualifications included a captain's commission and a decade of code work with the State Department. By reviewing an available British code and studying battlefield conditions, Barnes and his associates compiled the *American Trench Code*.

Developed to be issued along the U.S. Army's chain of command to companies on the front lines, the code was only distributed as far as regimental headquarters due to concerns that it would be compromised by capture. In the spring of 1918, one thousand copies of the *American Trench Code* were produced and printed in a breast-pocket, paper-bound format measuring 4½ by 7 inches.

Three thousand copies of the *Front-Line Code* were also made and distributed to the Army at the company level for security maintenance during the crucial early weeks of American Expeditionary Force (AEF) action on the Western Front.

The *American Trench Code* provided two types of codegroups, in either three-letter or four-digit arrangements. Although mixing character and letter groups was possible, the instructions advised restricting coding to one form. For example, the dispatch "patrol reports attack" could be encoded in two ways:

PLAINTEXT:	patrol	reports	attack
CODEWORD:	RAL	SAN	DIT
CODENUMBER:	2307	2408	1447

If a phrase was to be repeated a number of times in the same communiqué, it was to be encoded differently each time it was used by breaking it up into separate words. The phrase "gas attack" was encoded by the codeword *KOT*. If the words appeared separately in a dispatch as "gas" and later, "attack," their separate replacement terms were *KOR* (gas) and *DIT* (attack).

Although later security concerns forbade the sending of unencoded numerals, at the time of this code's printing numerals were still being sent in clear. When the message included numerals, the rules called for sending the preceding codegroup 2370 (*RUF*), which meant, "Read next group in clear."

The codebook was made up of 32 pages; pairs of opposing pages formed a cryptographic page. At the top of each were page numbers and consonants which ran from *B* to *Z*, with *J, Q, W* and *X* deleted. Each sheet contained 100 lines of digits arranged from 01 to 00 with combinations of vowel-consonant digraphs arranged beside each dinome (number pair). The codemaker built codegroups with either four-digit or three-letter combinations by joining the page symbols with the line dinomes or digraphs. Numeral codes ranged from 1200 to 2700 and literal groups ran from *BAB* to *ZYZ*.

The codenumbers 1201, derived from page 12 and digraph 01, to 1331, page 13 and digraph 31, referred to a list of equivalents for decimal points, weekdays and months. Alphabet and word equivalents began with the letter *A*, which was encoded by 1332 (*CEW*) and continued to 2795 (*ZYR*) which was the codeword for "zone(s)." To encode the letter *A* in numeric form, the encoder took the 13 from the top left page number and the 32 from the dinome group beside *A*. To encode *A* in letter form, the encoder used the letter *C* from the consonant pagination (beside the numerals) and combined it with the vowel-consonant digraph *EW* beside the letter *A* in the codebook.

In the following example, the instruction "acknowledge activity of artillery" can be concealed by either the digits 1343 and 1348 or by codeletters *CIN* and *CIV*.

ENCODING

01	AB	Decimal Point
02	AC	1.2
03	AD	5.9
04	AF	9.2
05	AG	9.45
07	AL	100
08	AM	155
09	AN	240
10	AP	Sunday
11	AR	Monday
12	AS	Tuesday
13	AT	Wednesday
14	AV	Thursday
15	AW	Friday
16	AZ	Saturday
17	EB	January
18	EC	February
19	ED	March
20	EG	April
21	EG	May
22	EH	June
23	EK	July
24	EL	August
25	EM	September
26	EN	October
27	EP	November
28	ER	December
29	ES	A.M.
30	ET	P.M.
31	EV	o'clock

DECODING

32	EW	A
33	EZ	abandon
34	IB	abandon first line
35	IC	abandon second line
36	ID	able (to)
37	IF	about
38	IG	above
39	IK	-ac
40	IK	accident
41	IL	according (to)
42	IM	accurate
43	IN	acknowledge
44	IP	act
45	IR	action
46	IS	active
47	IT	activity
48	IV	activity of artillery
49	IW	adjust
50	IX	adjutant

The use of distortion tables added extra protection to the letter groups. These tables consisted of 30 simple monoalphabetic cipher alphabets that were used to superencipher codeletters. These tables were changed frequently to maintain security.

To superencipher the codewords of the above example, *CIN* and *CIV*, the encoder simply substituted the original codeletters for their cipher alphabet equivalents. The superencipherments are *mcf* and *mcb*.

ENCIPHER

A	B	C	D	E	F	G	H	I	K	L	M	N
O	P	R	S	T	U	V	W	Y	Z			

h	o	m	s	v	a	r	e	c	z	k	n	f
l	u	w	y	i	t	b	d	p	g			

DECIPHER

a	b	c	d	e	f	g	h	i	k	l	m	n
o	p	r	s	t	u	v	w	y	z			

F	V	I	W	H	N	Z	A	T	L	O	C	M
B	Y	G	D	U	P	E	R	S	K			

To encode and superencipher the order "abandon second line activity," the codenumerals would be 1335 and 1347 and codewords would be *CIC* and *CIT*. Superenciphered with the distortion table, these letters became *mcm* and *mci*.

To decode a message, the recipient simply reversed the encoding process using the same codebook. Using the same distortion table as the sender, he or she found the letters *mcf* in the lowercase row and located their equivalents in the uppercase letters beneath them. Recovered from superencipherment, the uppercase letters were used to find the codebook's page and plaintext equivalent. If the code was in numeric form, the 13 or other first numeral pair indicated the page number and the second pair indicated the letter or word on the page. If the codetext was alphabetic, the recipient first established whether a superencipherment had been used. The distortion table would inform him or her of the plaincode (when recovered from superencipherment) which was again *CIC* or *CIT*. The initial letter then indicated the page and the digraph the plaintext equivalent.

Although more secure than some previous military codebooks, the *American Trench Code* was deemed too weak for extended service. An American codebreaker named J. Rives Childs of the Army's G2-A6 Enemy Code Solving Section easily broke through the distortion tables and read some 40 test messages. After his accomplishment, the code had to be abandoned, far sooner than originally planned.

The River Series: Potomac

After the *Front-Line Code* proved insufficient for World War I use, the codemakers of the U.S. Code Compilation Section worked to increase code complexity. On June 24, 1918, the result was the *Potomac*, the first of the "River series" of codebooks.

The *Potomac* was a two-part code with an initial edition of 2,000 copies and was issued for use to the level of battalion. With a combined encoding and decoding length of 47 pages, Potomac code contained about 1,800 phrases and words and measured approximately 7¼ by 9¾ inches.

Potomac was composed of three-letter codegroups with 100 sets of codes and equivalents per page. The trigraphic (three-letter) codes had both standard and particular construction rules. Vowels and consonants were used in varied positions rather than in the standard alternating consonant-vowel form. There were particular rules governing letter placement: letters *E, H, I, T* and *U* were not placed in the first position; *D, H, Q, T* and *Z* were not used as the second letter; and *H, I, Q* and *Z* were never put in the third position. The combinations were arranged so that there were no repeating letters in any one trigraph.

Potomac's creators placed special emphasis on the inclusion of nulls. The code's rules stipulated that at least one null should be placed amid every 10 codegroups in a dispatch and especially between double letters. Thus, the word *boot* was encoded:

b	o	null	o	t
KVG	LOC	ASY	VYN	ASG

As with other codes, transmitters sent the number and hour of filing of *Potomac*-encoded messages in clear, but the point of origin and addressee were encoded. Long dispatches were divided and sent as two or more messages. Following are some examples from a decoding page:

ABE	falling back
ABF	heavy
ABG	message received
ABK	supply
ABM	have you received
ABO	bombardment
ABP	barrage
ABS	battalion
ABV	automatic
ABW	must be
APE	relief completed
APF	retire
APJ	premature
APN	impossible
APO	withdraw
APU	machine gun ammunition
APW	e
APX	remove
APY	moving
ASB	92

The River Series: Mohawk

Following the release of *Suwanee* and *Wabash*, the *Mohawk* was the fourth of the River series. Thirty-two hundred copies

of the two-part *Mohawk* code were issued to the level of battalion in August 1918. It differed from its predecessors in that
its codes were in four-digit groups rather than three-digit
groups. Like *Potomac*, *Mohawk* contained variations for frequently used letters or numerals. Some examples of the
encoding and decoding sequences are shown here.

ENCODING				DECODING	
0	4616	4585	3524	2589	require
1	3211	4753	4997	2590	-ing
2	3323	4794	3305	2591	repeat
3	3657	3517	3188	2592	100
4	4498	2661	4213	2593	mile(s)
5	3667	2522	4431	2594	sending up
6	3272	2875	4900	2595	23
7	3766	4650	2613	2596	ought not
8	4004	3009	2650	2597	no
9	4550	3137	4313	2598	on this
10	3248	2525	4887	2599	q

Mohawk and its successor, *Allegheny*, which was also a four-
digit code with variants, were captured by the Germans in
October 1918. Learning from this loss, the last of the River
series, *Colorado* (issued on September 24, 1918) had printed
on its cover, "MEMORIZE THIS GROUP: 'DAM—Code Lost.'"
Upon the theft or loss of a codebook, the encoded missive
(*DAM*) was to be sent through the chain of command to alert
superiors of the security breakdown.

Telephone Code

In March 1918 500 copies of the *Telephone Code* were produced and distributed. The code concealed the names of commanders' staff officers and organizations. Originally planned
for telephone exchanges only, it did get applied in other
types of communications as well.

The lists of women's names that served as concealments soon earned the *Telephone Code* the nickname the "Female code."

ENCODING		DECODING	
1st Army	Bertha	Agnes	1st Corps
2nd Army	Dolly	Alexandra	81st Division
3rd Army	Kate	Alice	9th Corps
4th Army	Vera	Alma	6th Army
5th Army	Maude	Clementine	107th Division
6th Army	Alma	Constance	100th Division
5th Corps	Daisy	Cornelia	91st Division
6th Corps	Carrie	Joan	2nd Division
7th Corps	Violet	Jocelyn	102nd Division
8th Corps	Gabrielle	Priscilla	79th Division
9th Corps	Alice	Peggy	12th Corps
10th Corps	Helen	Vera	4th Army
		Victoria	16th Division

The Lake Series: Champlain

The role of the AEF increased during World War I as the new Second Army entered the combat zones. Codes called the "Lake series" were constructed for the Second Army's communication security while the First Army continued to use the then-current issue of the River series, *Colorado*.

Champlain's title was printed in red ink on its cover to distinguish it and its successors from the River codes, which were printed in black ink. It was two-part and had trigraphic concealments. With a growing number of codes in use, a command decision was made to identify the particular code being applied with an unencoded trigraph at the start of each communiqué. Thus, for example, *COL* indicated the Colorado code and *HUD* stood for the *Hudson*.

The Lake Series: **Huron**

In October 1918, *Huron*, the second of the Lake series, was initiated. This two-part code was noteworthy because it was the first code in U.S. history designed to encode *entire* telephone exchanges. As telephones were a new aspect of wartime security and were particularly susceptible to wiretapping, words transmitted vocally were given a phonetic alphabet. Words were spelled out over the phone using the following codewords to conceal each letter:

A	able	J	jig	S	sail
B	boy	K	king	T	tare
C	cast	L	love	U	unit
D	dock	M	mike	V	vice
E	easy	N	nan	W	watch
F	fox	O	opal	X	x-ray
G	george	P	pub	Y	yoke
H	have	Q	quack	Z	zed
I	item	R	rush		

When these dispatches were conveyed they were identified as the *Huron* code's phonetic group by the spoken word "Huron." The order Dock, Easy, Fox, Easy, Nan, Dock (pause) Rush, Item, Dock, George, Easy (pause) Opal, Nan, Easy meant, "defend ridge one." In some cases, the little-used *X-ray* or *Zed* was used to indicate a period as an alternative to the internationally known "stop," which disclosed sentence endings to the Kaiser's wiretapping eavesdroppers.

Huron also included the first *Emergency Code List*. This was a separate collection of 56 digraphs in two-part arrangement on a single page designed for use by front-line troops near the "no man's land" between Allied and enemy trenches. Following are some examples:

"EMERGENCY CODE" ENCODING

About to advance	SP
Ammunition exhausted	BX
are advancing	XF
at	AG
Attack failed	FS
Attack successful	XA
Barrage wanted	BD
Be ready to attack	SM
Being relieved	ZB
captured	AX
Casualties heavy	BJ
Casualties light	SF

DECODING

AB	Gas is being released
AF	trenches
AG	at
AP	Objective reached
AV	Enemy fire has destroyed
AW	Relief being sent
AX	captured
AZ	Look out for signal
BD	Barrage wanted
BF	right
BJ	Casualties heavy
BM	Using gas shells

American Radio Service Code No. 1

The introduction of the radio into the battleground had a substantial impact in the First World War. Having no radio encryptions upon entering the conflict, the AEF had to rely on France's *French Radio Code*. Language differences were not only awkward, but very dangerous as confusion with exchanges and dispatches required repetition for clarity provided more clues for the Kaiser's analysts.

In October 1918 a U.S. radio code was issued. It consisted of about 1,000 words and phrases in two parts, and 2,000 copies were distributed to brigade-level and artillery units. The codegroups were trigraphs without variants. The encoding section had standard alphabetized words in six sections that included transmission phrases as well as vocabulary for radio network operations and radio apparatus.

Following is an example of the *American Radio Service Code No. 1*, of which the AEF's Code Compilation Section developed, printed and delivered 2,000 copies in just six days.

ENCODING

126	B
642	bad
704	bag
908	balloon
994	bandages
545	base
325	battalion
180	battery
607	be
962	bearings
824	been
052	before

DECODING

001	alone
002	oil
003	ies
004	infantry
005	officer
006	except
007	code
008	fault
009	ing
010	account
011	grouping
012	er

PHRASES USED IN TRANSMISSION

304	Antenna was damaged
450	I am obliged to stop sending until . . . o'clock
367	I have been calling you since . . . o'clock
513	I have increased my radiation
584	It is forbidden to transmit until
024	Send faster
169	Send slower
212	Stop sending. You are interfering
312	Transmitting set was damaged
669	Wait few minutes; am changing batteries
739	Was obliged to stop sending until . . . o'clock
811	Will call you at . . . o'clock
715	Your sending is bad

Choctaw "Codetalkers"

"Codetalkers" was the name for the Choctaw Indians, who sent messages for the AEF during World War I. In October 1918 Choctaws with the 142nd Infantry saw combat in the area of Chufilly and Chardeny during the Meuse-Argonne campaign; they used a telephone exchange to convey messages that completely fooled the Kaiser's most skilled eavesdroppers, with concealments such as:

MILITARY TERM	CHOCTAW
allies	apepoa
artillery	tanamp chito, tanamp hochito
talk	anumpa
tank	oka aialhto chito

The original Choctaw language did not have direct equivalents for all military phrases, and between the wars, few efforts were made to develop formal codes equating Indian and military terms. But World War II caused a widespread reassessment of all communications security and more precise terminology was developed.

Commercial Codes

Codes have been as important in peacetime as they have been in war. The advent of telegrams and cablegrams launched the development of commercial codes for business transactions. Companies quickly realized that they could shorten messages and save money on both methods by putting frequently used terms and even whole phrases into code. Businesses began to devise their own lists of terms to replace money amounts, phrases and even entire sentences. Simple words such as *hat* and *coat* became codewords that might have meant, for example, "Send me the contract immediately at this address."

Eventually, as commercial codes became popular and were used by companies around the world, various codes and their meanings were published in books. These codebooks, with their substantial collections of nonsecret code terms, were prepared and sold commercially.

Such codes flourished through the late 19th and early 20th centuries. After the First World War, international trade expanded to new heights. Within businesses, however, disputes arose over the pronunciation of the many contrived and artificial letter combinations of these codewords. Clerks at cable and telegraph offices had numerous problems transcribing notes onto company transmittal forms and talking to people whose dialects or different languages made their use of codewords all the more complex. When misunderstood words or garbled transmissions led to business losses the result was often costly litigation. These problems with pronunciation put limitations on the number of codewords. This in turn placed restrictions on the number of terms available and raised message costs for businessmen who had to send large parts of messages unencoded.

In 1932 the International Telegraph Union decided that codewords built upon any arrangement of five characters would be permitted for cablegrams and telegrams. This resulted in another increased code use and publications of large codelists.

Beginning in September 1939, the Second World War led to security-driven censorship and the disruption of peacetime commerce. As the war disrupted normal communications, wartime development of the teletype machine spelled a decline for the commercial code business. Teleprinter operators and equipment were not affected by pronunciations, nor were clerks needed for encoding and decoding purposes. This made teleprinting faster and less costly, superseding the more awkward telegrams and cablegrams. Improved global radio and then telephone links hastened the demise of the huge tomes of commercial codes.

The Slater Code

With the completion of the Atlantic cable in 1866 and the corresponding boost in transoceanic communications, the need for a money-saving commercial code was greater than ever. By 1874, Englishman William Clausen-Thue, a shipping manager, had developed the one-part *ABC Code*, which was capably arranged and had a huge vocabulary. It soon became the first public codebook to achieve broad sales and longevity.

In 1870 Robert Slater, secretary for the French Atlantic Telegraph Company, published a large and popular one-part codebook titled *Telegraphic Code, to Ensure Secresy [sic] in the Transmission of Telegrams*. Constructed with economy and security in mind, Slater included 24,000 vocabulary terms and 1,000 more words for geographical sites, heroes, surnames and Christian names. The code equivalents were groups of five digits.

A

a	00001	abbey	00021
aback	00002	abbot	00022
abacus	00003	abbreviate	00023
abaft	00004	abbreviated	00024
abalienate	00005	abbreviation	00025
abandon	00006	abbreviature	00026
abandoned	00007	abdicate	00027
abandonce	00008	abdicated	00028
abandoner	00009	abdicates	00029
abandonment	00010	abdicating	00030
abase	00011	abdication	00031
abasement	00012	abdomen	00032
abash	00013	abdominal	00033
abate	00014	abduce	00034
abated	00015	abduct	00035
abates	00016	abduction	00036
abattoir	00017	aberrance	00037
abbacy	00018	aberrant	00038
abbe	00019	abet	00039
abbess	00020	abetment	00040

Additional protection was provided by a system of adding, subtracting and changing the numbers in the codegroups.

Slater's methods of superenciphering the codegroups were considered advanced for his time. They included adding and subtracting digits; transposing the last three digits of the five-part groups (thereby equating the number group with a new codeword); adding and transposing; subtracting and transposing; altering the numerals (the original 5 numbers sent as 4): altering and transposing (4 numbers sent with only the left-most digit not being transposed); altering, transposing and adding numerals according to the number of words sent (for example, first word add 1, second word add 2); and transposing the message itself.

As an example, Slater used the phrase "The queen is the supreme power in the realm." His subtraction process and the receiver's recovery of the plaintext is shown here.

WORD RECEIVED	CODENUMBER	SUBRACT ADDITIVE (5555)	PLAINTEXT
ounteous	02868	22313	The
wedge	23650	18095	Queen
purifying	17925	12370	is
bounteous	02868	22313	the
biography	02505	21953	supreme
transparent	22611	17056	power
posed	16981	11426	in
bounteous	02868	22313	the
yoke	23974	18419	realm

The Acme Code

Commercial codes were being developed even after the end of the first World War. The Acme was a one-part business code of 100,000 codewords compiled by William J. Mitchel in the early 1920s. As wryly described by writer Jack Littlefield in a 1934 *New Yorker* article, the Acme's terms ran the gamut from the mundane to the humorous, and from the intriguing to the tragic. Some example codegroups and their plaintexts follow:

BIINC	What appliances have you for lifting heavy machinery?
URPXO	For what use was the mixing machine intended?
CHOOG	lard, in bladders
GAHGU	cod-liver oil
GNUEK	rubber, slightly moldy
HEHST	clammy condition
ZOKIX	unhealthy trees
ARPUK	The person is an adventurer . . .
BUKSI	Avoid arrest if possible
NARVO	Do not part with the documents
OBNYX	Escape at once
CULKE	Bad as possibly can be
LYADI	Arrived here with decks swept . . . encountered a hurricane
PYTUO	Collided with an iceberg
YBDIG	Plundered by natives

Murray Teleprinter Process

Commercial coding processes were significantly enhanced when inventor Donald Murray improved British post office systems around 1901. Murray's advancement was an adaptation fo the Baudot code for teleprinters. The Baudot process (See "Unbreakable Ciphers" in Cipher section) involved a device that made marks (holes) and spaces in paper tape that corresponded to alphabet letters, numerals and keyboard functions. Murray studied the Baudot system and streamlined it by assigning the combinations with the fewest holes to the alphabet letters and keyboard functions used most often in telegrams. Murray also introduced a "start" signals element at the beginning of each letter and a "stop" signal at its conclusion for improved transmission. To overcome transmission garbles, a seven-unit error-detecting code was used on busy circuits; an error could be detected when this three-mark-and-four-space sequence was altered.

To send a telegraphed message with the Murray code, the operator typed on a keyboard. When each key was depressed, five long bars moved to form the code for that letter. Spaces occurred when the bars were held in their places by tabs located on the letter keybar. Marks resulted when the bars moved to their left and protruded beyond the space bars. The extended bars closed electrical switches, which sent current through the line as a mechanism called a rotating cam linked a battery to each of the bars in their turn.

At the recipient's teletype, each mark caused an electromagnet to depress one of five bars and form the five-unit code that represented each of the arriving characters in turn. The depressed bars (marks) and unmoved bars (spaces) caused a corresponding cut latch to drop and stop a turning typehead. The back of the character was then hit with a hammer mechanism that printed the letter on paper.

The Murray code was improved with duplex, multiplex and time-division systems whereby messages moved in different directions and time intervals on the transmission circuit and proved adaptable to later alternate-current signaling and frequency-division techniques.

Q Code

The Q code was developed as a means to facilitate global commerce, travel and emergency efforts as its three-letter groups allowed communication between international radio operators. To regulate developments in radio, a series of meetings was called, including the Berlin International Radiotelegraph Conference (1906), Washington Radiotelegraph Convention (1927) and the Madrid Conference (1932). The Q code developed from these and other meetings that convened until the outbreak of the World War II. With the end of hostilities, other conferences resumed and have maintained international standards into the age of satellite communications.

The Q code covered such subjects as recognition between stations (exchange of names and positions), radio operation (order of telegrams, transit and charges), control of aircraft (transmissions about height, speed and visibility); airport locations, meteorological conditions and dangerous situations.

A series of abbreviations beginning with Q were used to convey such information. Some sample codeletters along with their meanings, which could be in the form of either a question or an answer, depending on the context, are shown below.

The Q code has been revised over the years to reflect the technological advances affecting radio communications, such as use of geostationary satellites, and changes in maritime mobile services, radiotelegraphy, and radiotelephony. Below is a sample of recent equivalents.

CODE	QUESTION
QAA	At what time do you expect to arrive at . . . ?
QTC	How many telegrams do you have to send?
QFS (b)	Please place the radiobeacon at . . . in operation.
QFB	Are fresh meterological observations required?
QTM (a)	Send radioelectric signals and submarine sound signals to enable me to fix my bearing.
QOA	Can you communicate by radiotelegraphy (500 kHz)?
QOB	Can you communicate by radiotelephony (2,181 khz)?
QOH	Shall I send a phasing signal for . . . seconds?
QOJ	Will you listen on . . . kHz (or Mhz) for signals of emergency position-indicating radiobeacons?

CODE	ANSWER
QAA	I expect to arrive at . . . at . . .
QTC	I have . . . telegrams for you.
QFS (b)	The radiobeacon at . . . will be in operation in . . . minutes.
QFB	Fresh meterological observations are required.
QTM (a)	I will send radioelectric signals and submarine sound signals.
QOA	I can communicate by radiotelegraphy (500 kHz).
QOB	I can communicate by radiotelephony (2,181 khz).
QOH	Send a phasing signal for . . . seconds.
QOJ	I am listening on . . . kHz (or Mhz) for signals of emergency position-indicating radiobeacons.

■ 364

PRE-WORLD WAR II CODES

Gray Code

The Gray code, named for the color of its binding, was a diplomatic code used by the State Department from March 1918 through the early 1940s. The Gray used five-letter groups to represent commonplace terms and phrases. For example, *nadad* was the frequently used equivalent for "period." The code was so widely used and so transparent that it was the subject of amusement by the embassy personnel of other nations. It is said that a State Department staffer's retirement speech in Shanghai given in Gray was understood in its entirety by the career diplomats in attendance.

It is also believed that Japanese cryptanalysts in the Tokumu Han department of the Imperial Navy broke messages covered by Gray after the *kempeitai* (military police) gathered scraps of telegram messages from State Department trash. Japanese secret agents even used a wax imprint of a key to enter the U.S. consulate in Kobe to remove codebooks from a safe and copy them. Suspicions were apparently not aroused until a black bag mission was bungled at the office of the U.S. naval attaché in Tokyo. The break-in was discovered and a new site and code security resulted.

Despite the demonstrated insecurity of the Gray code, it was still in use at other U.S. embassies in the late 1930s as war began to threaten Europe and Asia. Even such high-level communications as the transatlantic messages between Franklin D. Roosevelt and Winston Churchill were still being covered by the Gray in 1940.

Department of State Code A-1

In 1919 the State Department activated a new codebook entitled *Department of State Code A-1*, and the confusing word "cipher" was finally dropped from State codebook titles. A superenciphered version of this code was created in the early 1920s by government employees who were alarmed by breaches in diplomatic communications. The new tome had two parts, the *Department of State Code A-1* and the *Department of State De Code A-1*.

The *A-1* provided a larger collection of codewords than previous State codebooks. For example, the plaintext "and" had a total of 55 equivalents, including *BIKEG*, *HUHUB* and *MOPUY*. "The" could be replaced by letter combinations such as *KEFUM* or *LIKER*. Like many codes of the time, the *A-1* codes grouped vowels (v) and consonants (c) in a "cvcvc" pattern with a regularity that compromised its security. If an enemy succeeded in breaking part of the code, the cvcvc pattern could give away spellings of other codewords.

Security concerns were evident in the directions given to code clerks, who were told to choose word replacements randomly, and by the use of a two-part code. For encoding, *A-1*'s rules required that the first letter should be a consonant ranging from *B* to *P*.

ENCODING		DECODING	
diplomacy	BIGED	BACUB	diplomatic
diplomat	BUJOH	BEDAC	diplomatic corps
diplomatic	BACUB	BIGED	diplomacy
diplomatically	BOHIF	BOHIF	diplomatically
diplomatic corps	BEDAC	BUJOH	diplomat

Unlike many previous code systems, number and letter tables provided superencipherment capabilities. However, *A-1* was rarely used, and U.S. diplomats were shackled with antiquated codes throughout the 1920s and the early 1930s.

Brown Code

Developed to replace the very vulnerable Gray code, the Brown came into use on December 30, 1937, known as the *Department of State Brown Code*. It consisted of an encoding section of 954 pages and a decoding section of 938 pages.

Like the *A-1* and other newer codes, the Brown contained multiple variations for frequently used words. There were 20 varied groups for "and" and 20 variants for "the." Due to high transmission costs, economy in dispatch sending was a key issue. The Brown had dozens of phrases starting with "the" that were concealed by cost-saving, five-unit groups.

The Brown also initiated a date group system that was new for the latter 1930s. Months were represented by the first letter, days by the second and third, hours by the fourth, and morning or afternoon were signified by the fifth letter. For example:

March 12, noon	XBAEK
May 15, 8 A.M.	IMUNF
July 20, 2 P.M.	LSZYS
November 26, midnight	ESJEX

Although the Brown was stolen from the U.S. consulate in Zagreb, Yugoslavia, in 1939, it continued to be used even during the onset of World War II. Eventually, growing security concerns caused the adoption of the M-138 strip cipher system which offered more protection for medium-security messages.

WORLD WAR II CODES

Black Code

Named for the color of its codebook's binding, the Black code was a U.S. military attaché code. It was considered invulnerable in the early 1940s as it was given extra security by cipher tables. In fact, it was compromised numerous times early in the war, having been broken by German radio interception and cryptanalysis and photographed by agents working for Fascist Italy.

Especially damaging to the Allied cause were the compromised messages sent by Colonel Frank Fellers, the U.S. military attaché in Cairo. Colonel Fellers's detailed reports about British Middle East operations were sent to Washington in the Black code. In the latter part of 1941, German radio listening stations began snaring the transmissions. Their valu-

able contents were decoded and sent to the "Desert Fox,"
Field Marshall Erwin Rommel.

Fortunately for the Allies, a surprise attack on a German radio
outpost in the desert at Tel-el-Eisa revealed the fact that the
Black code had been broken. The Allies then began to deceive
Rommel with false information sent in the same Black code.

Merchant Ships' Code

A code-broadcasting network known as BAMS (Broadcasting
for Allied Merchant Ships), was pivotal to organizing Allied
supply shipping for the British Isles during the Second World
War. The networks' dispatches were protected by methods
like the Merchant Ships' code, a two-part code with added
superencipherment.

When the Royal Navy's transmission stations sent out a message encoded in the Merchant Ships' code to a convoy captain, his radio operator would turn to the decoding section of
the ship's copy of the codebook to decrypt the missive. An
example of decoding pages is shown here:

36979	LPCR	island
980	DQ	locate
985	EP	port
990	FN	signal
996	GM	escort
37002	LQHL	bay
010	IK	dock
015	KJ	coast
021	LH	wind
027	MG	supply
38361	LRBP	convoy
376	CO	tons
383	DN	oil
392	EM	direct
39420	LSFL	current
422	GK	officer(s)

431	HJ	surface
440	IG	fleet
447	JE	buoy
453	KD	channel
456	LC	100 degrees

Even as this code was being used, it had been compromised by German cryptanalysts in the Navy's B-Dienst. They had read the Merchant Ships' Code firsthand when the Nazi raider ship, *Atlantis*, a fast freighter with camouflaged heavy armament, had taken copies from ships captured in the summer and autumn of 1940.

The "J" Series

The "J series" was the U.S. name for codes transmitted by the Japanese Foreign Ministry, and its embassies and consulates in the 1930s and 1940s. The J-coded exchanges were given additional concealment by a transposition that was arranged by a keynumber. The cryptosystem had a prearranged sequence of blanks that changed the standard length of the column segments and provided extra protection against enemy cryptanalysts. These transpositions and spaces were named "K" plus a numeral, for example, K-3, by U.S. analysts. U.S. codesolvers gave the J group numbers in sequences along with K transposition identifiers such as J17K6, J18K8 and J19K9.

With spaces represented by asterisks, the keynumber 4153267, and the message "all east asia co-prosperity spheres," an abbreviated example of the method follows:

4	1	5	3	2	6	7
a	•	l	l	e	•	a
s	t	a	•	s	i	•
a	c	o	•	p	r	o
•	s	p	e	r	•	i
t	•	y	s	•	p	h
e	r	•	e	•	s	•

The encryption was made by writing the message in columns according to the keynumber, and bypassing the blanks to read in order of keynumber, *tcsr espr lese asate laopy irps aoih*. These were placed in five-digit groups along with designators for the method and key before being sent by radio.

The J codes did not include the Imperial Navy's encryption method for flag officers, a four-character code with a transposition superencipherment, nor was the high-level JN25 fleet cryptographic system a part of this series.

JN25

The JN25 was a U.S. designation for a primary Japanese navy encryption method initiated in June 1939. The name JN25 was chosen because it was the 25th Japanese navy cryptosystem studied by U.S. cryptanalysts. As Japanese conquests in Asia began to threaten American interests in the region, special attention was focused on the Japanese fleets' activities. The solution of Imperial Navy message concealments became a primary goal of U.S. Navy intelligence.

Military and general history texts have variously described the JN25 as a code or a cipher. The consensus of cryptology experts is that it was two-part code containing 33,333 five-digit groups wrapped in an additive superencipherment. Every group was arranged to be divisible by three, such as 36876, to make it easier to detect transmission errors.

Agnes Driscoll of OP-20-G is credited with breaking through the additive to find the primary codenumbers and their plaintext equivalents. This important achievement received an apparent setback in December 1940, when the Japanese introduced a new concealment named JN25b by U.S. analysts. The JN25b was a larger enciphered code with more additives than its predecessor, but U.S. cryptanalysts received a lucky break when the Imperial Navy shortsightedly kept JN25 additives in service as the new ones were initiated. The solved JN25 additives gave OP-20-G a wedge with which to break into the new shield and portions of JN25b began to be understood in the winter of 1941.

The JN25 discoveries and the partial recovery of JN25b data did not reveal enough valid intelligence to avert the attack on Pearl Harbor in December 1941. However, the JN25 decryptions and subsequent solutions of other Japanese cryptosystems provided real dividends for the United States at the Coral Sea and Midway battles in May and June 1942.

Native American "Codetalkers"

During World War II Native Americans from the Chippewa, Comanche, Kiowa, Navajo, Pawnee, Menominee and Hopi nations served as "codetalkers" in regions from the Pacific to North Africa and Europe. After months of experimentation and testing in military camps, the official list of code equivalents was set at approximately 210 words. They included combinations of military lexicon, artificially created names for enemy-held sites and an alphabet equated with birds and animals to spell extra or unusual phrases. The following brief examples demonstrate why this simple code was never broken:

MILITARY TERM	COMANCHE	PRONUNCIATION
enemy	enemy	tu-wa-ho-na
machine gun	sewing machine	techa-keena
soldier	red belly	ex-sha-bah-nah
tank	tank	wah-ke-ray

MILITARY TERM	NAVAJO TERM	PRONUNCIATION
advance	ahead	nas-sey
camouflage	hid	di-nes-ih
defense	defense	ah-kin-cil-toh
grenade	potatoes	ni-ma-si
howitzer	short big gun	be-el-don-tso-quodi
a	ant	woll-a-che
scout plane	owl	nay-as-jah
battleship	whale	low-tso

Quadratic Code

The Quadratic code was developed by General Leslie Groves, who was in charge of U.S. production of the atom bomb, to protect his and others' telephone calls. It was applied by Groves and officials like Lieutenant Colonel Peer da Silva, chief of security at Los Alamos, New Mexico. While Groves gave versions of the Quadratic code to other important persons, he was the only one who held all the variations.

In spite of Groves's name for it, this "code" actually functioned more like a cipher. The code was based on a 10-by-10 checkerboard consisting of a typewritten square of about 3½ by 4 inches. The checkerboard contained a series of 86 alphabet letters mixed with 14 blank spaces. All the standard English letters were used at least once and some repeated letters were given subset numerals. The grid was bordered by the numbers 1 through 0 at its top and side. Concealed dispatches were formed by using the numerals at the end of the row and at the top of the column of each message letter. A pair of digits thus covered each letter. Able to be carried discreetly in a wallet or pocket, its loss was to be reported immediately.

The Quadratic code had enough alphabet letters to deter frequency counts of typical English letters by enemy interceptors, and by all accounts, the Groves system was not compromised. However, in spite of many precautions, the atomic research and development program was infiltrated by Soviet agents.

CODES IN EVERYDAY LIFE

Codes also have practical applications outside the demands of military and security. The following codes are nonsecret systems which continue to be used today. These are used primarily for communicative rather than cryptographic purposes. Like the commercial codes of the early 20th century, the function of these codes is to transmit generally low-security information in a concise form.

Universal Product Code (UPC)

The Universal Product Code is a nonsecret code that assists in the sale and distribution of consumer items. Product coding in the United States began as early as the 1930s, but the process did not become widespread until the 1970s. With the development of large supermarkets, stores needed to improve checkout speed and the processing of sales data, stock counts and inventory records.

After much study, an 11-digit code plus a check digit was chosen in April 1971 by an oversight committee representing grocery manufacturing, wholesaling and retailing. Still more study was necessary to choose the standard machine-readable symbol to accompany these numerals. After extensive tests, the UPC symbol design was revealed on April 3, 1973.

During the code and symbol development phrases, plans were made for an organization to oversee and manage the application and supervision of the Universal Product Code. In early 1972, the Uniform Code Council (UCC) was set up to regulate overall activity. Since then, the UPC system, popularly known as the "bar code," has been used in virtually every type of retail business.

The first digit of the code, the number system character, has a function similar to a cryptographic key, in that it determines the meaning and category of each of the other numbers. There are seven categories of the number system:

0   assigned to all nationally branded products except the following:

2   random weight items such as meat, cheese, and poultry

3   drug and certain health-linked products

4   products marked by the retailer for sale only in his store or stores

5   coupons

6,7   valid for assignment to retail products
since the first quarter of 1990. As of
1993, industrial applications are now
included here.

The number system character, *A* in the illustration below, is
included within a six-part numeral series (*B*) that identifies
the manufacturer. This number is assigned by the Uniform
Code Council, Inc. The next five-digit sequence (*C*) is a product code number. Controlled by the manufacturing company,
this numeral is unique for each of the manufacturer's products and includes color, size, flavor and other qualities.

The final number (*D*) is the check digit, which helps to reveal
error in the other numbers.

0   12345   67890   5

A   B   C   D

At the supermarket, the checker passes the UPC-labeled item
over an optical scanner, which reads the UPC symbol,
decodes it into the UPC code digits and sends these digits
to a small computer. The computer records the product's sales
data and transmits the item's description and price to the
checkout stand for visual display and printing on a receipt
tape.

Through the 1980s and 1990s the UPC has become ubiquitous on all types of packages, containers, forms and labels.

Zip Code

The zip code was originally a five-digit code designed to fa-
cilitate the processing of U.S. mail. On November 28, 1962,
Postmaster General J. Edward Day revealed the Post Office
Department's new sorting and distribution system called the
Zone Improvement Plan, or "zip" code. The new process was
formally initiated on July 1, 1963.

At first zip codes were intended to be used mainly by large-
volume mailers to help speed mail delivery, but the response
of the general public was so positive that the system was

expanded to include other business and personal usage as well.

In the beginning, the numbers had slightly different meanings for rural and urban areas. In less populated regions, the first three numerals of the code identified the central focal points of air, highway and railroad transportation. The last two digits identified the post office or delivery station. In cities that had local postal zones, the first three numbers identified the city and the last two designated the local zone number.

Sample zip code 54321 provides a sequence of information for the mail-delivery system. The first digit, 5, designates one of a series of national service areas. The second number, 4, identifies the service area subdivision. The third numeral, 3, signifies the city post office. The fourth and fifth digits, 2 and 1, are the station from which the mail is delivered. The recently activated "zip + 4" system is an elaboration of the location designator digit. The four additional numerals classify the sector and then the segment of the route.

Codes remain directly linked to the future of postal service. By 1998 a very rapid sorting and distribution system using scanner mechanisms and barcode type addresses had begun in several large U.S. cities.

ASCII

Among the nonsecret data exchange-enhancing codes is the American Standard Code for Information Interchange (ASCII, pronounced "ask-key"). It was developed in 1964 by computer experts as a method of transmitting facts in a standardized form. ASCII has become one of the most frequently applied codes in the United States and international variations are used worldwide.

Its groups of seven ones and zeros represent characters, numerals and punctuation in the binary pulse-and-no-pulse computer language. Such digits may represent numerous facts depending on how the computer is programmed. Following is a partial example of the code:

"	0	1	0	0	0	1	0		.	0	1	0	1	1	1	0
#	0	1	0	0	0	1	1		/	0	1	0	1	1	1	1
$	0	1	0	0	1	0	0		0	0	1	1	0	0	0	0
%	0	1	0	0	1	0	1		1	0	1	1	0	0	0	1
&	0	1	0	0	1	1	0		2	0	1	1	0	0	1	0
+	0	1	0	1	0	1	1		3	0	1	1	0	0	1	1
'	0	1	0	1	1	0	0		4	0	1	1	0	1	0	0
–	0	1	0	1	1	0	1		5	0	1	1	0	1	0	1

A check digit, an additional one or zero, is added to the ASCII's end or as part of the seven-bit encoding string as a mechanism for checking transmissions and catching mistakes.

This check digit, or parity bit, works on the basis of even and odd numerals. If the message total of ones is uneven, the transmission's parity bit is 1. If the total of ones is even, the parity bit is 0. This check digit is conveyed at the end of the communiqué. If the number of ones received fails to match the check digit, this mismatch signifies a sending error or errors and the transmission must be sent again.

When used with sections of dispatches, the parity plan functions in a type of self-correction process. Using a message part 0110, the first parity digit checks the first three numerals. The first three numbers would be 0 since the total of ones is even. A second check digit, for example, a fifth digit, can be set up to give the parity of the first, second and fourth numerals, which is 1 since the total of ones is odd in this group. If, however, more than one miscue occurs in the preparation or sending processes, the parity check may not catch the disruption and the check process would be nullified.

Banking Security Codes

These encryptions increase the security of banking transactions at automatic terminal machines (ATMs). This is accomplished with (1) a bank card containing a magnetically recorded primary key or account number and (2) a secondary key or personal identification number (PIN). A typical transaction follows these steps:

1. When the bank's card is placed in the terminal's card reader, the account number is read.

2. The cardholder then enters a personal identification number at the terminal's keyboard.

3. The card owner next makes a transaction request such as: Dispense $100 and debit checking account.

The security system checks to see that the account is a legitimate one and that the cardholder's PIN correctly corresponds with the account.

The PIN serves as a secondary key because it is intended to deny a segment of the private information to a potential thief. It usually consists of four to six alphanumeric characters. PIN numbers should be kept secret because, as with all codes, the discovery of the key would subvert the system.

The protection of such processes blend code and cipher functions. While the personal PIN is called an identity code (numbers equated with a person) the ATM's parent bank has its financial dealings protected by computers. They in turn are often secured by the DES algorithm, a product block cipher (See "Computer Cryptography" in Cipher section).

Genetic Code

Perhaps the least-understood and most complicated code is the genetic code, through which genetic information is transferred from the genes to the proteins that make up an organism. The chromosomes in the cell nucleus contain genes, which in turn hold the deoxyribonucleic acid (DNA that determines the genetic makeup of an organism. Like all nucleic acids, DNA is a long molecular chain made up of three types of molecules that are linked.

```
base base

sugar phosphate sugar phosphate

base

sugar phosphate . . .
```

The DNA sugar is deoxyribose and its bases are adenine (A), guanine (G), thymine (T) and cytosine (C). The complete DNA molecule consists of two molecular chains wound around each other in a form called a double helix:

```
A-    T-    T-    G-    C-    A-    A-    C

.     .     .     .     .     .     .     .

.     .     .     .     .     .     .     .

.     .     .     .     .     .     .     .

T-    A-    A-    C-    G-    T-    T-    G
```

The lines between the bases indicate hydrogen bonds, the weak chemical bonds that hold the two chains together. Opposite every A base in one chain is a T base in the other chain. The same is true for G and C. The order of the bases in the helix determines the message conveyed by that gene.

There is a one-to-one relationship between genes and proteins. The genes order the arrangement of the amino acids of proteins, and these orders are the genetic code.

The sequence in a DNA chain could be considered one alphabet of the four bases A, T, G and C. The amino acid pattern would then be a second alphabet of 20 letters (representing the 20 varied amino acids). The coding is not one-to-one, and the bases of the DNA chain set a single amino acid in the protein chain. Depending on the coding, the protein will perform different functions.

```
DNA:        GTA  /  ATG  /  CCA  /  GGA  /  . . .

Protein:     v  –    m  –    p  –    g  –
```

Every cluster of three bases (which indicates an amino acid) is identified as a "codon," the term for a coding of an amino acid. With 64 types of codons (4 x 4 x 4) there are enough to code the 20 kinds of amino acids. To signify the start and the conclusion of a "message" in these alphabets, three "blank" codons are placed before and after the identified chain.

The actual specification procedure is a complex chemical process. The DNA is transcribed into a major nucleic acid called messenger RNA (ribonucleic acid). One base of DNA specifies one RNA base. The RNA takes this genetic information and as a messenger, leaves the cell's nucleus, moves into the cytoplasm, and is translated into protein in that part of the cell.

Today many foundations and private companies are funding extensive research into gene structure as scientists search for cures for fatal diseases. As scientists learn to manipulate encoding processes to correct faulty protein sequences safely and effectively, this new "gene therapy" may offer hope for the millions of people with genetically encoded diseases, and begin a new era in science and medicine.

Codebreaking

As with cipher analysts, codebreakers cannot always be certain at first that the random groups of characters in a transmission form a code. Codebreakers must consider the possible sources of the particular message and guess at the intentions behind the communiqués. Particular nations and their personnel are often targets, as was the case during the Cold War, so clues can be garnered from the dates, times, origin sites, methods of transmission, languages of the exchanging parties or specific persons from whom the cryptic data was obtained.

Analysis usually begins with a search of letter frequency, often-used terms or typical pairs of letters. When these result do not seem to indicate a cipher system, however, an analyst may decide that a code has been used.

Codebreaking is a daunting task, as analysts must deal with hundreds and perhaps thousands of elements. Some patterns, however, hold true for codes in every language.

A sentence-based message must have a basic order to make sense to the recipient; it has a natural beginning, middle and conclusion. Codebreakers often begin their attack at the end of the missive, as certain codegroups represent a period or "stop." To an analyst's delight, enemy codesenders are sometimes lax about checking the codebooks new terms for "stop" and repeat the same codeword a number of times.

If this codeword can be identified, codesolvers can try to work back through the codegroups. Knowledge of the language used is essential because languages have sentence structures with nouns, verbs and adjectives in different positions. If the order of nouns and verbs can be determined, this may give clues to the code's construction. If, in the transmission language, a verb often ends sentences, verbs can be attacked at other phrase conclusions in the body of the text.

As in cipher solving, the beginnings of dispatches can also yield clues. In codebreaking, common terms are especially helpful as one typical term can be an easy bridge to another. When languages have gender-based word endings, the discovery of these codegroups could expose syllables, which can be built into words and phrases.

Some of the techniques mentioned in the cipher cryptanalysis section, such as repetition analysis, can also be applied to codebreaking. Yet perhaps more important is the solver's understanding of the language used. Knowledge of jargon and popular slogans are sometimes helpful as they may be used as codewords.

The context of the code can also give important information. For example, if a third party is intercepting codes from an arms manufacturer, frequently repeated codewords may represent words such as "ammunition" or "guns."

An aspect of languages called *pattern words* is also an essential tool for codebreakers. Pattern words are found in all established language. The word *eye* has a pattern of 121, while the word *afternoon* has the pattern 123456776. In these examples, the letter *e* repeats as do the *n*'s and *o*'s in *afternoon*.

1	2	1	•	1	2	3	4	5	6	7	7	6
e	y	e	•	a	f	t	e	r	n	o	o	n

Nonpattern words contain no repeating letters and range from the single letter *a* to the longer *something*. (For more pattern and nonpattern words, see Appendix).

Knowledge of pattern and nonpattern words gives the codebreaker an angle with which to rebuild full words from scattered letters and digraphs that may emerge. The analyst compares these fragments of terms with the codelists to see if combinations of letters might form a sensible word. If repeats, such as double *e*'s, *l*'s and *o*'s, seem to be occurring in the codetext, they can be compared for pattern word matches. In some cases, however, these may be nulls which only draw the analyst down false paths.

As the codesolver finds some possible identifications, he or she begins the process known as "building up a repertory." For many years possible plaintext matches were written in "pencil groups." When they were verified as correct they were listed as "ink groups." Although the means of recording such matches has changed with technology, the idea remains the same.

From a series of more fully decrypted intercepts, clues from careless codegroup repeats and many other searches, the codebreaker begins the arduous task known as "book building." This involves the compilation of the syllables, words and phrases of the enemy's codebook. The difficulties involved in these efforts were greatly compounded when rival cryptographers added numbers to codenumbers to create enciphered codes.

BREAKING ENCIPHERED CODES

Codes are often superenciphered with additives to ensure secrecy. When trying to solve such encicodes, codebreakers use what is commonly called the "difference method" to find the placode (plain code) behind the additives.

From studying message origin points, recipients, times and patterns of the numbers in the transmissions among other factors, analysts try to find sections where the same encicode may have been applied. This can range from brief groups of numbers that have similarities to a longer key of additive numerals where similar digits also appear in concentrated points.

The first group of digits sent in a message is called an indicator. These are usually the starting point where the enemy began the additive process. An indicator such as 4527 could mean that the addition was begun on page 45, line 2 and column 7. These numbers and other repeated numbers in the body of a cryptogram are good places to make tentative comparisons. When possible, a series of dispatches sent on the same day are also compared.

The numbers believed to be indicators are placed in columns. The analyst then puts the transmitted digits underneath their respective indicator to check for numbers containing similar digits or digits with like arrangements.

INDICATOR	A	B	C	D	E	
1	4527	9534	3291	6792	4317	2253
2	3816	2785	. . .	. . .	. . .	. . .
3	5329	1275	. . .	. . .	. . .	. . .
4	8291	6211	. . .	. . .	. . .	. . .
5	4527	5297	. . .	. . .	. . .	. . .
6	9161	4820	. . .	. . .	. . .	. . .
7	2834	1739	. . .	. . .	. . .	. . .

The cryptanalyst hopes to find that the same additive has been used to cover more than one codenumber group. As these number groups are studied, they are placed in increasingly specific comparative arrangements.

Suppose that a series of these numbers are gathered according to their relationship to an indicator and their positions in a dispatch. They are aligned in columns and the differences between them are found by nonborrowing subtraction. A brief example follows:

1	2	3
6795	5254	3668
5827	4558	1502
3013	2127	8796

To check whether 6795 from the first column and 5254 from the second column could have the same placode, the smaller is subtracted from the larger, resulting in 1541. The difference between 5827 and 4558 in the second row is 1379, which is not the same result. When 2127 in the second column is subtracted from 3668 in the third column, however, its result is also 1541.

A large series of codegroups requires many more comparisons and subtractions. The numbers of each column are subtracted from every other column until more foundation placodes are found. Tables of the differences are prepared to check for patterns. These subtraction results are compiled in what is called a "difference book." The goal is to identify as many of the placode numbers as possible.

This is not the conclusion of cryptanalysis, as the first codelevel still exists, but these can then be attacked as basic codes using the techniques mentioned above.

During wars when large numbers of transmissions were sent, analysts were required to perform thousands of these subtractions. Both the subtraction process and the compilation of the difference books were greatly aided by the introduction of mechanized tabulation known as "punch card machines," which IBM supplied to the U.S. military before and during World War II. The machines took on the tedious task of counting and sorting totals, allowing the analysts to concentrate on attacking the code itself.

Today analysts are aided by computers programmed with special algorithms designed to attack modern encicodes, yet the basic techniques of codebreaking remain the same.

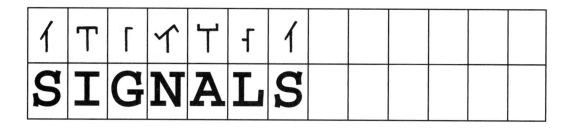

SIGNALS

Signals, from the Latin *signum*, meaning "sign," have existed from the origins of human existence. Although their meanings can be secretive, signals have relied more upon their efficiency over long distance than on their cryptic nature. Signals have made use of various mediums, from general motions such as waving to fire signals and sounding warnings using conch shells. From general hand gestures intended to transmit information or give orders, signals gradually became more complex as hunter-gatherer peoples settled in agricultural areas and developed towns and cities.

Very few records of ancient signaling systems have survived to the present day. The ancient Egyptians used horns for combat orders and Alexander the Great signaled his army using horns, torches and smoke. Some manuscripts also indicate that the famous lighthouse at Alexandria signaled ships by reflecting sunlight from polished metal objects.

Mediterranean sea traders, such as the Phoenicians and the Greeks, developed very basic forms of fire or horn signals to ensure safe passage among the many hazardous island and coastal passageways. The Greeks also developed more elaborate pyrotechnic communications across their islands and among their city-states.

It was not until the rise of the Roman empire that organized signals played a significant role in military communications. Both the Greeks and Romans used early trumpets made of wood or goat horns, which were used on land and sea to signal brief battle orders. The Romans also used a horn called the *cornu*. This was a large curved instrument with a fairly high tone that was used in conjunction with the deeper notes of a type of tuba. The combination of tones passed along orders about advances or retreats. Around the camps, a smaller horn called the *buccina* gave the wake-up calls, signified divisions of the day's routine and times and sounded "torches out."

Fire signals were another of the many Greek practices adopted by the Romans as they expanded their territory. Under the Caesars, the Romans erected an elaborate system of some 3,000 towers for torch communications throughout their vast empire. With the fall of Rome in 476 C.E., signaling systems fell into disuse.

During the Renaissance's resurgence of interest in cryptography, various signaling methods were also revived. The growth of commerce between Mediterranean city-states required the development of more complicated messages to be sent from ship to shore and between ships in the form of torch, trumpet or drum signals.

During the latter 1500s and the 1600s, individual inventors proposed visual signaling mechanisms from banner systems to barrel-and-flame signals. Although few of these passed the test of time, many of these overly complex and sometimes amusing methods were the basis for later systems.

The development of visual signaling systems was given an important boost by the invention of the telescope. Prior to the telescope, there had been few sight-oriented methods as poor visibility limited the effectiveness of the signal.

Though scientists have understood the principles of lenses since ancient times, it was not until the beginning of the 17th century that telescopes began to be sold in European cities, mostly as items for amusement. The government of the United Netherlands, however, saw possible military applications, and bought several telescopes from Dutch inventor Hans Lippershey beginning in 1608. The increase in distant vision created a revolution in signaling systems. Encouraged by the telescope's visual aid, the military adapted primitive torch-waving methods to the new technology from the mid-1600s onward.

The invention of mechanical telegraphs in the 18th century led to the development of other semaphores, shutter signals and related methods throughout Europe. These new networks proved so critical to military communication that other systems soon emerged, from arrangements of lanterns hoisted on masts to groups of colored flags representing numbers and letters.

In the late 18th and early 19th centuries the British Royal Navy finally recognized the need for better organized signal and codebook combinations and developed instruction manuals and guide texts. The rival French navy and Britain's rebellious American colonies were not far behind. In the late 1700s England introduced a flag-based system of signals for the Royal Navy, which was eventually also used by merchant ships. In 1817 an international code using colored flags to indicate numbers that had been assigned to words was published in book form. The flag-based system was maintained for ship-to-ship needs into the 1800s, but by the mid-19th century the mechanical telegraph had been rendered obsolete by the electric telegraph and transoceanic cables.

The technology developed during the Industrial Revolution had a profound impact on the progress of signaling systems, as electricity provided once unimaginable speed to communications. The telegraph, telephone, electric light bulbs and radio initiated a plethora of systems for land, sea and air transmissions. On land, the sun-reflecting heliograph was replaced first by telegraphy and later by radio. At sea, pyrotechnic rockets and lanterns were superseded by electric lights provided by ships' own power sources.

Signal makers kept pace with technological improvements, adapting as many processes as possible to the ubiquitous dot-and-dash of Morse code. Telephone technologies slowly replaced radio signals for personal communications as microwaves opened the possibility of global networks. By the 1970s, telephones and radios were joined by satellites that conveyed messages through new voice-coding processes and brought signal transfer speeds to new levels.

These developments have profoundly affected warfare technologies in the 1990s. Wars are fought on electronic battlefields as much as real ones, and new signals security and signals intelligence consist of elements that would only have been dreams to the first signal makers.

Hand and Arm Signals

Hand configurations are probably the oldest true signaling method. Before they developed waving objects or portable sunlight-reflecting surfaces, men and women used their hands and arms to create basic signals. If the gestures were noticed, the first link in a communication had been forged.

Formal signals probably began when small groups banded together to hunt and forage. Cooperative planning required prearranged ways to communicate silently at long distances, and gestures probably indicated everything from "game near" and "fish here" to "danger" and " danger gone." As these groups formed permanent bonds, built villages, and raised families, their waving, directional and recognition gestures became standardized.

Finger and hand gestures were part of daily life in ancient cities. In thieves' dens, recognition signs signaled membership and kept outsiders at bay. In the bazaars, merchants gestured, bartered and traded, often using special finger combinations to signal, "friends, charge less," or "wealthy, charge more." In his text *Tactics*, the Eastern Roman Emperor Leo VI described using various hand gestures at sea to avoid errors with fire or banner systems.

Many hand-based methods were still being developed during the 20th century, despite the mechanical advances over the past century. Two specific examples of hand signals include a U.S. Navy harbor code, and a method used by American prisoners in North Vietnam.

Both of the following examples of harbor and prisoner-of-war signs are variations of earlier signals for the hearing- and speech-impaired. The first example has been applied by Navy personnel since the 1950s for guiding harbor tugboats. The signs mean: (1) "Half speed ahead or astern"; (2) "Tug use right rudder"; (3) "Full speed"; (4) "Tug use left rudder"; (5) "Dead slow"; (6) "Tug to rudder amidships"; (7) "Stop"; (8) "Cast off, stand clear."

Harbor tugboat hand code

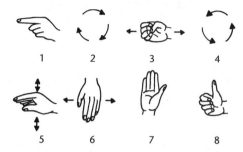

The following numerals were used by American service personnel to exchange silent messages in the infamous North Vietnamese prison camps. Hand signs have been used by U.S. military personnel in many situations in which silence was essential for survival.

Hand code used by prisoners of war

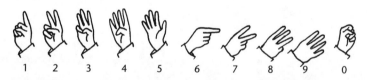

Although the number of arm-and-hand signaling methods declined through both world wars, some forms have remained active into the current era. The U.S. Army, Air Force and Marine Corps have used various land combat signal systems similar to those used over the previous century. One example is the signal for a forward march. A squad leader faces in the direction of march. Extending his arm and hand to a full vertical position, he lowers it toward the direction of desired movement.

Combat personnel signals undergo periodic evaluation and change. As surprising as it may seem, in the 1960s and 1970s a variety of training manual styles required two-hand or two-arm gestures which did not take into consideration the fact that troops under fire often do not have both hands free. As recently as the mid 1960s, Marine Captain Leon Cohan was calling for a standard signaling system. At that time, the Corps' gesture for "I did not hear you" and "I do not understand" was not the cupping of a hand behind an ear. Rather, signalers hid their eyes behind their hands—a certain way to miss the next signal.

Hand-and-arm signals for noncombat and security personnel still vary between U.S. sites and services. They range from silent signals made by sentries and recognition signs made by gate guards during emergency security tests to motions made by persons directing heavy vehicle traffic.

Hand signals are also used outside the military in situations which preclude spoken exchanges. One common example is the hand and body gestures used in sport to convey plays or umpire decisions. Using certain finger signals, a baseball catcher indicates a fastball, curve or slider to pitchers. Managers instruct batters and baserunners to swing away, bunt or steal a base through a series of gestures such as adjusting their caps, folding or unfolding their arms or touching their hands. These signals are relayed to the third base coach who, using a different set of motions, informs the batter of his manager's decisions. Whatever the sport—baseball, basketball, football, hockey, soccer—many hidden meanings are conveyed in these ways.

Naval Flaghoist Signals

Naval flaghoist signals are flags of various colors and designs that are raised on ropes or cables at the captain's bridge, at mastheads or from the yardarms of a mast.

The first recorded naval signal, attributed to the Greeks, was something like a flaghoist signal. In 480 B.C.E., at the Battle of Salamis in the Saronic Gulf near Athens, a Greek fleet won a crucial victory over their arch enemies, the Persians. The victorious Greek commander, Themistocles, broke with the naval tradition of fighting in parallel lines and gave the order to his fleet to turn and ram the Persian vessels. The order was conveyed when he waved his cloak and then threw it over-board.

Many years of single flag use preceded the development of multiple flags. William the Conqueror, who ruled England from 1066 to 1087, flew a white banner with a blue border from his command ship, known as his flagship. Each ship in his squadron flew the personal flag of the knight in command.

Records of naval signals during the Middle Ages are vague, as none of the great seafaring nations at this time had formally standardized their signals. Sea captains gave their own signs and the varied meanings to a few trusted subordinates, rarely committing these signals to paper.

In 1530, however, the Royal Navy ordered all admirals' vessels to raise a recognizable flag to their masthead. These banners were to be standardized, and every supporting craft was instructed to defer to orders from this ship. This decision created the command ship, or flagship, which continues to be a part of naval tradition today.

Great Britain rose to naval predominance after the historic defeat of the Spanish Armada in 1588. Before a major naval expedition in 1596, Queen Elizabeth I had her secretaries prepare instructions for her fleet commanders. To ensure secrecy, her orders were not opened until the ships reached a planned latitude. When they were unsealed, her instructions included the display of certain flags to indicate a meeting on the flagship or the pursuit of an unknown vessel. Some historians consider Elizabeth's directions to be the first formal set of signal orders ever given to the leaders of one of England's fleets.

The courtier and adventurer Sir Walter Raleigh is also credited for his advances to naval signals. He helped finance and direct naval expeditions to the New World, and in 1617 compiled a set of signal directions for use between vessels. In addition to flag signals, Raleigh's instructions included a combination of raised and lowered sails, gunfire and cannon shots to convey intentions to other captains. His method of combining some flags with other types of signaling methods was continued for many years.

In the mid-17th century, Admiral Sir William Penn, the father of Pennsylvania's founder, made some important signaling changes. Penn had served under the duke of York, who had greatly improved England's battle formations before conflicts with the Dutch occurred in the 1660s. Influenced by the duke's reforms, Penn used a series of differently colored signal flags as a meeting calling system when he commanded a fleet in the West Indies in 1655. A red flag called the captains of the admiral's squadron to board the admiral's ship, a white flag signaled the vice-admiral and the captains of his squadron, and a blue flag meant that the rear-admiral and the captains of his squadron were to board the admiral's ship. In addition to meeting signs, Penn also added flags indicating tactical maneuvers and designations for chasing the enemy.

Through the 15th and 16th centuries the Royal Navy made gradual changes in the positions of flags, designating certain masts for peacetime sailing and others for wartime instructions. Some naval historians credit the duke of York, later James II, with first codifying naval flag signals. When a certain banner flew from his mizzen mast, the aft-most mast, all other ships' commanders raised the same flag on their masts and moved into the preplanned battle order. In 1673, the Duke issued England's first signal book, which consisted of signals with 15 flags. This first version of the *System of Sailing and Fighting Instructions* lasted until 1705 when a new code of some 20 instructions was published.

In 1738, a French captain named Mahe de la Bourdonnais invented the first numerical flag code. He used ten variously colored pennants each equated with the numerals 0 to 9. Three sets of the flags, a total of 30 flags, represented ones, tens and hundreds, allowing him to signify the numbers 1 to 999. The large number of variations made it possible to equate a large vocabulary of words or phrases with the flags. However, the system was restricted to conveying prearranged dispatches and needed 30 special pennants to complete the planned words, so this apparently revolutionary breakthrough was overlooked by the world's other naval powers for decades thereafter.

By the 1700s, the American Navy also had their share of signals, generally lanterns mixed with banners and occasional cannon fire. These were used primarily by privateers, privately owned vessels, whose crews fought against the powerful British squadrons during the American Revolution. Shipowners first applied for a privateering commission and Letters of Marque from a Colonial governor, who had vice-admiralty powers in wartime. The Letters of Marque were similar to licenses and authorized the arming of a vessel for the purpose of capturing the ships and property of an enemy nation. By April 1776 the Continental Congress started to grant these commissions as well and they eventually became the sole purveyor of these credentials.

During the war, the privateers accrued quite an impressive record. They accounted for the loss of 16 Royal Navy warships and more than 2,900 royalist merchantmen, in addition to captured goods with a combined value of nearly $50 million.

Privateers carried a number of national flags, as switching banners to trick a foe was considered an acceptable *ruse de guerre*. When two rebel vessels used this scheme on each other, however, the potential losses often outweighed the gains. For this reason, as the war continued and more Colonial ships operated in pairs or groups, more effective flag and lantern codes were developed.

If the commanding officer wanted to give an order for "full sail with first wind," he would place a European flag at the peak of his ship's mainmast:

"Full sail with first wind"

If the enemy were sighted and the commander wished to pursue, he would order "begin chase" by placing an English ensign on the ensign staff at his vessel's stern. He would also hoist a pennant to his fore-topgallant masthead:

"Begin chase"

The command to "begin engaging the enemy" could be given by lowering the English ensign from its staff and raising a European flag in its place:

"Begin engaging the enemy"

When actually ready to fire opening salvos, captains raised the flag of their own colony.

Despite the loss of the American colonies, the Royal Navy continued to pioneer in naval communications throughout the 18th and 19th centuries. During the late 18th century flag colors became more standardized, generally being combinations of blue, black, red and yellow. Signal books also became more detailed and tables of squares were devised to coordinate the flag combinations. One was as large as 16-by-16 squares, permitting an arrangement of 16 signal flags vertically at the side and 16 banners across the top.

In the 1780s, Admiral Richard Kempenfelt produced the first scientific naval signal book; his flag system was adapted for general use. Familiar with the neglected numeric code of Bourdonnais, Kempenfelt developed a flaghoist method with 12 banners which represented the numbers 0 through 9, as well as a first and second repeater flag. A repeater flag indicated that the numeral just displayed was to be used again, allowing repeating numbers such as 11 or 222 to be sent without confusion. Although Kempenfelt may not have invented this system, he was the first to document the process.

During 1780–81, Kempenfelt also produced a signal book which aligned ten banners in a ten-by-ten table, ten down and the same ten banners across. Each flag in the grid could then be equated with any one of the 100 numbers in the ten-by-ten square. The numerals were linked with various naval terms and their meanings were supplied to commanders before sailing. When flags were raised on the admiral's ship, signalmen in the accompanying squadrons could check their signal books to find the signaled order.

In 1803 Admiral Sir Home Popham published his *Telegraphic Signals or Marine Vocabulary*. Popham's codes included nearly 3,000 numeric signals that represented words, phrases and entire sentences. Between 1803 and 1812, Popham also made improvements in the banner system using combinations of numerical flags. These numerical flags were equated with an alphabetic word list, and Popham's combinations of three and four flaghoists led to an expanded vocabulary of 30,000 words.

In 1817, British Captain Frederick Marryat wrote the first book of international codes. In his *Code of Signals for the Merchant Service*, colored banners signified 9,000 different numbers that were equated with words. It included standard naval terminology and was especially popular among international sailors as its 1851 edition contained specific listings for nations' warships, merchant vessels, lighthouses, ports, headlands, shoals and reefs. It remained in service until the latter 1870s, a remarkable lifespan for a signal book.

Despite these advances, military competition, diplomatic distrust and business rivalries hindered the search for a uniform international flaghoist system. In 1857 the British Board of Trade made one of the better attempts at formulating a universal maritime communication system. Since the most practical use of the open seas required cooperation between noncombatants, the need for mutually understood flaghoist signals became urgent. While lanterns, lamps, whistles and horns were now used only for night and fog warning mechanisms, flags maintained their primary role for day signals.

The board decided upon a standard system of 18 colored flags. These banners, which represented all the consonants except *x* and *z*, were to be used openly by English merchant vessels. Combinations of these banners provided a code of 70,000 signals, which were published in a text entitled *The Commercial Code of Signals*. Not to be outdone by their rivals, the French responded the next year with their *Code Commercial des Signaux*, and disagreements regarding signals continued between different seafaring nations.

In the latter half of the 19th century, many changes occurred both in individual nations' flag signals and in international revisions. The illustration shown here generally exemplifies how flags became equated with the alphabet:

International maritime standard

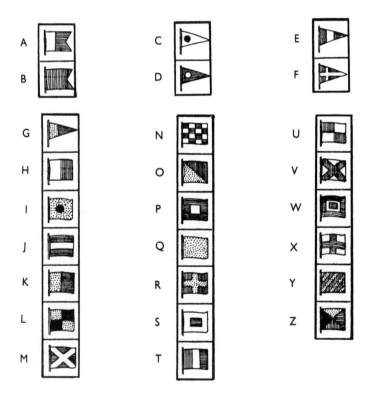

In 1870 the *International Code of Signals* was published, which still carries this title today. Other signaling procedures were also formalized through the 1870s and 1880s and signals were created with from one to four flags representing alphabet letters.

On January 1, 1902, a new international code was activated. The number of banners was raised from 18 to 26, restoring vowels and the letters *x* and *z* to make a full series of equivalents for the English alphabet. The new system was followed until 1914 when the First World War affected most naval procedures.

Flaghoist signals had greatly increased in complexity during the 19th century. In 1805, at the Battle of Trafalgar, in which Admiral Horatio Nelson's ships defeated a combined French and Spanish fleet, Nelson ordered only two flag signals sent. In May 1916 at the Battle of Jutland between Britain's Grand Fleet and Germany's High Seas Fleet, the Royal Navy communicated with nearly 260 flag signals.

After World War I and the invention of the radio, the International Radio-Telegraph Conference was convened in 1927. Reflecting the many differences that still existed among nations, the new *International Code* was not adopted until January 1, 1934. Published in various languages including English, French, German, Japanese and Spanish, it contained two volumes, one for visual signals, including flags and lights, and one for radio.

Through the 1930s and the Second World War, other practices became standardized. Individual nations' code and cipher books were entirely separated from the International Code system. For example, NATO, founded in 1949, had a series of secret naval signals for military purposes. The merchant ships of NATO nations applied the international non secret signals when no security needs were involved, but used their own signals when secrecy was required.

After many attempts at standardization, the International Code was finally developed. The following list illustrates single-letter signals:

A I am undergoing a speed trial.

B I am taking in or discharging explosives.

C Yes (affirmative).

D Keep clear of me. I am maneuvering with difficulty.

E I am directing my course to starboard.

F I am disabled. Communicate with me.

G I require a pilot.

H I have a pilot on board.

I I am directing my course to port.

J I am going to send a message by semaphore.

K You should stop your vessel instantly.

L You should stop. I have something important to communicate.

M I have a doctor on board.

N No (negative).

O Man overboard.

P (Displayed in harbor) All persons are to repair on board as the vessel is about to proceed to sea.

   (Displayed at sea) Your lights are out or burning badly.

Q My vessel is healthy and I request free pratique.

R The way is off my ship; you may feel your way past me.

S My engines are going full speed astern.

T Do not pass ahead of me.

U You are standing into danger.

V I require assistance.

W I require medical assistance.

X Stop carrying out your intentions and watch my signals.

Y I am carrying mails.

Z Calling a shore station.

The international flags are usually square, although the flags for *A* and *B* have an angle on their left side. The designs of the square banners vary with patterns of diagonals, stripes and colors. There are ten numeral pennants representing the numerals 0 to 9 and three repeater or substitute pennants. These send four types of signals: position of longitude and latitude (prefixed with the letter *P*), time (prefixed by *I*), position in degrees from true north (prefixed by *X*) and course signals (no prefix).

Three-letter flaghoists signal information such as compass points, wind direction, degrees to starboard or port and time zones. Four-letter signals beginning with the letter *A* represent around 11,600 geographic locations. Four-letter groups beginning with letters other than A are found in the "Beme List," which compiles signals for ships and their radio call signs. The first letter of the first letter pair often signifies the nationality of the ship such as Britain (GM) and New Zealand (ZK).

Hand-Held Signaling Systems: Flags, Disks and Torches

While flaghoist signaling systems were progressing at sea, extensions of the arm-and-hand methods using flags, disks and torches were being developed on land. The first "flags" in the Western world were actually carved wooded images attached to poles. Known as standards, these symbols were also used in the Middle East as early as 670 B.C.E. and from there spread to Crete, Greece and Rome.

During the decline of the Roman empire, their poles with varied insignia gave way to standards bearing a cross and by the 1100s the French had their "royal vexillum," a square-shaped piece of material suspended from a traverse bar atop a staff. Later French noblemen developed the "oriflamme," a banner split at one end forming flame-like streamers, which was in use from about 1124 to 1415. One of the earliest uses of flags to convey signals was around 1225, when regimental commanders' in the Persian army signaled tactical orders with black-and-white flags.

One of the first written documentations of a hand-held signaling system was a small text from around 1803–1816 titled *Signals to be made by One Man*. The text described a formal code that contained 61 different signal positions holding a hat, handkerchief or lanterns. The varied postures indicated the standard ten digits and 26 alphabet letters, along with 11 indicators such as "Affirmative," "Annul," and "Finish."

Signals to be made by one man

16   14   Finish

In 1809 Knight Spencer presented a signaling system to the secretary of the Royal Society of Arts. His proposal for the "Anthropo-Telegraph" was a numeric method with two round disks bearing two-toned "bull's-eyes." In addition to the ten numerals, four information signals were included: "Attention," "Period or space," "Calling-up" and "Error or annul." The numerals were arranged so that they could be linked with Popham's number-word code. While the Royal Society of Arts gave Spencer a silver medal for his creation, no record exists of his system being put into practice.

Similar methods followed with increasingly more involved combinations of rods, handkerchiefs, swords and flags. In 1817, several particularly complex processes were proposed by Lieutenant Colonel John Macdonald. One, his own version of the "Anthropo-Telegraph," involved seven sergeants who stood at predetermined distances holding white flags with red borders. The process required two groups of three sergeants who were positioned 20 feet from each other and stood to the left and to the right of a center point. Signaling involved raising but not waving the flag. A seventh sergeant with a red banner provided additional movements when rare

Spencer's
Anthropo-
Telegraph

words or special code was needed for a particular dispatch. An officer gave orders to the group while a second officer used a telescope to observe the replies from the next squad of men.

The right side's flag represented tens, the second flag hundreds and the third signified thousands. The left trio's banners represented letters, words or phrases. The numeric or alphabet equivalents were to be drawn from Macdonald's publication, *A Treatise Explanatory of a New System of Naval, Military and Political Telegraph Communication of General Application.* In addition to a 1,000-page dictionary for specific signaling equivalents, this text also contained nearly 80 pages commenting upon previous message methods.

Unfortunately for Macdonald, the seven sergeants and two officers needed at each communicating point and the overly detailed signal books rendered the system too complex for military use. Although the colonel gained personal acclaim in the military, his signaling system was never used widely.

Another signaling system was developed in 1850 by Commander A. P. Eardley-Wilmot. His huge *Dictionary of Signals* was written for fleets as well as flotillas and individual yachts. In addition to a flag system, he proposed man-formed signals that used rods and handkerchiefs, caps, banners and shipboard objects raised in formations on masts. These single-man systems gave way to team efforts as different objects and more complicated motions were needed. While Eardley-Wilmot rose through the naval ranks, his system proved to be too complex for naval use. His system was also too difficult for military bases and railroads as he had complicated his formations with too many shipboard tools or structures.

In 1852 Captain Robert W. Jenks of Boston compiled *The Brachial Telegraph* for land and sea communications. Designed more specifically for signaling at sea, Jenks's method included signals to be used when weather conditions, such as thunder and high winds, precluded vocal communications. While still retaining the old, awkward stick-and-flag signals, he recommended a "voice-enhancing trumpet," which was probably an early megaphone. On the following page is an example of Jenks's spelling system.

Although his system was not particularly ingenious in itself, many of Jenks's recommendations for signalers were ahead of their time. He suggested signaling from clearly seen vantage points, having brief intervals between transmissions and writing received signals on a slate to ensure accurate translation into the alphabet. Although no nation adopted this advice at the time, variations later became part of accepted civilian and military signaling procedures.

Just three years later, a single-man arm-and-objects method was accepted as a part of a larger communications system.

Jenk's Brachial
Telegraph system

The creator of this process was Captain Charles de Reynold-Chauvancy, whose *Télégraphic Nautique Polyglotte* became better known as *Reynold's Code for Maritime Flags*. Primarily a large compilation of naval banner combinations, it also included distress signals using a flag and a hat on a 4-foot rod to signal from ship to shore. While it was the banner code, rather than the individual signals, that enjoyed wide usage, some historians suggest that the success of the flag code carried the individual signals along with them. This signaling system may have had an influence on the later acceptance of the two-arm semaphore process.

During the American Civil War, the Union navy developed the *Flotilla and Boat Squadron Signals* to aid communication as it strove to blockade the Confederacy's long coastline. Within the publication was a "Plan of Homograph" that consisted of ten digits and six specific signs (see facing page). The method had sporadic applications for ship-to-ship contacts in the early months of the conflict until it was superseded by Albert Myer's wigwag.

A doctor who had proposed a communications system for the hearing impaired, Myer also created a method for single-flag

Reynold's distress signals

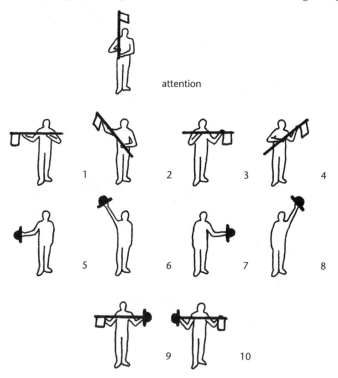

attention

1

2

3

4

5

6

7

8

9

10

dispatches. While serving with the U.S. Army in Texas in the 1850s, he developed the system that he called flag telegraphy. Although it was approved for use in 1860, the year Myer became chief signal officer, the system was not fully appreciated until the outbreak of the Civil War in 1861. By this time, it had come to be called "wigwag" because of its distinctive signaling motions.

First successfully used by the Confederacy, wigwag was a system of positioning a flag at various angles that corresponded to the 26 letters of the alphabet. The flags used by the communicants measured 2, 4 or 6 feet square. They were usually red or black with white square centers or white banners with red square centers, both versions of which provided adequate

U.S. Navy's homograph signals

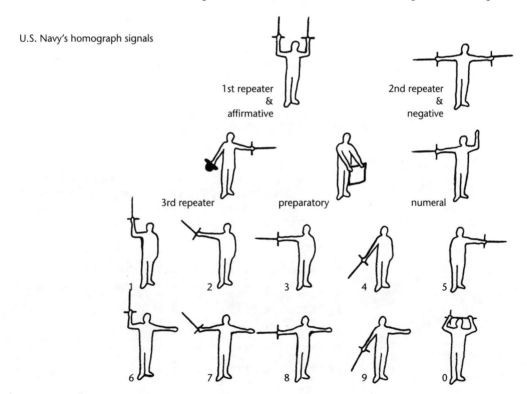

1st repeater
&
affirmative

2nd repeater
&
negative

3rd repeater   preparatory   numeral

1   2   3   4   5

6   7   8   9   0

contrast against standard field backgrounds. The flagstaff itself was usually made of a strong wood such as hickory and was built in sections that could be joined for extension if required. At night, torches made of a metal canister filled with a flammable liquid attached to a staff were substituted for the flags. A second "foot torch" was placed on the ground in front of the signaler as a fixed point of reference, making it easier for the recipient to follow the torch's movements.

The method, as outlined in *Myer's Manual of Signals* (1872), required a combination of three basic motions for each letter. All sequences were initiated with the signaler holding his device vertically and motionless above his head. The first motion was made by bringing the device downward on the signaler's right side and then quickly returning it to its upright position. Motion 2 involved bringing the device downward on the signaler's left side and then returning it to the starting position. Motion 3 lowered the device in front of the signaler, then returned it to its vertical position.

Myer's wigwag with flags

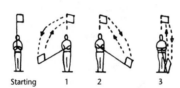

Starting   1   2   3

The chart below indicates how letters and directions were conveyed. For example, 112 would be signaled by motion 1, motion 1 and motion 2 in rapid succession. The periods signify a pause in a sequence of movements.

a	112	h	312	o	223	v	222
b	121	i	213	p	313	w	311
c	211	j	232	q	131	x	321
d	212	k	323	r	331	y	111
e	221	l	231	s	332	z	113
f	122	m	132	t	133		
g	123	n	322	u	233		

MYER'S SIGNAL DIRECTIONS

3	End of a word
33	End of a sentence
333	End of a message
22.22.22.3	Signal of assent: "I understand," or general affirmative
22.22.22.333	Cease signaling
121.121.121.3	Repeat
211.211.211.3	Move a little to the right
221.221.221.3	Move a little to the left
212121.3	Error

In 1864 Myer developed a similar daytime system using standardized disks. Somewhat easier to handle than the flags, they measured twelve to eighteen inches in diameter and were made of wood or metal frames with canvas stretched across them. The disks had handles and were used in the series of motions shown below:

Myer's wigwag with disks

Starting   1   2   3

Lanterns were also substituted for torches at nighttime. One lantern was mounted on a pole in a fixed position, while a second lantern was moved above and to either side of it as shown below:

Myer's wigwag with lanterns

Myer's flag method was also used in both the Federal and Confederate navies. Despite the efficiency of the system, changes in the signals' meanings as the war progressed often caused problems for commanders. One instance was the changing of the right and left motions for positions one and two and their alphabet equivalents in the latter stages of the struggle.

Signals were always subject to interception by experienced observers on both sides. Always security conscious, Albert Myer applied for a patent for a cipher disk intended to accompany the wigwag system. The disk used the numerals 1 and 8 in various combinations on its outer ring and a randomly placed alphabet on its movable inner ring. Both the sender and the receiver each required a disk, coordinating it with the flags or other visual signals. Myer's goal was to give better protection to wigwag by sending signals with numbers that could only be decrypted with the disks.

As the Civil War wore on, Myer increased the wigwag motions to four, enabling more specialized words and abbreviations to be used. His signaling system was used extensively by both the Confederate and the Union armies throughout the Civil War, directly affecting the outcome of battles such as Bull Run.

Another Union success in August 1864, Admiral David Farragut's naval victory at Mobile Bay, Alabama, also benefited from signals exchanged between Northern ships and land forces. Several reports indicate that a flagman conveyed Farragut's legendary command, "Damn the torpedoes, full speed ahead!" using Myer's wigwag system.

Another flag signaling system that was proposed during the Civil War was Swaim's signals. James Swaim was a U.S. Navy man whose service prior to the U.S. Civil War had exposed him to various methods of visual communication. Hoping to be awarded a war contract, he brought his signal system to the Washington, D.C. in November 1862, and gained the attention of two telegraphers of the War Department, David Bates and Charles Tinker.

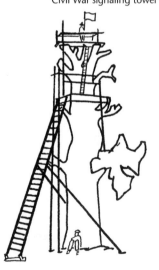

Civil War signaling tower

Swaim presented Bates and Tinker with a codebook of several thousand phrases, words and numerals that were represented by combinations of as many as four digits, which in turn were conveyed by six visual signals using a flag or a torch. In this system, which was similar to the one devised by Albert Myer, the flag or torch was moved through different positions to form a nonsecret code. It was secure as long as the true meanings of the movements were kept secret.

The signalman held the flag or torch up and to the right to indicate the number 1, straight out to the right for 2, and to the right and down for 3. The numbers 4, 5, and 6 were indicated by the same movements performed to the left side of the signalman. These positions were equated with number pairs that indicated alphabet letters and digits:

a	g	m	s	y	5
11	21	31	41	51	61
b	h	n	t	z	6
12	22	32	42	52	62
c	i	o	u	1	7
13	23	33	43	53	63
d	j	p	v	2	8
14	24	34	44	54	64
e	k	q	w	3	9
15	25	35	45	55	65
f	l	r	x	4	0
16	26	36	46	56	66

Swaim proposed a mnemonic device to help signalmen remember the key to the alphabet. His formula asked, "Who is the inventor?" The answer was found by repeating the first four horizontal letters of the chart rapidly, so that *agms* sounded like "a Jeems" (or "James," Swaim's first name). The next question was "Why?" (*Y*). The letters *a, g, m, s, y* corresponded to the first letter of each column.

Although Bates and Tinker found the method relatively easy to learn and Abraham Lincoln was encouraging, Swaim's proposal never gained favor with the War Department hierarchy or the field commanders.

After the Civil War, fewer manual signaling systems were developed. In 1872, however, the prolific Myer, now the Signal Corps' commander, proposed a new arm code among the various methods defined in his *Manual of Signals*. His ten-element form was especially applicable for use on slanting ship decks, as it substituted the often-unwieldy flags for disks from a 12 to 18 inches in diameter, with handles for holding in different configurations.

Despite Myer's advances, though, the telegraph with its Morse Code became the predominant system for long-distance communication. Inevitably, officials created Morse code versions of wigwag. The banner movements were tried in both a right-and-left and a left-and-right order equating banner motions with Morse's dots and dashes. For example, a flag motion to the right was a dot and one to the left was a dash. Flag signalers could use Morse signals to exchange dispatches with men waving other objects or using the light-reflecting heliograph to make long and short flashes. In these often confusing times, different signaling languages might be used on different days by the same unit.

In 1886 the U.S. Army made International Morse its official system for electric telegraphy, wigwag and the heliograph. Many signalers found Morse more difficult to apply with flag signals, however, and in 1896 the familiar Myer flag motions were restored to wigwag and the heliograph, and the Army returned to American Morse for telegraphy.

In England, the Royal Army used a one-flag style to convey the digits 1 through 0 using all white or all black flags. The English system signaled in long and short waves which were both broader in motion and longer in duration than those of Myer. Each signal also had different motion and number equivalent; a short wave represented 1, a long wave was 6, and two short and one long signified 9.

Flag waving continued into the 20th century with the publication of the *International Code of 1901*. In the United States, Myer's method was still used with flag and torch signals and the military had expanded wigwag to include hand lanterns and searchlights. Great Britain's flag signalers had switched from their long-and-short waves to dots and dashes. A dot was shown by an overhead motion, while a dash was a broader sweep of motion from a 45° angle to 225° as depicted here:

British flag signals

dot   dash

A   B

In 1908, Lieutenant Colonel R. B. Dietz of England's Seventh Dragoon Guards created a clever and uncomplicated process which depended solely on a disk held by pieces of wood. One side of the disk was colored black or light brown and the other was white. Dashes and dots were created by flipping the disk from side to side. A longer-held color represented a dash, and the quicker-held one indicated a dot.

A similar version of this disk was brought into action in France in 1915 during World War I as trench warfare became standard practice. However, the First World War came to mark the end of land-based waving signals, as enemy marksmen eventually became too accurate for even the bravest signalers to transmit safely.

One-flag waving continued to be applied sporadically in the Royal Navy through both World Wars. Single flags were used in daylight or in unusual circumstances. During radio silence, or if electric blinker lights were considered a security risk, the flagmen could transfer silent orders from the stern to the next ship's bow. For brief dispatches, naval officers preferred to use a single flag rather than hoisting a group of banners on ropes or halyards. Single flag methods were only officially discontinued in March 1946, finally superseded by electric systems.

Two- or three-flag systems existed in U.S. Army armored and mechanized units from the 1920s through the 1970s, before the widespread use of computers and satellites. For many years, a green and an orange or red banner were combined to give attack orders, maneuver directions and warnings of enemies. A truly standard code for such flags never emerged, and eventually radio and radiotelephone links outdated even informal tank codes.

Some arm-with-flag or disk versions are still in use today. One form, a U.S. Navy signaling system, is called high-line semaphore. During resupplying and refueling at sea, a connecting cable called a highline aids the transfer of personnel, supplies and fuel. During such procedures, a strong cable is fired from the vessel needing resupply to the replenishing ship, often not more than 50 yards apart. Another cable (the highline) is also transferred to the supply ship and attached to the first line at mutual connection points on both vessels. These

Aircraft carrier
landing signals

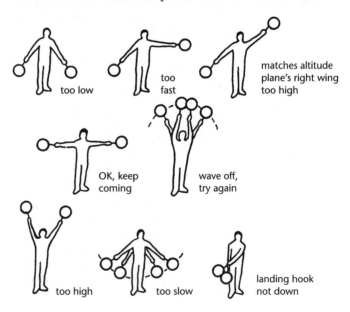

too low

too
fast

matches altitude
plane's right wing
too high

OK, keep
coming

wave off,
try again

too high

too slow

landing hook
not down

cables support chairs or box-filled nets that move along the lines by pulleys. For refueling, a special hose is passed along the cables and connected with the fuel tanks. On the recipient ship, the person in charge of signaling about the condition of the vessel-connecting lines must be constantly ready to warn of any problems during the transfers.

Though largely supplanted by electronic systems, basic waving techniques are used as a backup communications system during aircraft carrier operations. On deck, the landing signal officer helps direct incoming aircraft by using a pair of red or yellow paddlelike disks. Through various disk positions, he or she alerts the pilot about the aircraft's height, speed and wing alignment. Following are some examples most frequently used from the 1940s through the 1970s.

Light Signals

Light signals include constant illumination systems from groups of torches to multiple long-burning lights. The second category, intermittent lights, include flashing dots and dashes, and chronosemic (time-governed) signals with pyrotechnics.

CONSTANT ILLUMINATION SIGNALS

The history of single-light systems combines legends, myths and practices from around the world. Greek myths frequently mention communication with fire signals, and two lovers communicate with flame signs in the play *The Supplicants* by Aeschylus. In 1194 B.C.E. a fire signal chain of nine or ten sites conveyed the news of the defeat of Troy some 60 miles to Agamemnon's palace at Mycenae, and the chroniclers Homer, Julius Africanus, Livy, Vegitus and Plutarch also described the military use of smoke signals by day and fire by night.

The Macedonian rulers and territorial conquerors Philip and his son Alexander used fire signals in the 4th century B.C.E., but the more organized methods of Aeneas the Tactician and

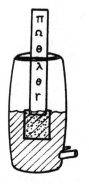

Aeneas's torch-and-barrel apparatus

Polybius did not develop until the period from 340 to 170 B.C.E.

Both Aeneas and Polybius were Greek scholars, but Aeneas's works have only survived through Polybius's later analysis of his writings. In one such text, Polybius described one of Aeneas' systems, created around 350–345 B.C.E., that combined water and fire. Two groups of communicants, consisting of at least two men each, first waved torches to attract the recipients' attention and indicate that they were ready. At each site was a central apparatus, which consisted of 4-foot-deep earthen vessels filled with water with a valvelike aperture at the base. A long rod with military phrases carved in sections along it floated on a piece of cork.

When a torch was waved in response to the attention light, both lights were covered by a screen. The valves of both containers were opened, permitting the water and cork to descend. When the desired term on the rod reached the upper rim of the barrel, the sender again displayed his torch to halt the flow of water. The receivers stopped the valve and read the words at the graduated mark. If the timing was accurate, the words at each site would be identical.

Although an awkward method by modern standards, Polybius's writings suggest that the device may have been used in actual sieges. He was so intrigued by such accounts that he applied himself to devising a signaling system of his own.

Aeneas's signals

Recorded in his *Universal History* (of which only 5 volumes out of 40 survive), the torch system for which Polybius is best remembered is partially credited to Cleoxenes and Democritus. Polybius's system divided the Greek alphabet into a five-by-five grid. Below is one of the most complete illustrations of the system:

Polybius's torch system

Translations of Polybius's writings indicate that he suggested using torches to convey the positions of letters on the tablets. For example, torches held to the left of the tablet may have indicated a column, while a torch on the right may have signified its rows. A letter in column 1 and row 2 of the first tablet would then be indicated by one torch raised on the left and two on the right. Because each letter required several torch movements, screens were set up to conceal the torches in between signals.

Since Polybius was spelling individual words with his torches, this method is considered a true alphabetic signal. Although there is no known account of his signaling style being applied, Polybius's checkerboard, as it came to be known, became the basis of many types of ciphers.

In 65 B.C.E.. the Greek biographer Plutarch recorded the use of fire signals from towers by raiders around the Mediterranean Sea, and in Julius Caesar's reign, a series of masonry towers were constructed to transmit fire signs.

Like much of cryptology, knowledge about fire signals was lost with the fall of Rome and the onset of the Dark Ages and signals took several centuries to reemerge.

By the late 16th century, fire signs were beginning to be used as flagship locators at sea. The Spanish Armada had a very basic light pattern called a station-keeping system, in which each vessel moved past the flagship to receive the evening's watchword and the next course. A metal container filled with combustibles burned at the stern of the flagship to guide the ships. If the weather conditions were inappropriate for fire, a lanternlike arrangement with four covered lamps was placed in the stern.

Records of later signaling systems are vague. In his *Histoire de la telegraphie*, Ignace Chappe described a type of mirror for sun and moon reflections. It seems to date from the late 1100s as English scholar Roger Bacon reportedly used a version of it in the 1200s.

In 1616 Franz Kessler published *Unterschiedliche bisshero mehrern Theils Secreta oder Verborgene, Geheime Kunste* (Varied, until now mostly hidden, secret arts), in which he described a light signal combined with telescopes. A lead-lined barrel containing a flame was placed on its side with an opening facing the recipient. A shutter or trap door covered the opening and the shutter was opened and closed to signal letters. As this meant numerous signals to send letters further in the alphabet, Kessler reportedly developed a code that squeezed the alphabet into 15 letters. The numerals 1 to 15 could then be equated with different letters to preserve secrecy.

Great Britain used similar shipboard fires which served during periods of fog and as a locator for complex course-changing maneuvers. By 1617, Sir Walter Raleigh was guiding his subordinate commanders using a fire signal aboard his ship.

During the American Revolution privateers used lanterns to communicate with friendly ships. A seaman would report sightings of an unfamiliar sail configuration or a possible enemy vessel by raising a lantern at the ensign staff, lowering it and raising it again in a preset signal pattern. To reduce the chance of detection, canvas was used to shroud the brightness:

Lantern signals

Battles at times of poor visibility also required light signals. Lanterns were often hung fore and aft or along the rail in alignment with different masts as signs that friends were near:

Lantern signals

During this rebellion the always sea-conscious British learned from the lantern tricks used by rebel privateers and adopted similar techniques for their own use. From 1775 to 1783 British admirals began to use arrangements of lanterns as standards to signal daily operations such as sailing speed and directions. While these were basic configurations such as three lights in a row or in a triangular form, they were the foundation for future light signaling systems.

One early system was Admiral Lord George Rodney's "Fighting Instructions and Signals by Night" (1779). The Rodney method used varied clusters of lanterns mounted with ropes and attached to yardarms in a hoist apparatus. A formal list conveyed meanings for night engagements.

Detailed code dispatches were not yet possible, however, and only preplanned communiqués could be sent. For example, a signal could indicate that an enemy ship had been sighted, but could not give details of the approaching squadron's numbers, speed or armaments.

Elaborate rituals for lantern procedures eventually developed. Groups of lanterns were mounted on frameworks and hoisted to different masts for varied types of signals. Some lanterns used color-tinted glass of red and occasionally green hues. A metal panel operated with ropes was attached to each lantern and could be applied to create combinations of visible and covered signals,making more varied signals possible.

Coding possibilities expanded when light signals were combined with other implements. According to the Royal Navy's *Night Signal Book for the Ships of War* (1799), lanterns were equated with the digits 1 through 4, one rocket represented 5, one cannon was 10 and two cannons were 20. Numbers up to 79 could be signaled. As the cannons could not signal when they were involved in combat, lanterns in triangular or square frames could also signify multiples of ten. These numbers were equated with orders listed in signal books:

```
1-4    1 to 4 lanterns
5      1 rocket
6      1 rocket, 1 lantern
7      1 rocket, 2 lanterns
8      1 rocket, 3 lanterns
9      1 rocket, 4 lanterns
10     1 cannon
11     1 cannon, 1 lantern
15     1 cannon, 1 rocket
16     1 cannon, 1 rocket, 1 lantern
20     2 cannons
25     2 cannons, 1 rocket
30     3 cannons
```

In 1803 a light system aided the fortunes of Captain Edward Preble, commander of the U.S. frigate *Essex*. Captain Preble had served in the Mediterranean during the U.S. conflict with the four Barbary States, Algiers, Morocco, Tripoli and Tunis. The seamen of these North African kingdoms had become buccaneers who attacked others' merchant ships.

U.S. ships had also been attacked and President Jefferson ordered the U.S. Navy to take action. While guarding U.S. ships and needing night signals, Preble used a series of lanterns to send numbers equated with signal book orders. Adapted from British fleet signals, these allowed him to send more efficient ship-to-ship exchanges.

Preble's lanterns

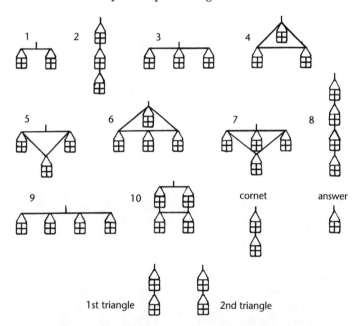

When he rose to the rank of commodore in 1803, Preble initiated attacks on the vessels of the Barbary States, and by 1805, only the ruler of Algiers remained defiant. He finally capitulated to U.S. Naval forces in 1815.

Amid a number of rudimentary codes, the signals of the French vice-admiral Edward Burgues, comte de Missiessy offered a more standardized system. His method combined flags by day, torches by night and cannons for inclement weather. In 1826 he published his *Tactiques et signaux—de jour, de nuit et de brume, a l'ancre et la voile* (Tactics and Signals used at anchor or on the waves, by day, by night, and in the fog). Incorporating time intervals, his torch version indicated numbers with a base sequence of one to five, then a timed pause, followed by different additions of torches according to a set pattern.

Following is the comte de Missiessy's number code.

Missiessy's torches

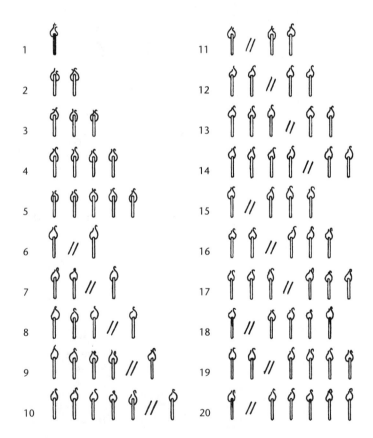

In this chart, dual diagonals (//) represent a minute interval. Similar to Polybius's checkerboard, the motions were also accompanied by a grid that mirrored the numeric equivalents. The first five digits were equated with the corresponding number of torches. The pattern for numerals 6 to 20 were determined from the grid. The numbers on the outside were read from the top first, followed by the side. Number 6 was the intersection of 1 at the top and 1 at the side, indicating one torch motion, a minute interval, then another torch motion. The numeral 10 was 5 torches, an interval, and one torch; and 15 was one torch, a minute pause and then three torches.

FOR ONE TO FIVE: 1 2 3 4 5

FOR SIX TO TWENTY:

	1	2	3	4	5
	(one-minute interval)				
1	6	7	8	9	10
2	11	12	13	14	
3	15	16	17		
4	18	19			
5	20				

The same system could be used for flags and cannons. For daytime use, flags signaled that flag motions would follow. A rocket was fired if the message was to be conveyed by cannon while gunshots indicated that torches were to be used.

The signals' numerals were equated with lists of nautical terms and sailing directions. This type of nonsecret signal code was adopted by other navies and was used until it was superseded by multiple lights and wigwag systems.

As the 19th century progressed, color increasingly entered the signaling process with mixed results. Depending on the glass-making skills of a nation's craftsmen, glass often distorted the light. Tinted glass also had varying degrees of transparency; if it was too thick it dulled the light and if it was too thin it appeared as white light to observers.

Different maritime nations debated whether to cover the light with materials of different hues or to tint the glass during manufacture. The thick substances required to withstand weather conditions at sea often obscured the light so that no illumination could be seen at even short distances.

One color system was developed in the late 1850s and 1860s by William Henry Ward, who conducted experiments at the Washington Navy Yard near Washington, D.C. Ward's proposals influenced the trend away from hanging groups of lights in squares or triangles to arranging them in sets of vertical lines.

Because blue and green tints were often mistaken for each other at a distance, Ward settled upon a pattern of white, red or no light, a system popular among naval signalmen in the 1850s. He used movable cylinder lenses to change the colors of the lanterns from inside them and applied metal or canvas screens to cover the lanterns. His lantern arrangements were adopted by nations as diverse as Norway, Egypt and Turkey and remained popular into the 1880s.

In the late 19th century, more advanced light codes were influenced by Myer's wigwag and the electric telegraph. A rather unusual multilight method is credited to the Confederate Navy during the Civil War. For communications between land and sea, the South assigned army signal corpsmen to the swift vessels that tried to break through the Union's naval blockade. They set up a system of wigwag flags and a pair of long tubes with shielded lanterns at one end. The lights could only be seen in the direction at which the shipboard tubes were pointed. A red light indicated the wigwag 1 and green was wigwag 2; the wigwag alphabet was used to decode the light patterns. These tube-sent transmissions were considered safer than flag- or torch-sent dispatches.

Electric-based light signals joined other signaling systems after Thomas Edison invented the electric lightbulb in 1879. Electric signals were possible after a series of U.S. Navy experiments and the development of electric power sources on board ships. The introduction of electricity profoundly improved naval communications. When electricity could be generated aboard ship, whale-oil lanterns were replaced by electric lamps with green, red and white glass covers that indicated alphabet letters and numbers. In the late 1800s, a new generation of electric codes were introduced, including the Conz, the Ducretet, the Sellner and the Ardois.

In 1892 U.S. Navy officer Albert P. Niblack gave a detailed description of electric light-based mechanisms in the U.S. Naval Institute's publication *Proceedings*. Some examples included:

1. **Berg** Tested by the German navy, this process involved three electric or oil lanterns with green, red and white shades powered by electromagnets. A total of 27 three-light signals could be made.

2. **Conz** A German invention with three double lanterns that conveyed red and white light.

3. **De Meritens** A French naval method, the De Meritens had 11 lamps. The two upper and two lower lights were red, and seven white lights filled its middle section.

4. **Ducretet** Used in the French navy, this system had only four signal lanterns, capable of transmitting 15 different signals. They could only be seen in one direction, however, and the system was quickly superseded by others.

5. **Kasolowski** Incorporated by the Italian navy, this method involved four double lamps supported by a bronze frame hanging from a yardarm.

6. **Massari** An Italian apparatus, this had four double lamps and was supported on a yardarm like the Kasolowski method. The Massari system also had a Morse printing device which recorded the signal on paper.

7. **Ardois** A French system that had five lamps supported on a frame and yardarm arrangement.

Developed in the 1880s, Ardois became the U.S. Navy's official lighting system in 1891. Originally consisting of five lights, it was changed after being adopted by the U.S. Navy, resulting in a process that used four double lamps. Each lamp contained one 32-candlepower incandescent light in its upper half and another in its lower half. Cover shades for the glass globes provided the red for the lower half and white for the upper half, a combination that became standard for this method in the United States.

Clusters of lights were arranged to make a code and hoisted-up masts with current-supplying cables were attached and controlled by a keyboard device. The lamps provided much clearer and more varied messages than the lanterns of past eras. Using red and white lights, the Ardois system conveyed messages in the familiar dots and dashes of International Morse code. This made the illumination of alphabet letters and numerals more practical.

Signals were first set up according to the Myer wigwag system, with red representing motion 1 and white motion 2. International Morse later became the standard, with red signaling a dot and white a dash. Ardois continued to be used as better generators and onboard electricity sources became more common. The following alphabet chart is a representation of Ardois.

Ardois system

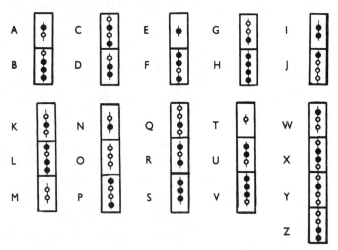

key: red = dot (in the International Morse); white = dash.

Aspects of multipart light methods continued into the 20th century. In the First World War, pilots used an air-to-ground method that incorporated green, red and white lights to communicate with persons on land. Informal signals equated International Morse with certain combinations of the colors. For example, a green light mounted on the right wing would signify a dot and a red light on the left represented a dash. Brief messages would range from "enemy sighted" to "fuel low" or "need to land." The introduction of radio soon superseded these limited flight messages.

As single lights were replaced by multiple types, constant illuminations were rivaled and often surpassed by pyrotechnics and sun-based processes.

Constant illumination lights are still useful in identifying aircraft and water-borne vessels, such as green lights for starboard and red for port. Naval craft also have signaling distinctions for international, coastal, harbor, lake, river and at-anchor procedures.

PYROTECHNIC AND CHRONOSEMIC SYSTEMS

The word *pyrotechnic* is derived from the Greek *pyros*, meaning "fire" and *technè*, "art." Pyrotechnic devices have included firecrackers, flares, rockets, smoke bombs and tracers and have been applied for private communications as well as warfare for centuries. Made of elements that consume themselves, pyrotechnic devices have been used to attract attention, attack enemies or send messages to subordinates and allies. Because of their irregular burning qualities, they have also been known as periodic or intermittent systems.

The Chinese developed forms of explosives around 1161 C.E. During the latter 1100s and 1200s Mongol cavalry used a type of explosive made with naphtha, an inflammable oily liquid. Gunpowder as a propellant for weapons was not used until the mid-1300s in Europe.

The first pyrotechnic mechanism of some lasting value was the rocket developed by Sir William Congreve. When launched, the device exploded into light, then emitted a parachute that enabled it to float to the ground over several minutes. This became a signal to initiate or halt military action, and also served to illuminate battle areas.

Improvements in its launch height, illumination power and duration led to the adoption of the Congreve rocket by the U.S. Navy as its illumination could serve as a warning or emergency signal. The rocket was soon adapted to carry a short cord to provide a lifeline to victims in the water or a link to other watercraft. Some larger devices, fired by a mortar, could also carry a line to the shore.

Although not yet applied as a formal code system, it was the Congreve device which created the "rockets' red glare" recorded by Francis Scott Key during the attack on Fort McHenry. In September 1814 Key, a lawyer practicing in Georgetown, near Washington D.C., had been asked to intercede on behalf of a doctor held by the British. The release was successfully arranged, but during the night of September 13, Key himself was detained on a ship by the British. Their bombardment targets included a key location, Fort McHenry. On seeing the American flag still flying above the ramparts at daybreak, Key wrote a poem whose words, combined with a popular English song, became the American national anthem.

By 1849 Brevet Major A. Mordecai had compiled an official ordnance manual for the U.S. Army which contained accounts of rocket construction. Made by hand, the rockets had thick paper cases. A powder called "stars" was composed of alcohol, antimony, gum arabic, niter and sulphur. U.S. war rockets, like the Congreve model, were built from sheet iron and had a long wooden stick that functioned as a directional tailpiece. When the fuse was lit, the rocket emitted colored smoke that was visible from several miles away.

The next advance with periodic light systems was made by the U.S. Navy in 1858. Benjamin F. Coston (sometimes spelled Costen) improved the composition material to give it much more illumination, but his real innovation was making the device hand-held, which allowed it to be ignited by hitting it on a hard surface. This was the first flare, a device well ahead of its time.

In 1861 Albert J. Myer took a direct interest in the process. Although Coston was deceased by this time, with the permission of his widow Myer had the system manufactured as a standard piece of U.S. Army Signal Corps equipment. Known as "Coston's Lights," the green, red and white flares could be fired by hand percussion or with a percussion cap in a signal pistol.

After more experiments, the rockets were mounted in a holding device. The containers were numbered according to the colors they emitted, and were set off by igniting a match rope. Also known as the Coston Composition Fire System, the method was easily applied to code formation. The digits 1, 2, and 3 could be represented by the standard colors red, white and green. A number such as 726 which might mean "regroup" was sent by lighting rocket number 7, which gave off a green illumination. Immediately upon its going out, number 2 was lit, which glowed with a white and then a red flame. This was then followed by number 6, which burned red and green. After no more than three-quarters of a minute, another series of digits was sent. If the signaler waited beyond this time, the recipient could begin his response.

Following are some other signal codes possible with this system:

1. Ready?
2. I am ready.
3. I am not ready.
11. The enemy are at our front.
12. The enemy are on our left.
13. The enemy are on our right.
111. Shell the trees on our left.
112. Shell the trees on our right.
113. Shells falling too far.

One red Coston device signified the inquiry "ready?" The combination of 1 and 2 (red and white) meant "the enemy are on our left" while 1, 1 and 3 (red, red and green) signaled "shells falling too far."

The Congreve rocket and Coston method influenced the development of a new type of system called *chronosemic* (Greek: *chronos*, "time," and *sema*, "sign"). The brainchild of Franklin Greene, a civil engineer and inventor, he proposed his plan to Union officials during the Civil War. Greene hoped to bring together the standard signal codes in a more effective arrangement for use during foggy conditions.

Greene incorporated an array of processes including rockets, lanterns and flashes of gunpowder. He included sound signals using cannons, steam whistles, human- and steam-powered trumpets, bugle calls and drum rolls. These were to be conducted under the command of three officers. A signal officer directed the procedures, kept the signal book and wrote the content of the dispatches. A second officer directed the crewmen who made the signs and a third officer handled all the necessary timekeeping aspects.

In order to time intervals between signals, Greene recommended using watches, deck clocks, hourglasses, metronomes or a seconds pendulum. The latter was made of twine and a small weight, and when suspended between the forefinger and thumb, it provided a surprisingly accurate way to count the passing of seconds.

This organization of time and methods made synchronized codes more practical. When followed carefully, a cross section of methods could be combined. Lanterns could be raised before cannons were primed, but after steam was built up for fog horns. Rockets' launches had been standardized so that a quick-match fire burned in 12 seconds. These actions were governed by Greene's deck clocks, metronomes and hourglasses.

This precise timing expanded code options, as instructions could be signaled not only with combinations of colors, but the same colors used in different sequences and at specific times could represent different meanings.

PRE-CHRONOSEMIC

```
green = 1

green and red = 2

green, red, and white = 3

green, green, red, white = 4

green, green, red, red, white = 5

green, green, red, red, white, white = 6
```

CHRONOSEMIC

```
green = 1

  (time interval, e.g. 20 seconds)

green (appearing after a set time,
e.g. 10 seconds) = 2

  (time interval, e.g. 25 seconds)

green (appearing after a set time,
e.g. 15 seconds) = 3

  (time interval, e.g. 30 seconds)

green and red = 4
```

Albert Myer was so impressed that he included a detailed description of Greene's facts in his own *Manual of Signals* (1872). This influential work led to more pyrotechnic and chronosemic applications in regions such as the western United States. Scouting units of the U.S. Army were aided by Coston's lights, rocket signals or better-timed volleys of gunfire as links across the vast expanse.

One of the most practical devices for safety and secrecy was the Very light. Born in 1845, Edward Very served in the Navy during the Civil War. Later, at the Washington Navy Yard, he began experiments with weapons and ammunition. In 1877 he introduced a code for day and night messages and a year later he patented a signal cartridge. His method was first used in 1878 by the U.S. Navy.

His Very signal consisted of a pistol designed to take a 10-gauge, center-fire, brass-headed paper shotgun shell. The original pistol was single-loading and had a steel barrel around nine inches long that was tapered at the muzzle. To operate it, one pressed the barrel catch and broke open the barrel by pushing it down. The cartridge then was inserted and the barrel closed and locked.

The cartridge was similar to a standard shotgun shell, with a primer and a firing charge of about 25 grains of musket powder. At night the cartridge propelled a "star" skyward and in daylight the shell produced a trail of smoke. These stars, of red, green or white, were made of cylinders packed with pyrotechnic material. Each was reinforced with wire and wrapped with a quick match, and one end was primed with a small portion of black powder to ensure its ignition at full illumination just before it reached its maximum height of 200 to 400 feet.

Each star color had a distinctively shaped top for easier handling at night. In comparison with earlier pyrotechnic methods such as the Coston system, which had 18 to 20 different accessories, the Very was a less awkward nonsecret code.

The Very code was similar to telegraphic systems. Resembling the alternating dots and dashes of Morse code, the Very system signaled numbers with different four-part sequences of red and green. The digits 1 through 0 were indicated by the following patterns of red (*R*) and green (*G*) bursts:

1	R	R	R	R
2	G	G	G	G
3	R	R	R	G
4	G	G	G	R
5	R	R	G	G
6	G	G	R	R
7	R	G	G	G
8	G	R	R	R
9	R	G	G	R
0	G	R	R	G

The alphabetic code used alphabet letters and corresponding alphabet positions, for example, *A* = 1, *B* = 2 and so on. The letter *A*, for instance, was represented by a single star. To prevent confusion between numbers and letters, a color variation was used to indicate that numbers would follow.

The Very system was not without its problems, however, as unlike the Greene process, it had no planned time intervals. The correlation between number of stars and alphabet position also caused problems with the number of shots needed to convey the letters later in the alphabet.

The convenience of the pistol compared with other systems seems to have been a primary reason for the acceptance of the method by many of the world's navies and armies. Albert Niblack suggested that letters be represented by different color combinations rather than numbers of signal bursts (the system shared previously), and greatly improved the efficiency of the system.

The Very system brought a much needed order to night signaling. It was included in the U.S. Navy *General Signal Book*, and the U.S. Army soon adopted the Very method as well.

During the 1890s, Very's signals were tried with Myer's alphabet and numeral equivalents. With the advance of telegraphy and Morse code, some attempts were also made to translate the Very system into Morse. Although by the time of World War I the Very was considered ill-suited to Morse code, a system did exist for translating its colored flashes into dots and dashes:

NIGHT

```
red = dot
white or green = dash
```

DAY (COLOR NOT DISTINGUISHABLE)

```
one shot = dot
two shots (fired in quick succession) = dash
```

Many Very signal codes were impromptu arrangements during battle. These included quick messages such as "mission succeeded" or "begin supporting advance" rather than long sequences of star bursts amid bomb hits and artillery fire. While the process was obviously nonsecret, Very did create a system for dispatch security. For encoding, extra star bursts could be added at certain points in the firing sequence to make the illuminations indiscernible even to an experienced observer.

Although by World War I, pyrotechnic and chronosemic devices had been supplanted by electric telegraphy and radio as vehicles for Morse code, they did play a large part in a different role. Both sides employed huge numbers of illuminations including large "searchlight" shells, many forms of color bursts and smoke-emitting cartridges, locator flares, tracer bullets with flaming chemicals, flares that hissed to warn of poison gas attacks, and some experimental rockets that carried secret communiqués.

Although military and special forces groups still have flares for particular operational signals such as alerting aircraft for supply drops, pyrotechnics in general are now a thing of the past.

REFLECTED AND FLASHED LIGHT

The use of reflected sunlight to send messages has existed since ancient times. Pythagoras described reflecting the sun's beams with a burnished surface and Herodotus's texts contain accounts of Persians communicating with polished shields. Around 300 B.C.E. a mirror was placed in the lighthouse at Alexandria, Egypt, which was used to convey messages. Plutarch's chronicles of Greek naval conflicts also mention shield-formed signals. The Roman emperor Tiberius, who ruled from C.E. 14 to 37, sent reflected-light communiqués from the Isle of Capri using metal implements.

These various reflecting devices were precursors of the heliograph. The actual mechanism apparently developed from advances in surveying methods. In the early 1820s, a Royal Engineers colonel named Colby developed a sun-reflecting instrument made of polished tin. The reflections helped to make measurement of angles and degrees of distances between locations.

Other British pioneers of heliography included Sir George Everest and a Royal Navy captain named Sheriff. The former consisted of a sun reflector used for surveying, called a heliotrope, after the plant that turns toward the sun's beams. Around 1851, the scientist Charles Babbage presented his sun-flashing "occulting telegraph" to Arthur Wellesley, the duke of Wellington, although his invention was apparently not put into practice.

While these methods may have been forerunners of the heliograph, a more direct antecedent was the heliostat, a surveying mechanism created by in the early 19th century by German scientist Carl Gauss (who some historians also credit with inventing the heliotrope). His device had a pair of mirrors (one coated with silver and the other without) placed at right angles to one another. The surveyor measured distances by manipulating the mirrors and directing the light toward certain sites. An unknown inventor added a shutter to the heliostat, which allowed flashes to be sent using the apparatus.

A heliograph

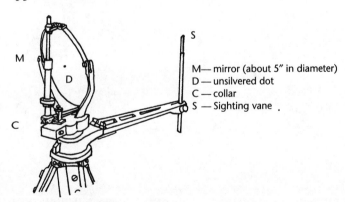

M— mirror (about 5" in diameter)
D — unsilvered dot
C — collar
S — Sighting vane .

At some point in the late 1860s, a practical heliograph became operational. It was developed by Henry C. Mance, a telegraph employee in Baluchistan, an area in western Pakistan. In that vast rugged terrain, Mance changed the heliostat by adding what was called an oscillating mirror. This made it possible to create the short and long flashes that could transmit Morse's dots (short flashes) and dashes (long flashes).

The Royal Navy showed an interest in Mance's method before the Army, and it was not until 1877–78 during a British Army mission in what is now northern Pakistan, that the heliograph was actually used in active operations. Known as the Jowaki expedition, the mission involved the heliograph in general dispatch exchanges as well as security reports from scouting parties. While the Jowaki expedition's primary duties included surveying and locating potential base sites and water sources, units that patrolled dangerous areas were accompanied by signalers. The heliograph helped to send scouts' reports and permitted the main body spend less time on guard duty.

A heliograph team

From about 1870 to the early 1900s, the heliograph was the British military's main frontier transmissions system prior to the introduction of radio. In late 19th-century America, the U.S. Army frequently used the heliograph to counter the Native American smoke signals between the Mississippi and the Missouri Rivers and the Rio Grande, as wigwag was limited by terrain and weather and telegraph wires and poles were vulnerable targets.

U.S. Army Brigadier General Nelson Miles also used the British heliograph in the 1880s through wide stretches of Arizona and New Mexico. Flashing Morse signals were conveyed between mountain peaks from 25 to 30 miles apart. The heliograph teams became so skilled that they were able to traverse distances of up to 800 miles in under four hours.

Varying in design according to the needs of the particular region, the version used by the Plains army units consisted of a mirror and a sighting vane. Some versions also had two mirrors, both mounted on a tripod.

The heliograph required bright sunshine for optimum efficiency. Reflected light was sent in Morse's dot-dash pattern using a screen or shutter attached to a second tripod. A longer shutter application signaled a dash and a shorter one represented a dot. A change in position relative to the sun required a different configuration of the mirrors. With the sun in front of the sender, the light was reflected directly to the receiving station as in line with the sighting vane. If the sun was behind the sender, the rays were reflected with the two-mirror arrangement.

Weather, terrain, and mirror size all placed some limits on the heliograph, but on a good day, flashes could be seen without a telescope from a distance of 30 miles. Visibility could be further enhanced by using field glasses or telescopes.

Heliographs did not fare well generally in navies as they required clear skies and immobile sending and receiving points. During World War I the Royal Army did use the heliograph in some engagements in Palestine and Mesopotamia where weather conditions were much clearer than near the trench lines in Europe. While the U.S. military's interest in heliographs quickly faded, the Royal Army continued to use them until 1942.

On both land and sea, heliographs were rivaled by flashing lamps and blinkers. Flashing signals with lamps were first developed in the 1820s when scientists discovered that pure lime gave off an intense light when it came into contact with a controlled supply of oxygen. Applied to illuminate theaters (the legendary limelight) and some lighthouses, the process soon drew the attention of military signalers as well.

In the 1850s Lieutenant Phillip H. Colomb of the Royal Navy, experimented with different lamps fueled with whale oil and kerosene, which were used with lights raised on yardarms. In the early 1860s he worked with Royal Army Captain Francis John Bolton to develop a portable lamp fueled with lime and compressed oxygen and hydrogen in copper cylinders. When used with a metal shutter, the lamp could be made into a signaling system which sent long-and-short Morse flashes.

Bolton and Colomb also created codes for their flashing systems that were used by the Royal Army and Navy. Colomb's version was a more efficient form, as its flash combinations were based on numbers and could be equated with letters. While some of Bolton's numbers and letters were similar to Colomb's, Bolton's method of spelling the alphabet meant a longer transmission time.

Colomb also patented his own "Flashing Night Signals" in 1862 which combined his code with apparatus appropriate for naval and land forces. In this system the letter *A* was represented by the number 5, *B* by 6 and so on until *Z* was transmitted by the number 30.

For seafaring signals, Colomb designed a larger suspending lamp that first ran on lime and later on electricity. Its shutter

Bolton's signals

Colomb's signals

1
2
3
4
5
6
7
8
9
0

was operated by a long cylinder known as a barrel organ, which was activated by turning a crank. Metal wedges called keys were placed in openings in the cylinder. They controlled a cable that opened or closed the shutter while the cranks turned the cylinder to complete one full message sequence. The same system was converted into a smaller, more portable Royal Army version which was mounted on walls or on wagons for quick mobility.

The U.S. Navy had begun tests with incandescent lights in 1875, and by 1878 Lieutenant W. N. Wood had developed a light that could be seen at some distance at night. Thomas Edison's invention of the lightbulb then spurred renewed efforts.

Among the incandescent forms developed at this time was the arc searchlight, named for the sparks that moved between two closely placed electrodes when current leaped the gap from one to the other. As the equipment was improved, early signals were made by turning the light from side to side creating a pattern of light–no light. After this development, interest in signaling systems faded for a few years.

In the 1880s a series of kerosene, oil, lime and magnesium lamps with shutters made their way into the signaling repertories of most European nations. An American inventor named Bradley Fiske and a British admiral named Sir Percy Scott proposed that lights should be covered with shutters. The better-known Scott promoted the use of a Venetian blind--type shutter on an arc searchlight. The blind was built into a circular casing that could be placed over the circumference of the light. A handle was pressed to turn the blind's slat flat or on its edge. This created the ordered flashes for Morse code and Sir Percy was credited, rightly or not, for the development

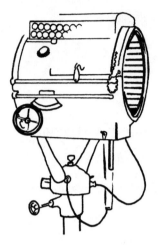

A "blinker" light

of the "Scott shutter." From these motions and the eyelike appearance of the searchlight, the nickname "blinker" arose.

A brief summary of the legacy of flashing lights should include the flashing mirror signals taught to Second World War pilots in case they were shot down. Using the sun or moonlight, they were able to signal rescue planes or friendly ground units and resistance groups. In addition to mirrors, modern pilots are now aided by battery-powered radio beacons which help pinpoint their location.

A descendent of some of the early light systems was the British Aldis signal light, a small hand-held illumination mechanism with a pistol grip. Morse dots and dashes were made by pulling a triggerlike control that tilted a parabolic mirror that reflected the light made by an incandescent lamp. The United States also developed versions of this method in the 1890s. Eventually, it was made mobile by battery power, enabling signaling from more than one shipboard position. Types of these "blinker guns" are still in use today.

Infrared light systems were also affected by the various pyrotechnic and flashing light experiments. During World War II, an infrared method nicknamed "Nancy" or "Nancy gear" was developed for the U.S. Navy. This was a nighttime process in which specially filtered infrared beams could be seen only by a certain receiver on the recipient ship. The infrared transmitting lamp was fitted with a hood which filtered out all the wavelengths of the visible spectrum, permitting only infrared rays to pass through. Crewmen on the recipient vessel read the invisible transmissions with a converter coated with a radioactive phosphorus substance that made it sensitive to infrared. Optical amplification with a mechanism known as a Schmidt lens made reading the patterns of the infrared rays possible.

Infrared systems continue to be applied today. With the many developments in laser beam research, however, that technology is now considered at the cutting edge of flashing-light telegraphy. Lasers are used for weapons guidance systems, and are also being developed for communications between surface ships and submarines or aircraft carriers and aircraft.

Mechanical signaling systems

The word *semaphore*, derived from the Greek *sema*, "sign," and *phero*, "to carry" encompasses a number of signing systems and is applied both to hand and mechanical shape-based signaling methods.

One of the first predecessors to the mechanical semaphore was developed in England in the mid-17th century. Edward Somerset, the second Marquis of Worcester created a day-and-night signal process that consisted mainly of large letters and symbols cut into squares that could be seen at some distances on clear days. By night, torches illuminated the cutout portions. The letters provided an alphabet and each symbol had a prearranged meaning.

A similar method was introduced in May of 1680 when the English scientist Robert Hooke gave a lecture to the Royal Society in London titled, "On Showing a Way How To Communicate One's Mind at Great Distances." Like Somerset's system, Hooke's method did not contain standardized equivalents for each shape. By agreement of the exchangers, the shapes could signify any word or letter the signalers desired.

The shapes were moved from behind a screen with the aid of ropes.

Hooke advised signalers to plan transmission times and suggested using pendulum clocks to ensure accuracy. Early forms

Hooke's apparatus

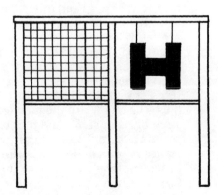

Hooke's shapes

| ─ ＼ ／ ┒ ┚ ┖ ┗ ┳ ＋ ┴ ┛ ⊏ ⊐

Ｈ □ ⌐ ⌐ Ｖ ∧ Ｘ ▽ △ ◇

of telescopes were to be used for viewing signals at distances, and torches were advised for night illumination.

Although Hooke's lecture was well accepted, no one seems to have financed his signaling system. Although never used, Hooke's proposal was the first to describe control codes for better two-way exchanges. These codes included signals meaning "ready to transmit," "not ready to send or receive (delay)," "show again (error code)," and "not so fast (rate adjustment)." He was the first to understand the importance of operational commands that are now found in every international signaling system.

While many others tried to build upon Hooke's system, the first shape arrangement to achieve widespread use was Claude Chappe's "aerial telegraph." Born in Brûlon, France in 1763, Chappe's early interest in science was encouraged by an uncle, Jean Chappe d'Auteroche, who was himself an astronomer and traveler. Having studied varied communications methods, Claude was apparently influenced by Hooke's imitators, if not Hooke himself.

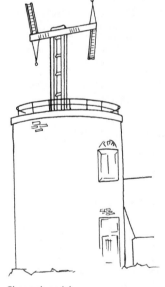

Chappe's aerial telegraph

One of five brothers, Chappe had struggled to have his invention both noticed and funded. His opportunity came when his eldest brother, Ignace, won election to the Legislative Assembly. Ignace eventually helped Claude secure the monetary support he required to build a few towers for a test between small towns. In the 1790s, Claude Chappe developed his aerial telegraph. This semaphore-type system consisted of signaling devices placed on top of structures and hills. The rapidity of the message exchange impressed the assembly enough to fund the construction of sixteen towers covering 140 miles between Paris and strategic Lille on France's northern border. Except during heavy fog and in inclement weather, Chappe's invention could convey messages north from Paris to Lille in 2 minutes and south from

Paris to Toulon in 20 minutes. This remarkable achievement gave real speed to long-distance dispatches for the first time in history.

The signaling device consisted of an iron support column between 15 and 30 feet high, with a moveable wooden cross-piece, or regulator, connected to the top. The regulator measured 14 feet long, 13 inches wide and 11/2–2 inches thick.

At each end of the regulator was an indicator, which was made of a brass grating to reduce wind resistance. Each was 6 feet long, 1 foot wide and 1 inch thick. Steel rods with lead counterweights were connected to the indicators' bases as a way to balance them. A series of pulleys at the base of the center post enabled operators to move the regulator and indicators to varied positions. A watcher at the next tower studied the shapes with a telescope and recorded the signal, repeated the pattern to the next tower and so on, until the message had traveled the entire distance.

The first aerial telegraphs were painted black due to early concerns about visibility, although later telegraphs were also painted in other colors. Following are the signals representing the alphabet and the numerals one through ten. (The letters *i* and *j* are combined):

Chappe's signal alphabet

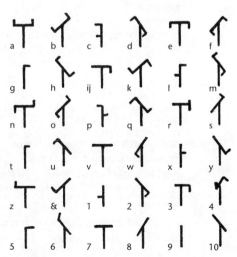

The regulator was flexible enough to be set in four positions: vertical, horizontal, a 45° tilt left and a 45° tilt right. The indicators had seven varied positions that formed a basic alphabet which was eventually expanded into a larger code list. With the help of his skilled associate, Léon Delauney, Chappe developed a codebook of some 9,999 entries. Using this codebook, signs could be equated with words, phrases or a number representing a word or phrase. This greatly increased transmission speed as entire messages did not have to be spelled out in full.

Another procedure designed to enhance speed was the sending of only one word, phrase or digit through a network at one time. Station 1 would initiate a word. Station 2 would see the sign and transmit it to station 3 and so forth through the chains of towers. In this way, the message moved constantly through the system rather than waiting for each full communiqué to be completed at each sending site.

"Semaphore fever" spread to other inventors eager to find a faster, less costly method of their own. An Anglo-Irish politician named Richard L. Edgeworth experimented with what he eventually called a "tellograph," his version of the standard telegraph, from 1767 to 1794.

Spurred on by news from the Continent about Chappe's invention, Edgeworth put a presentation together for the speaker of the House of Commons. The demonstration mechanism was a large triangle attached to a swivel that was manually turned with handles. It stood 14 feet high with an accompanying index, on which were drawn eight different parts of a circle. These segments stood for eight numbers, 0 through 7, which could be equated with different letters or words.

Edgeworth's presentation failed to impress Irish or English authorities, however, and the system was never put into practice. Some scholars believe that Britain's naval leaders blocked the acceptance because they had already committed to a rival system.

During this time, Edgeworth had developed a friendship with another signal inventor, Abraham N. Clewberg, later Edel-

Edgeworth's
Tellograph

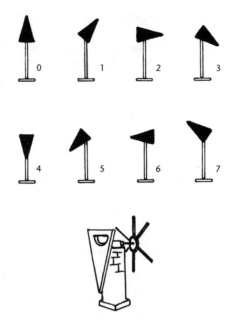

crantz. Clewberg was born in Åbo, Sweden, into a wealthy family which encouraged his interest in science. He won the favor of King Gustav III and soon became his personal librarian. When news of Chappe's work arrived in Sweden in 1794, Edelcrantz (who had reached the peerage and had changed his name) began his own experiments with royal backing. After discarding a method similar to Chappe's, he settled on what he called a shutter telegraph. This consisted of an arrangement of three-by-three rectangular panels or shutters placed in a supporting framework that looked like a large grid. A tenth, larger shutter (A) was mounted on top of the middle column. When these shutters were closed in a vertical position they were visible as rectangles at the recipient site. A horizontal plane, signaling "open," revealed the background above and below the shutter.

Each shutter was given a digit for easy reference. Beginning at the top of each column and reading down, the shutters were numbered 1, 2 and 4. By adding the number of the closed shutters in each column, a code was created from the three totals. The combinations of the three columns could produce 512 varied signals for numerals, letters and syllables.

A code such as 361 was set up as with the largest top shutter (A) open; the first and second shutters in the left column closed; the second and third shutters in the middle column closed; and the top shutter in the right column closed.

(A)	–	
1	–	1
2	2	–
–	4	–
3	6	1

When shutter A was included, the arrangements could produce 1024 different combinations for coding purposes.

Using combinations of connecting wires, handcranks and pedals, a skilled operator could manipulate the panels at an impressive speed. Tests in 1794 were successful and Edelcrantz began his network in Stockholm in January 1795. As his network grew Edelcrantz met inventors with similar interests, one of whom was Richard Edgeworth.

During visits to see Edgeworth, Edelcrantz met his daughter Maria and fell in love with her. Edgeworth approved of the potential match, and Edelcrantz proposed to her. Maria lost interest in him, however, and Edgeworth returned to his message network. Neither ever married.

Meanwhile, Edgeworth's competitor system, the Murray shutter, had won the Royal Admiralty's approval. Lord George Murray's six-shutter system, a direct variation of Edelcrantz's design, could create 64 different codes using two columns of six rectangular or octagonal shutters.

In September 1795 the Admiralty officially selected Murray's method for a new line of signaling stations. The first chain consisted of 15 sites constructed from London to an eastern coastal town called Deal, with a branch heading north to Sheerness. Messages traversing the distance from London to

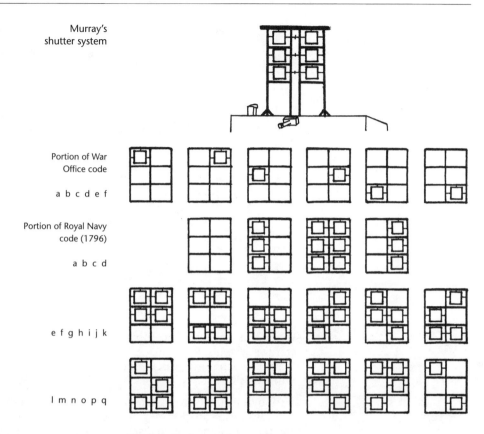

Murray's
shutter system

Portion of War
Office code

a b c d e f

Portion of Royal Navy
code (1796)

a b c d

e f g h i j k

l m n o p q

Deal took about 60 seconds. Extensions were added in the ensuing years, and 65 sites were in use by 1808.

The Murray system remained the predominant method until 1816. The shutters were hampered by a fixed, direct line-of-sight arrangement and the often inclement weather across the British Isles caused disruptions in signaling.

In 1815 Sir Home Popham, an admiral who had already instated a flag signaling system, presented a two-arm semaphore-like apparatus which could be used both on land and at sea. Its arms could be arranged in seven configurations to create 48 different signals.

Popham's two-arm semaphore

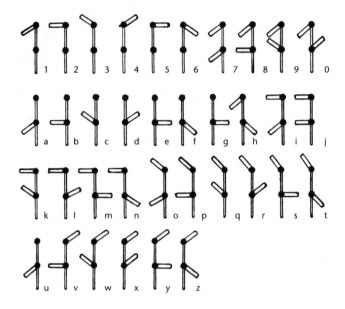

While this was limited compared with Chappe and Edelcrantz's systems, Popham's Admiralty connections ensured that his mechanism was accepted, and soon it supplanted Murray's shutters.

Although it functioned more like a semaphore, Popham's creation was also called a telegraph, as this had become a general term for a mechanized signaling system. The telegraph was constructed of a hollow wooden mast some 30 feet high with two wooden arms 8 feet long and 16 inches wide. The arms pivoted at the 12-foot mark and at the top of the mast to form the configurations. When inactive, the arms folded inside the pole.

Having learned from the immobility problems of the Murray shutter, Popham designed his telegraph so that the entire mechanism could be pivoted to turn a full circle. This allowed messages to be sent to passing vessels by gradually turning the telegraph to keep the arms aligned with the ship. Different arm positions and flags were combined to distinguish between sea and land signals and to accommodate prevailing navy flag procedures.

During this time, England's combat forces developed the first interchangeable signaling systems. During the Peninsular War in Spain (1808–14), Wellington confronted Napoleon's French armies with the aid of the Royal Navy. On land, Wellington used mobile semaphores in wagons. The wagon-transported signalmen sent messages along a long defensive line near Lisbon called Torres Vedras. Thirty miles of fortifications had been built to cover the region from the Tagus River to the ocean; in 1810–11, Wellington's troops successfully conveyed dispatches with the semaphore along this defense.

Supporting Wellington from a location in the River Tagus, the Royal Navy maintained water-to-land links using a "visual telegraph." The system consisted of a flag, a pennant and from one to five large balls hoisted upon the mast and yardarm. Each combination was equated with numerals that corresponded with codes in a codebook. The numerals 1 to 99 were signified with different balls suspended below the yardarm. The numbers 100 to 900 were indicated by combinations of the flag, pennant and a single ball hung in different combinations above the yardarm. Another flag and pennant enabled numbers up to 10,000 to be transmitted, although these were not used during the war.

Wellington adapted the naval signaling system to connect Lisbon with a key town called Badajoz 130 miles away. Poles

British sea-land telegraph

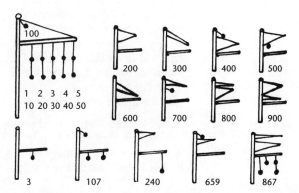

replaced masts and the naval code dictionary was slightly altered for army applications, but the basic signals remained the same.

At the turn of the 18th century, many countries had still not adopted more advanced signaling systems. Although the Dutch government was aware of Chappe's invention, in 1794 their plans for a message system still included proposals for cannons and beacon fires. After an invasion by France in 1795, a small series of flag signals was introduced and in 1798, the Dutch developed a sequence of disks combined with triangular projections resembling weathervanes. The Dutch used signaling for the same kinds of messages as other nations did, but with one addition local concern: the Dutch disks served as a flood-warning system for the lowlands. Variations on the disks were used into the 1830s, when the government adopted a six-shutter system similar to Murray's method.

Semaphore's popularity spread from France to nearby Spain and as far away as Australia. Spain's version imitated Chappe's style, and by the mid 1840s signal chains connected cities such as Madrid, Barcelona and Cadiz. In Australia, Great Britain's influence affected the signaling choices, and in 1827 a replica of Popham's 1816 telegraph was constructed in Sydney. Originally used to announce ship arrivals and departures from the bustling port, the system later connected sites farther inland and was used to send messages across the country.

In Russia, a full semaphore network was not set up until Nicholas I came to the throne in 1825. When construction finally began, extremely tall Chappe-style towers were constructed from Warsaw to St. Petersburg and Moscow. The Warsaw-to-Petersburg branch alone had 220 sending sites which employed more than 1,300 men. It remained in operation for about 25 years.

In the United States, fixed semaphore signaling lagged behind. Neither the Chappe, Edelcrantz or Murray systems won favor among American military, government and business leaders. The first real signaling network was designed by

Jonathan Grout in Massachusetts, whose system was active from 1801 to 1807. Modeled on Chappe's method, it connected Boston and towns like Hull and Plymouth with Martha's Vineyard.

In the 1830s, a three-armed apparatus linked Manhattan and Staten Island, and in the 1840s, a broker who wished to keep abreast of the stock market financed a link between New York and Philadelphia.

From 1849 to 1853 the West Coast boom town of San Francisco used versions of Popham's two-arm apparatus to signal ship arrivals and departures. Announcements were conveyed from Point Lobos, the Presidio and Telegraph Hill.

Although some networks were still being built, in general the construction of land-based semaphore dropped sharply in the 1840s. Backed by military and commercial leaders, the new communication marvel, the electric telegraph, eventually replaced all of the manual land-based systems. The electric telegraph could not yet operate at sea, however, and mechanical semaphore continued to be used in shore-to-ship and ship-to-shore transmissions.

Systems similar to early 19th-century mobile semaphores were still used into the 20th century. The 1916 U.S. Army *Signal Book* describes a semaphore with two vanelike arms adapted to hold electric lights for night dispatches. Its alphabet corresponded to the Army's two-arm flag semaphore code. This semaphore was presumably at least semimobile since fixed semaphore sites had disappeared by this time.

Popham's Sea Telegraph, which was really a semaphore, was the first shape method widely used by the Royal Navy. First installed on vessels in 1816, it was adapted as two 12-foot poles that could be viewed from both port and starboard. While Popham's design was quickly accepted, it only lasted until about 1827. After this time, with the exception of the London-to-Portsmouth line, Popham's Sea Telegraph was supplanted by a system designed by Colonel Charles W. Pasley.

Pasley was a Royal Engineer and a badly wounded war veteran who financed his experiments by clearing wrecked ships

British two-arm
semaphore

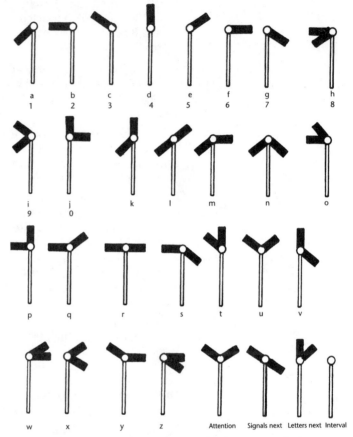

from channels. His proposal for a single mast with two arms
was finally accepted by the Admiralty and was used until the
mid-19th century.

In the 1850s the French navy developed a shore-based system
similar to the method used by Wellington during the Peninsular War. Reynold's 1855, the *Télégraphie Nautique Polyglotte*
also contained a daytime signaling process that combined a
square red flag, a half-red and half-yellow pennant and a
large ball. It sent ten numerals in a code that equated numer-

Reynold's day signals

0 1 2 3 4

5 6 7 8 9

als with words and phrases. This system was also used in some inland stations until 1863, when it proved inadequate for the demands of land communication. The French navy kept the system until the latter 1800s.

By the late 1870s naval authorities were briefly interested again in the mechanical semaphore. By the end of the decade, however, a hand-held two-flag semaphore had won their favor. Masthead semaphores were also tested during the turn of the century. A number of these systems consisted of movable sheet-metal arms operated by handles at the base of the mast. These ship-to-shore systems were kept in service until the invention of the radio rendered them obsolete.

Hand-Held Semaphore

Hand-held semaphore, or two-flag semaphore, was quickly adopted by the Royal Navy as it was well-suited for shipboard use. Based upon its mechanical predecessors, two-flag manual semaphore first came into use in the 1880s.

Two-flag semaphore differed from other two-flag signaling systems in that flag positions were equated with alphabet letters. Most arm-hand systems can convey specific predesignated signals, but they are unable to send orders which do not already exist in the signal's vocabulary. By contrast, semaphore has alphabetic codes, thereby enabling unfamiliar commands to be spelled out in full. A more rapid technique than Albert Myer's wigwag, semaphore was ideal for deck-to-deck communications.

Signal historians suggest that human semaphore began when seamen imitated the arms of Pasley's mechanical semaphore. In the absence of mechanical semaphores, signalmen stepped up to high-vantage points and gestured with their arms to send messages. Seamen also used banners to enhance visibility over long distances and during inclement weather.

In 1875 the Royal Engineers, the precursors of the British Signal Corps, discontinued the two-flag semaphore, but in 1896 the British Army reinstated a type of semaphore. Adapting the Royal Navy's two-flag system, which used a 1-foot red-and-yellow square flag, the Army's version consisted of a black-and-white 20-inch square flag that was used for about 40 years. In 1939 the British army finally discontinued flag semaphore training and nearly 30 years later, in 1966, the Royal Navy followed suit.

Influenced by the British system, human semaphore also developed in the United States. Although Albert Myer's one-flag wigwag system was the predominant signaling system in the Civil War, it does not seem to have directly inspired two-flag semaphore. While an 1884 American edition of the *Inter-*

U.S. Navy semaphore

national Code of Signals described mechanical semaphores, it did not contain any reference to flag signals.

In a 1903 article in the *Proceedings* of the U.S. Naval Institute, Lieutenant Commander Albert P. Niblack noted "fad" experimenting with British hand semaphore. By 1905, the U.S. Navy *Boat-Book* recognized the official use of two-arm semaphore.

The U.S. Army was slower to accept two-flag systems. Between 1900 and 1910, Signal Corps personnel were required to learn three code systems: Albert Myer's wigwag, American Morse for regular telegraphy, and Continental or International Morse for cable and radio transmissions.

After years of dissatisfaction with wigwag, the Army officially adopted two-arm semaphore in 1914. Although wigwag was occasionally used in World War I and in the western United States before wire or radio links were completed, it had been replaced by two-arm semaphore by the end of the war.

At the beginning of the 20th century, the *International Code of Signals* still included some fixed two-arm semaphores that were descendants of Pasley's mechanism. It also contained a standard shore arrangement for rescue signals called the International Distant Signals code. The standardized alphabet equivalents could be sent by mechanical semaphore or with flags of varied shapes and sizes.

Flag semaphore has remained a fixture in the U.S. and British navies into the late 20th century. Although not able to signal over long distances or cut through fog, semaphore has its benefits. Semaphore is often used in conjunction with standard flaghoist signals as the hand-held flags are more mobile than flaghoist banners.

Semaphore is particularly suited for transmitting information between ships in single formation, as dispatches are sometimes sent from the flagship's stern to the first support vessel's bow and then back through the squadron. Two-flag semaphore is also useful in times when radio is unavailable or impractical, or when blackouts are enforced, as the fluorescent flags are still visible over short distances. Semaphore is still used for some administrative and maintenance purposes between ships when they are anchored at sea or in port.

Technical Advancements

Any history of signaling should include the highly influential technical developments that accompanied advances in communications and which ultimately changed the course of signaling. Through the 18th century undependable static electricity was the only electric power available for experiments, but in 1800 Italian physicist Allesandro Volta produced the first battery, making the first central, manageable source of electric current.

This development was immediately applied to tests of the newly invented telegraph, but there was still no way to detect the arrival of the current at the recipient's site. In 1832, a Russian diplomat named Paul Schilling demonstrated that a compass needle could serve as a telegraphic instrument. He showed that when electricity entered a coil of wire around the compass, the needle moved or deflected.

Scientists such as Carl Gauss, Wilhelm Weber and William Cooke set out to turn this discovery into a practical communications medium. After numerous setbacks, Cooke sought the guidance of Charles Wheatstone, who was at that time a professor at King's College in London.

In 1837 Cooke and Wheatstone built the first practical telegraphic system. A message letter was conveyed by sending an electric current through two of five wires between a transmitter and a receiver. The current moved two of five compass needles on the receiver, and the pair of deflected needles pointed to the transmitted letter. Cooke-Wheatstone telegraphs became popular in England, but were later supplanted by the system developed by Samuel Morse in the 1840s.

Morse's code-and-telegraph system was arguably one of the most significant developments in communications during the 19th century. It allowed speedy and reliable message sending over vast distances and was used extensively by both military

An early Morse
telegraph

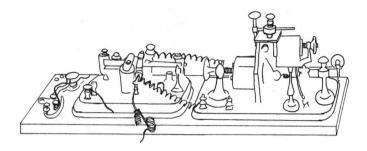

and civilian groups throughout the 19th century until the mid-20th century.

Conceived by the American artist and inventor Samuel F. B. Morse, this early electromagnetic telegraph device underwent a series of developmental stages in the 1830s, as did the methods for the messages sent on it. An early model had 10 symbols denoting 10 numerals, and Morse created a vocabulary of numbered words in order to send phrases with these digits.

In February 1838 Morse sent the world's first Morse message, "Attention, the Universe, by Kingdom's Right Wheel," and by 1844 Morse's familiar dot-and-dash system was established. On May 24 of that year, over the 40 miles between Washington, and Baltimore, Morse sent his better-known words, "What hath God wrought."

In 1851, as the use of the telegraph spread, a convention of nations met and agreed upon an International Morse code with several variations from the United States Morse code.

A number of word lists such as the *Telegraph Dictionary* and the *Secret Corresponding Vocabulary* appeared soon afterwards to help businessmen improve the economy of their telegraph messages. During the U.S. Civil War, Morse code was used to transmit encrypted military dispatches by telegraph.

Morse code could be used to send encoded or enciphered messages, and the military often used telegrams to transmit information across battle lines. Experienced clerks were nec-

INTERNATIONAL MORSE CODE	AMERICAN MORSE CODE
ALPHABET	

A chart showing the International Morse Code and American Morse Code for the Alphabet (A–Z), Numerals (1–0), and Punctuation (Period, Comma, Interrogation, Colon, Semicolon, Hyphen, Slash, Quotation marks).

Morse code

essary, however, as a misplaced dot or dash could easily garble the message.

The next great communications advance, the telephone, is credited to inventor Alexander Graham Bell in 1876. Bell's fame was based largely upon his ability to turn the telephone into a commercial success, as other scientists—Faraday, Volta, Oersted, Ohm and other electrical pioneers—had already laid the foundation for his invention.

While the telephone quickly became popular among civilians, it took longer to be accepted by the military. Although official approval was slow in coming, the U.S. Army Signal Corps experimented with phone connections as early as 1877. Chief Signal Officer Albert Myer in Washington, D.C., frequently called officers at Fort Whipple (later Fort Myer) in Virginia.

Some British military scholars claim that the men of the Jowaki expedition in 1877–78, where the heliograph was active, also used a few crude phones and phone lines. These were supposedly built from descriptions of Bell's phone in magazine articles. Other historians have suggested that the more trustworthy Morse signals were tapped on the diaphragms of phonelike transmitters which were then heard by recipients using a receiving "telephone." In Britain, however, telephones were frequently installed in British military bases by this time.

The U.S. Army used the telephone for the first time in combat during the controversial Spanish-American War of 1898. In addition to standard telegraph practices, some 13 miles of phone lines were suspended on trees, walls and shacks. Although often garbled, the phone messages guided U.S. naval gunners during the assault on the Spanish fortress at Santiago, Cuba. Spotters called in shell hits and misses to a shoreline wigwag signal station and from there the news was conveyed to a flagman aboard a U.S. Navy ship. These reports are credited with supporting the resulting American victory at Santiago.

In Italy during this period, Guglielmo Marconi established a business, known as the Wireless Telegraph and Signal Company, that would change communications. In building his wireless set, Marconi had built upon the electromagnetic wave research conducted by scientists such as Heinrich Hertz, who discovered the waves, Oliver Lodge, Alexander Muirhead, Reginald Fessenden, Nikola Tesla and Augusto Righi.

These scientists discovered that electromagnetic waves were similar to light and heat waves but were generally much lower in frequency. Some of these waves, called "carrier waves," could be modulated or varied by the electrical signals. Messages were transformed into Morse code's signal waves, which were then transmitted from one site's antenna and captured by a recipient antenna. At the recipient site, special equipment separated the signal from the carrier wave and turned the signal into recognizable sounds. This was the basis of the wireless telegraph that came to be known as radio.

With Righi's help, Marconi successfully transmitted and received signals on his family estate in Bologna in 1895. By 1898, he had the business expertise and the technical help to become the leader of the new radio industry.

While radio was in its fledgling stages, new types of telephones were also being developed. Not sure that voices could be clearly transmitted over long distances, civilian and military leaders encouraged the development of phonelike mechanisms for Morse applications.

One such device was the "buzzer" phone, also known as the "sounder" or "vibrator" phone. Created first by the U.S. Army Signal Corps, the buzzer phones were really telegraphic mechanisms that sent Morse with a hand key. Military men were especially concerned about wiretapping, so the buzzer phone was often used with additional codes or cipher covers.

In 1911 Englishman Algernon C. Fuller suggested that a carrier frequency could be adapted to send more than one communication over a single wire. After talks with Marconi and various physicists, Fuller spent a substantial personal sum to produce the Fullerphone. The Fullerphone had a direct current that was difficult for an enemy to overhear. The current was purposely broken at the recipient phone where the transmitted sounds became audible once more. The breaking was done multiple times a second by an electronically-driven interrupter. The electric current moving from the line to the receiving instrument was alternately directed into the Fuller-phone's condensers and then into its receiver. As this current oscillation occurred, the sounds of Morse dashes and dots became discernible. The Fullerphone was accepted by the British army and was standard equipment by 1914.

Radio was accepted much more quickly by military officials than Bell's telephone, as American and British naval officers had already been conducting experiments with wireless signals. Bradley Fiske of the U.S. Navy made tests in the early 1890s with electromagnetism, succeeding in launching a torpedo with wireless signals.

In 1899, with Marconi equipment and other instruments developed by British Commander Henry Jackson, the Royal Navy conducted successful ship-to-ship radio exchanges. By 1902 signalers on the U.S.S. *Prairie* and *Topeka* had conducted experiments with different European-made radio sets. Over the next two years, radio connections were made over longer distances and by 1904, some 24 ships and 19 coastal installations had radio equipment. The new International Morse code was chosen in 1905 for radio transmissions.

The U.S. Army had also become interested in wireless communications and had developed the first viable portable radio in 1906. In 1908 the Army collaborated with the Navy on ground-to-air Morse wireless messages when Navy stations in Washington, D.C., and Annapolis, Maryland exchanged communiqués with signal corpsmen in the basket of an Army balloon.

More experiments were initiated on land and sea, and in 1911 a ground signaling station succeeded in transmitting wireless Morse signals to U.S. Army planes. In 1912 a U.S. Navy aircraft transmitted the first aerial-sent signals to ships and ground sites. In this same year the U.S. Navy officially replaced the word wireless with radio, and the U.S. Army eventually followed suit.

In August 1914 the First World War began, and signaling systems were pushed to the fore. Signal communications and their security were critical to both sides and fledgling radio technology had a direct role in both particular battles and in overall strategies. At Tannenburg in eastern Prussia, for example, the Russians' failure to encrypt their radio dispatches led to a crushing defeat. Royal Navy distrust of radio interceptions of German High Seas Fleet dispatches during the naval Battle of Jutland in 1916 lost them the chance for a decisive victory. In 1918 interceptions and solutions made by the French were essential in halting the last offensives of the Kaiser's forces.

With military radio communications came the development of direction finding. Antennas were angled toward the direction of enemy messages and could be turned until the signals were heard clearly. When two direction-finding antennae fixed their bearings on the origin site, the control center tracked and coordinated their findings. The point at which their lines of direction crossed on a map was the location of the enemy's transmitter. This process was eventually adapted to track ships at sea, where continued direction finding could chart a vessel's direction and speed.

Direction finding became the foundation for a new type of military intelligence gathering called "traffic analysis." This

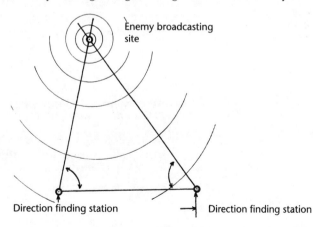

Two direction finding stations tracking an enemy broadcasting site

Enemy broadcasting site

Direction finding station

Direction finding station

was the study of communication sites' locations, sending times, amounts and types of dispatches, and their recipient points. Tracking traffic could often tell a signaler the location of a control center, and whether the enemy were advancing or heightening security in preparation for an attack.

In 1915 the radiotelephone, or voice radio, was introduced. A group of Bell Telephone engineers succeeded in broadcasting voices that were heard from Hawaii to Panama and France. This advance was made possible by the development of triodes, electron tubes with three electrodes and better vacuum tubes, that transmitted voices more clearly and over longer distances. By 1917, radiotelephone systems known as SCR-67 (ground based) and SCR-68 (airborne) sent two-way voice reception during tests at Langley Field, Virginia, in July and in Dayton, Ohio, in December. By mid-1918, radiotelephone sets began to arrive in France. Although they saw only limited action, the radiotelephones promised to revolutionize ground-to-air signaling.

In the years before radio, pilots had dropped messages in weighted bags. A few planes had wing-mounted lights that flashed Morse dots and dashes and some pilots also practiced wing-wagging in an informal short-and-long Morse style.

Observation panel systems for ground-to-air communication were used during the early years of the airplane. These panels, known as "strips," consisted of pieces of cloth placed on the ground in various configurations to convey messages to pilots. For ease of handling, the sizes and shapes of the cloth became standardized over time and white cloth was preferred because of its superior visibility. In time, the infantrymen who arranged the panels and their aerial observers developed more numerous and more systematic configurations, and the panels were formalized into true codes.

In World War I, panels were used to designate friendly camps, to direct aerial observation and to transmit news near the front. With 12-by-1-foot strips of white cloth, coded messages were sent using a series of letters and symbols:

Observation panel code

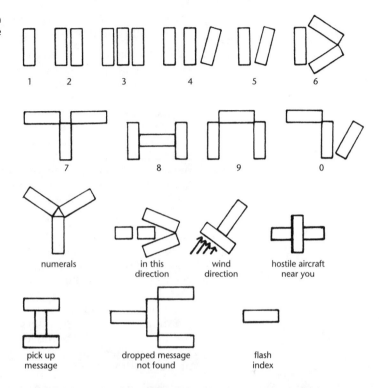

The use of strips diminished after the development of ground-to-air radio, but they continued to be used in special cases in which radio was susceptible to interception.

Once wireless exchanges became a primary part of military operations, means of protecting transmissions (communications security) or intercepting and solving them (communications intelligence) also increased. Radios were also used in a type of subterfuge known as *Funkspiels* (radio games) in which an enemy's radio station was captured or compromised. Because an outright capture had to be swift to snare the operator, equipment and ciphers intact, blackmailing or bribing an enemy radio agent was often easier than a full-scale raid. If the game was successful, the compromised station could be used to transmit false information to enemy operations.

In 1940 a highly productive U.S. radio game was accomplished by the FBI with the help of a double agent named William Sebold. A native of Germany who had served under the Kaiser in World War I, Sebold had returned from the United States to his homeland in 1939 to visit his mother. His immigration card alerted the Gestapo to his previous employment with a U.S. aircraft manufacturer, and they learned that Sebold had done some jail time as a young man and had not mentioned this when applying for citizenship in the United States. Blackmailed by this past sin, Sebold was persuaded to join the espionage training center at Hamburg, but he remained loyal to his adopted country.

Sebold returned to the United States in February 1940, a full-fledged operative with radio skills and postage-stamp-sized microphotos of documents, including the names of German agents and transmission instructions, hidden in his watch-case. Following orders to set up a front, he established the Diesel Research Company in the Knickerbocker Building in Manhattan. He also began broadcasts of German agents' information from his home in Long Island. Sebold appeared so efficient that a growing number of Abwehr spies sent their messages through him.

This funkspiel was a major success for the FBI, whose agents transmitted both screened and false data to Hamburg. Furthermore, the Manhattan Diesel Research office was equipped with cameras, one-way mirrors and audio devices to record the spies who visited Sebold. Using this survelliance, the FBI conducted a series of raids in June 1941 and broke a major espionage network.

Counterintelligence groups of the Third Reich also used radio games. One of the most successful funkspiels is credited to an Abwehr branch in the Netherlands called Section IIIF. Beginning in January 1942, its NORDPOL radio game led to the compromise of the underground movement in Holland. Until concluded in April 1944, the operation accounted for a huge cache of captured money and materiél meant for Resistance groups and resulted in the demise of many Allied agents.

Communications security devices created to protect radiotele-phone transmissions were the A-3 and the SIGSALY. Frequently used in World War II message transmissions, the A-3 was a band splitter first put into operation in December 1937 on the circuit between the Mutual Telephone Company in Honolulu and Radio Corporation of America in San Fran-cisco.

Band splitting involved splitting a signal, such as the human voice, into smaller frequency bands and interchanging them. When a voice signal entered a band splitter the signal was divided into bands by filters. These were then altered by mod-ulators and sent through other band filters, an encoding device and other modulators. Finally, at a concluding filtering point, a scrambled signal emerged.

On September 1, 1939, the A-3 carried the momentous news from Paris to Washington, D.C., that World War II had begun. Throughout the war, President Roosevelt used phones equipped with better scrambling systems to communicate with Churchill and other Allied leaders.

In June 1943 the SIGSALY, a designation of the U.S. Army Signal Corps, promised to provide fully protected radio-telephone transmissions for Roosevelt and Churchill. Created by AT&T's Bell Telephone laboratories, SIGSALY was also known as the "X Project" and was nicknamed the "Green Hornet." The latter sobriquet was derived from the advanced mechanism's control tones which sounded like the theme of a popular 1940s radio program.

The SIGSALY was composed of 7-foot-high futuristic-looking cabinets. The sending equipment was arranged in a semicircle placed tip-to-tip with the semicircular receiving equipment. The combined apparatus housed vacuum tube and gas tube logic circuits, synthesizers and a very sensitive 24-hour clock. The latter's exact frequency was maintained by a crystal oscil-lator kept in a controlled temperature oven.

The SIGSALY split and synthesized the voices of Roosevelt and Churchill and the result was covered by a sequence of random sound. The random noise was created using special vinyl disks with recorded thermal noise which were played over the speaker's voice. Staff persons at the send-and-receive locations controlled SIGSALY mechanisms that removed the noise shields and restored the distorted voices. By 1945 there were a dozen SIGSALY terminals aiding the Allies' top-secret communications.

With the increased use of the television in the 1950s, means of encrypting images became important as television confronted subscription viewing for the first time. Devices to scramble TV signals to nonsubscribers rapidly advanced with the expanding technology. In the 1950s fiber optics, which provided direct delivery of an image to the recipient, were developed. By electronically shuffling the thousands of tiny points of dark and light that compose a fiber-optics image, the sender could encrypt a visual message. Decryption was achieved by sending the jumbled image back through an identical cluster of fibers, thereby reversing the mixing process.

The Electronic War Zone: Signals Security and Signals Intelligence

During the aftermath of World War II and through the Cold War, U.S. communications security and intelligence were renamed "signals security" and "signals intelligence." Although top-secret dispatches are still carried by courier or sent through radioteletype encrypted by a one-time tape, most information is carried by electronic signals which are subject to interception and decryption. With the improvements of electronic equipment available to both the civilians and the military, the military have been forced to increase protection of equipment, records and Internet connections.

Signals security has two major branches, communications security (COMSEC) and electronics security. COMSEC is the protection of U.S. national security–related telecommunications. It also involves measures taken to ensure the authenticity of telecommunications. COMSEC includes encrypted messages or broadcasts, encrypted fax and phone transmissions, steganography and traffic security such as radio/telephone site call-sign changes, fake dispatches and periodic radio silence.

Electronics security includes the protection of radar emissions by changing frequencies, sending out false signals from radio beacon sites and periodic radar and radio beacon silence. These procedures make it difficult for an enemy nation's pilots or satellites to locate and electronically analyze the radars and beacons. An added feature of radar security is counter-countermeasures, such as "seeing through" jammed radar.

Signals security's rival, signals intelligence (SIGINT), also has two major branches: communications intelligence (COMINT) and electronics intelligence (ELINT). COMINT is one of the primary categories of signals and involves the interception and processing of foreign communications passed by radio, wire or other electromagnetic means, and the processing of foreign encrypted communications. More generally, COMINT includes cryptanalysis, traffic analysis, and interception and direction finding.

ELINT includes electronic reconnaissance such as eavesdropping on radar emissions, and countermeasures for undermining enemy radar systems. ELINT is defined as "the collection (observation and recording) and the processing . . . of information derived from foreign, noncommunications, electromagnetic radiations." ELINT can include data acquired by spy planes eavesdropping on radar bases, satellites' interception of telemetry signals from missiles in flight and the location of radio beacons used for navigation. This information is combined with human intelligence sources to assist cryptanalysis in various ways. For example, an encrypted message from a military site to a radar center might contain specific orders for the radar crews. If observation of activities at the radar station can be matched with of the encryption, a wedge could be driven into the cryptic concealment.

Within electronic intelligence are the divisions of RADINT (radar intelligence) and TELINT (telemetry intelligence). RADINT involves finding and recording a radar site's location and its antenna beam motions, frequency, beam width and pulse repetition rates (the number of pulses or "bundles" of radio waves it emits per second). These are known as a radar site's "fingerprints." These signals are recorded by satellites and sent back to ground reception stations by encrypted superfast burst transmissions. Rival nations with these technologies study their opponents' radar systems in order to develop countermeasures.

Telemetry, from the Greek *tele*, "far off," and *metron*, "measure" refers to the signals sent by sensors on a missile or flight test vehicle which diagnose onboard equipment, guidance systems, acceleration rate, temperatures at different points on the vehicle, fuel flow rates and similar data. These facts are transmitted with encrypted signals to engineers at ground stations for evaluation of the test flight. This information, which includes the rocket's onboard instrumentation and in-flight status, is a primary target for telemetry intelligence (TELINT), which gathers information using special sensors mounted on satellites.

After intercepting the microwave signals that carry telemetry, the observer satellites return their captured encryptions to their control center in rapid spurts that make them difficult to intercept. The ground-based analysts then try to break these encryptions to learn more about a particular missile's technology.

Following is a summary of the basic functions of electronics security and electronics intelligence:

ELECTRONICS SECURITY

`missile telemetry (encrypt signals)`

`radar (shift frequencies)`

`radio beacons (emit false signals and falsify sites, maintain periodic silence)`

ELECTRONICS INTELLIGENCE (ELINT)

```
satellites (intercept signals)

aircraft (record emissions)

aircraft (locate sites, record data)

radio stations (detect radio signals)
```

These interwoven technological competitions are called ECM (electronic countermeasures) and ECCM (electronic counter-countermeasures), known collectively as electronics warfare. The following hypothetical situation illustrates how advanced electronics warfare has become.

Nation A protects its radar emissions, such as early-warning systems or short-range tracking, by shifting its frequencies. These early-warning systems attract attention to aircraft that do not answer the radar's electronic interrogation, as a friendly plane transmits an electronic code that identifies it as belonging to a nonthreatening craft.

The shifting frequencies are intended to evade Nation B's electronic intelligence, which tries to find and interpret A's radar information. This data is valuable in developing ways for B to protect its own bombers and missiles from radar detection.

Nation A may further protect its missiles by developing electronic countermeasures such as radar jamming, ejecting chaff (metal strips) to throw off radar, creating false radar echo techniques and designing new weapons using stealth technology. B might then counter with electronic counter-countermeasures, including devices to "see through" jamming systems and instruments to find a radar's true echo pattern.

At other levels, B also tries to intercept and record all A's communications from missile telemetry to telephone calls. Because A will usually encrypt these data transmissions, B uses electronics intelligence to take signals from space or the earth's airwaves, and then tries to unwrap the encryption protections by using computers and human analysis.

Electronics warfare has also brought together a combination of ciphers and codes with signal communications to ensure security. On satellite links that transfer textual information, ASCII characters are used to encode the text for rapid transmission. If the message is an analog signal, such as the human voice, the speech must be converted into binary codes.

The binary codes can be encrypted with a method similar to that made for the Vernam/Baudot teletypewriter (marks and spaces with a key of random marks and spaces). In the modern version, binary numerals are added to a binary key generated by a computer. They can also be encrypted by putting the binary code digits through the data encryption standard.

The resulting cipher stream is then encoded again with error-correcting codes for smooth transmission flow before being sent from the land site to the orbiter. After transfer from the satellite to the recipient, the coding and ciphering stages are reversed to recover the original speech or text.

If the signal was analog, the error correction codes are checked for garbles. If these are correct, the DES or other key stream encryption is reversed and the original voice is restored from the deciphered binary code digits. If the message was textual, the ASCII form is converted to the plaintext equivalents.

Since humans first sent messages from the moon, we have generally taken these signals for granted. These communications, however, represent exponential advances that would have amazed the first inventors of signaling devices.

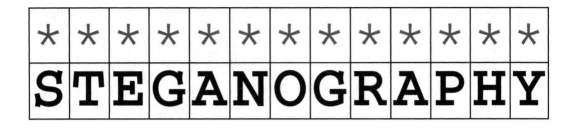

✱✱✱✱✱✱✱✱✱✱✱✱
STEGANOGRAPHY

The term *steganography* is derived from the Greek *steganos,* "covered," and *graphein,* "to write." In the past, it was used interchangeably with *cryptography*, but was revised in 1967 by historian David Kahn to describe processes that conceal the presence of a secret message, which may or may not be additionally protected by a cipher or a code. While codes and ciphers are susceptible to cryptanalysis because they are obviously an attempt at secret communication, missives covered by steganography are disguised to appear like an innocent message, thus hiding the presence of a message without necessarily altering its content. It can involve a literally physical means of transmission known as technical steganography. It can also refer to types of linguistic concealments that include two general forms: the semagram and the open code. Examples of open codes include null ciphers, geometric methods like the grille and jargon codes.

The physical or technical methods have often appeared in popular literature and on films as messages conveyed in secret compartments, hidden pockets, hollowed-out books and umbrella handles and disguised with invisible ink, among many other examples.

This type of steganography dates back to ancient times. Herodotus, for example, described a revolt of the fifth century B.C.E. against Persian rule that directly benefited from steganography. Two powerful regional leaders communicated secretly by shaving a slave's head and tattooing it with a secret message. After his hair grew back, the slave was sent to co-conspirators, who shaved his head again and read the message. Thanks to the information contained in this ingenious-

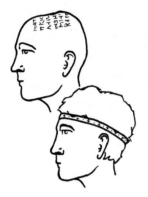

Tattooed missive

ly concealed missive, the Greeks succeeded in overthrowing the Persian oligarchy.

Herodotus also wrote of a steganographic warning that saved the Greeks from the Persians in 480 B.C.E. In that year, a vast Persian army under Xerxes I was on the march toward the Greek city-states. (Incidentally, the Persians themselves had a historically significant system for communicating over their vast territory, a system which helped Xerxes mount his invasion, the largest yet against the Greeks. The Persians had developed a full-fledged messenger service that traversed the realm, allowing rulers to send word to friends, warnings to enemies and notices to tax collectors. A very credible predecessor to the Pony Express, this system used fresh mounts and couriers located at sites not more than a day's riding distance apart. During times of impending conflict, these swift horsemen quickly roused the troops into action).

Hearing that Xerxes was on the march, a Greek named Demaratus alerted the Spartans, the most famous warriors in the Aegean region. Using a makeshift concealment device created by scraping wax from two wooden tablets, Demaratus inscribed what he knew of the Persians' intentions, then replaced the wax covering. The seemingly plain tablets were passed untouched to Spartan control.

In Sparta, Gorgo, the wife of King Leonidas, studied the tablets and discovered the hidden message. Gorgo's achievement was bittersweet. Her spouse, Leonidas, loyally rushed into action with his men, hastening by forced march to a crucial defensive position on the route of the Persian onslaught. The site was a pass called Thermopylae. Traitors and their knowledge of a hidden path led to the eventual slaughter of Leonidas and his 300 troops, but they held their position for three days, allowing the city-states time to prepare for battle and enriching the honor of the Spartan warriors.

During this same period, the Chinese used couriers who committed rulers' dispatches to memory. Their ideographic script, which puzzled Westerners for centuries, was difficult to encrypt, as the characters were extremely difficult to transpose or substitute in a methodical way. By 1000 C.E. Chinese military leaders were writing important communiqués on very thin paper or silk. These missives were then rolled up very tightly, covered with wax, and when cooled, they were hidden in the messenger's clothing, swallowed or inserted in a body orifice.

Steganographic methods of linguistic concealment are less well known. The term *semagram* refers to covering methods that use neither numbers nor letters to conceal the message. The concealment might be certain images on a deck of cards or figures in a painting designed to convey a meaning known only to the intended communicants. Other forms of linguistic steganography discussed in this chapter are geometric forms, null ciphers and jargon codes.

During the fourth century B.C.E. the Greek historian Aeneas the Tactician developed an astragal, the oldest known semagram. In his text about military science, entitled *On the Defense of Fortified Places,* Aeneas described a type of disk that was punctured with a series of holes representing the Greek alphabet. Yarn or string was drawn through the appropriate letter-apertures in the order of the message letters. The recipient unwound the string from the holes, marking down the letters sequentially to reveal the message.

During times of war counterespionage censors have been so concerned about the possible transmission of information through steganographic means that they have gone to sometimes extraordinary lengths to verify information. Such practices escalated in the U.S. during World War II. Wartime counterspy units changed the arrangements of the hands on clocks in deliveries meant for timepiece dealers. When the Office of Censorship was in full force after Pearl Harbor, it interdicted large groups of certain types of objects, newspaper articles and even comments about children's report cards. Other examples included loose stamps, which were replaced with new postage; reports of sports statistics, which were studied by sports experts; the number and frequency of x's and o's on lovers' letters; and even knitting directions.

Cablegrams also had outright prohibitions and close restrictions. Orders for flower deliveries with their many flower types, addresses and dates, were banned internationally by America, Britain and other Allies. Only U.S. territories, Canada and Mexico were given a waiver. Broadly used cable codes were also limited, with each company required to have special licenses to send communiqués with their own particular business codes.

Nor were the media spared. Radio station scripts were studied and broadcasts were monitored. Telephone requests for specific records and live, in-studio songs had to be denied. Newspapers were studied and cautioned about want ads, especially from new clients. Screenplays were not so closely watched, although many locations near military plants were closed and on-screen references to real U.S. weapons were prohibited.

Technical Methods of Concealment

The practices of physically concealing messages expanded through the ages. In some cases, the arrangement, color, shape, style or number of an object itself acted as the message. In one ancient example, two gold coins were passed to a contact person as a signal to begin his or her mission. More frequently however, clothing or items carried on one's person were used as conveyances for messages. These have ranged from the removable handles of daggers and extra layers in a cloak to false sides or bottoms in bags.

Concealing missives in secret compartments became such a common practice that the organized black chamber postal interception sites of the late 1600s and 1700s intercepted and examined mailed objects as well as personal and business mail. The chamber's well-orchestrated practices continued until the late 1840s. By the time of the world wars, postal censorship was again resumed by both the Allied and Axis powers.

One example of a simple physical steganographic technique uses a deck of playing cards. The pack is sorted into a pre-arranged order and the message is written on the side of the deck. The pack is then shuffled. In this disarranged state, the markings on the side are barely visible. The recipient then sorts the pack into the agreed-upon order so that the markings again form the message on the sides of the pack.

Another innovative example of physical concealment is a smoking pipe reportedly enjoyed by secret agents. In its regular use the pipe conceals a dispatch in the hollowed portion of the bowl and can be used to smoke tobacco in the exposed part of the bowl. If the spy was confronted and put at risk of discovery, he gave the inner bowl a slight turn, letting the paper missive fall into the burning tobacco.

Many types of technical steganography have survived through the ages. In his historical writings, Aenaes the Tactician described a technique by which tiny holes were poked in parchment above or beneath the existing writing to convey secret words. This method survived in England in the times before telegraphy as a way of avoiding the high costs of mail. While letters were very expensive to send long distances, old newspapers, which had already been stamped, could travel back and forth across the country. Many of those unable to afford postage placed dots above letters in these journals, thereby writing their own letters which were then delivered free of charge. The recipient simply wrote down each letter so marked in sequence until the entire message was spelled out.

Some systems very similar to these remained in use as a cryptographic method during both world wars.

INVISIBLE INKS

Although physical concealment in secret compartments does provide a measure of security against casual examiners, these simple tricks would not be proof against high security situations when all mail and unusual objects are carefully checked by censors. Those wishing to outwit organized censors or single sharp-eyed enemies have sought out more complex methods. A very early technical steganographic process was that of invisible, or secret, inks.

Invisible inks are special fluids or chemicals used to conceal the presence of writing. This method is considered a form of steganography, as the purpose of such substances is to render a message unseen by the unaided eye. The use of invisible inks has been recorded since Roman times. In the first century C.E., in his *Natural History,* the Roman writer Pliny the Elder described using the liquid of a tithymallus plant for secret writing. A Greek military scholar named Philo of Byzantium also wrote of a gall nut-based liquid that rendered script invisible. Arab scholars in the early 1400s mentioned some mixtures from plants native to their region, as did Renaissance writers such as Leon Battista Alberti and Giovanni Porta.

The inks are generally of two types: sympathetic and organic. The former are chemical solutions that become invisible as they dry. When other chemicals, called reagents, are applied to them, the hidden terms become visible. The organic group consists of commonly obtainable substances such as onion, lemon, milk and vinegar. These are generally made visible by careful applications of heat. Organic inks were used to write between the lines of an innocuous letter or on blank parchment in a stack of seemingly empty pages.

As early as C.E. 600, the Arabs were using invisible inks to communicate throughout their flourishing empire. Invisible inks were also used in Europe in the Middle Ages and were the subject of several treatises, ranging from the fanciful and heretical to the scholarly and scientific. During the Renaissance, Alberti's important texts about cipher making and solving included references to invisible inks. Later, the French satirist François Rabelais humorously discussed concealed writing in his classic *Pantagruel* (1532). Amid his witty comments about life, he described making invisible ink from such substances as the juice of white onions, ammonium chloride and alum.

By the late 1700s, knowledge of such methods had crossed the Atlantic. During the American Revolution, George Washington and his spies used inks that they called "stains." These inks were provided by a London physician named James Jay, brother of patriot John Jay. Some of the ingredients of Jay's "stains" were so well hidden that they are still unknown today. The agents, Benjamin Tallmadge and the so-called Culpers, Abraham Woodhull and Robert Townsend, passed important information about British troops, ships and armaments in occupied New York, writing their reports in a book on seemingly blank pages. At his headquarters Washington recovered the words by applying a "counterpart liquid" to the parchment.

British loyalists also had their invisible solutions. In April 1775, not long after the battles of Lexington and Concord, one Benjamin Thompson used secret ink in a letter to British officials in Boston. He had learned of rebel plans for military actions around the city and conveyed his secrets with gallotanic acid (made from the gall nuts mentioned by Philo of Byzantium). The British brought out his message by applying ferrous sulphate. Fortunately for the American colonies, Thompson's spy career faded before he could divulge any more crucial secrets.

Invisible inks were increasingly applied as advances in chemistry made them even more effective. The most sought-after substance was one that reacted with very few chemicals and was therefore designated a "specific" ink. Well-hidden messages were sometimes also encoded or enciphered to protect the message in case its invisible ink cover was compromised.

In the early 20th century, advances in chemistry provided additional counter-concealment tools. During both world wars, postal censors searched for secrets in letters using tests such as iodine vapor and ammonia fumes. The World War I spy Maria de Victoria was arrested in 1918 at a hotel on Long Island due largely to the work of British intelligence and the U.S. Military Intelligence Division's Codes and Ciphers Unit which discovered invisible inks in her correspondence from Berlin that exposed her plans for sabotage. Indicted but never brought to trial, she reportedly met her demise through drug addiction in 1920.

Fortunately for the United States, de Victoria did not hide her inked writings as well as others. One commonly used double concealment attributed to World War I spies was to draw a symbol such as a type of war weapon with an invisible liquid, let it dry, then cover it with an innocent postage stamp moistened only at the edges. This combination of invisible

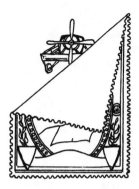

Symbol drawn in invisible ink

inks and physical concealment was a perfect example of technical and physical steganography.

The discovery of invisible ink also became evidence in the case of saboteur George Dasch during the Second World War. Put ashore with three other men from a U-boat near Long Island in June 1942, Dasch lost his nerve and turned himself in to the FBI in the New York area. A test with ammonia fumes was done on one of Dasch's handkerchiefs, which was shown to contain writing with a copper sulphate compound. The ammonia brought out incriminating names of mail addresses and contacts for Dasch and the other would-be saboteurs, three cohorts and four more agents who had landed in Florida on the night of June 16. Dasch and another spy named Ernest Burger were imprisoned and spared the execution the other six faced in July 1942.

The ammonia fume test that exposed the Nazi spies' connections was one of several advances applied by the censors and counterspies that included various chemical developers and infrared and ultraviolet light. They "striped" letters with developer on brushes in order to find the necessary reagent for a specific ink. Ultraviolet light was used to detect words spelled with starch, while infrared beams identified messages written with ink of the same color as the material (usually paper or cloth) on which it was placed.

Probably the biggest drawback to even more extensive applications of invisible inks was the fact that spies could not send enough of their information by this means. It was often difficult to slip letters past mail censors even if a very good secret ink was applied between the lines. Some agents wrote on blank pages that were placed amid a stack of new paper. Other spies tried the old technique of ink-dotting their enemies' newspapers, which were a large daily source of letters and phrases. However, this was often time consuming and awkward with no set character arrangement for reference, such as that found in a dictionary's columns or the lines of a novel. Added to these problems was the fact that the mailing of newsprint became an obvious target for counterespionage forces.

The need to convey even more detailed information with increasing speed eventually led to the virtual abandonment of invisible inks as a concealment technique. Although still used occasionally, invisible inks have been generally replaced by the ingenious product of microphotography, the microdot.

How to Make an Invisible Ink

Both citrus and onion juices make easily available invisible inks. Simply dip a thin brush into the juice and brush your message onto the paper. A brush is better than using a pen nib as it will not leave indentations on the paper which may give away the presence of a secret message. Fibrous paper is best as it will absorb the ink and will not shine where the ink has been brushed. These inks will become visible with gentle heating over a lightbulb, or with a heater, hairdryer or iron.

Milk can also be used as a secret ink. Brush your message on a piece of bond paper, and to reveal the message, rub a dark powder such as ash or powdered lead from a pencil over the message. The charcoal-like substance will glide over the surface of the milk, revealing a white message on a dark background.

An even simpler form of invisible writing makes use of the indentations left by writing hard on a piece of paper. Simply write the message on a pad of paper, pressing hard so the message can be seen on the next sheet of paper. Depending upon the amount of pressure you use, you may want to send the third or even fourth sheet of paper so that the indentations are not too obvious to the casual observer. The recipient can rub the paper with the side of a pencil point, or use a flashlight or table lamp turned so that the light shines sideways across the paper. Both techniques will reveal the indentations that are invisible under an overhead light.

CRYPTOPHOTOGRAPHIC TECHNIQUES

Cryptophotography comprises both image reduction and chemistry. The chemical processes include latent imaging, gelatin hardening and bleaching.

In the first of these techniques, the "blank" space of a seemingly innocent photograph is actually occupied by a latent image. To produce the image, the nonsecret negative is printed with ordinary chemicals. The secret negative is then placed in a white space within the other image and exposed. The photograph is then put into a fixing bath. The intended recipient places the photograph in the necessary developer, processing the photograph further until the hidden image is revealed. Before the invention of the microdot, this process was considered one of the better concealment systems.

Gelatin hardening, also known as tanning, takes advantage of the changeable properties of gelatin. Its concealment potential depends on skilled application of solutions and type and quality of paper in one process, or the use of fixed films or plates in a more complicated procedure. In the paper-based system, a message is written with a solution of formaldehyde on gelatin-covered, high-quality photographic paper. After being permitted to harden, the surface appears untouched and provides a level of concealment against the casual observer. The intended addressee recovers the message by placing the photograph in tepid water or ammonia. The gelatin swells up around the formaldehyde-formed characters which can then be seen by examining the contents under low-angle illumination.

In the bleaching method, an exposed but undeveloped photographic image (either negative or positive) is placed in a chemical solution which "bleaches out" the visible image. The image is recovered by the recipient with a developer that reverses the effects of this solution by reacting on the solution and causing the original image to be visible again. The following are examples of chemicals that are used in this process:

SOLUTION

copper sulphate and sodium chloride

potassium bichromate and hydrochloric acid

mercuric dichloride

mercuric chloride

DEVELOPER

`metol hydroquinone`

`metol hydroquinone`

`metol hydroquinone`

`sodium thiosulphate or ammonia`

Bleaching is not a very effective concealment; it often leaves suspicious traces of images, visible especially in sunlight. A fixing bath helps with the masking aspects but makes development and image recovery more difficult. For these reasons, the process never achieved widespread use as a cryptophotographic technique.

Microdots

The Germans were reportedly the first to find a solution to the age-old problem of concealing large quantities of pilfered data. They did so with what was then an amazing technological advancement, the microdot. Now a well-known tool of factual and fictional spy lore, the microdot had its precursors when experiments were conducted with reducing the size of pictures for advertising or for making groups of personal photos.

Since its development in the 19th century, photography has had a supportive role in providing physical concealments in cryptographic practices. The first instance of microphotography as a process of hidden communication appeared during the Franco-Prussian War of 1870–71. René Dagron, pioneer of miniature photos, and his associates left besieged Paris by balloon and set up a message service from Tours, in west central France. Dagron reduced military documents on film measuring 30 by 55 millimeters, and these miniaturized items were placed in tubes attached to homing pigeons'. The information generally reached Paris safely and, although the Prussians won the war, the value of microfilm was established.

The best known type of miniature photography is the microdot. Microdots are miniature copies of documents used for the secret transmission of information and were "the enemy's masterpiece of espionage" according to FBI director J. Edgar Hoover.

Using Dagron's miniature film technique, German scientists invented the microdot process itself in the 1920s. It was developed by the following series of steps: a secret communication was photographed and put through a series of reduction stages. After the negative was developed, a hypodermic needle was used to isolate and raise the microdot from the photographic emulsion. Finally, the dot was inserted into an innocent-looking text in place of a period or dot where it was held with an adhesive substance such as collodion. A number of the dots were made in very small film squares as the system progressed.

Microdot technology improved to the point where the process could be performed in a trunk-sized device. Additional advances enabled the image to be fixed but not developed. This kept the film clear and less noticeable on shiny surfaces such as the gummed edges of envelopes. Soon these improvements enabled Nazi spy rings to hide greater numbers of dots in telegrams and letters and minute film strips under postage stamps.

According to Hoover's account, the FBI received its first warning of the dots' existence in January 1940. It was not until August of 1941 that an example was sighted, when a lab technician happened to notice a glint of light on an envelope that had been taken from a suspected German agent. The gleam came from a microdot masquerading as a typewritten period. By late 1941, the counterespionage teams began to find dots in telegrams, letters and under postage stamps.

Allied postal censors, lab technicians and counterespionage agents combined their talents to break Nazi spy rings, succeeding even where the microdot messages were encrypted as well as reduced and concealed.

After the end of the war in 1945, microdots were used widely by Soviet agents operating worldwide. One such agent was Rudolf Abel, who relied on them in the 1950s to back up his espionage work in the New York City area. Abel made microdots by reducing 35-millimeter film negatives with a

The development of a microdot:

(1) An agent photographs a secret document;

(2) The photograph is photographed again;

(3) The dot is handled with a syringe;

(4) The dot is fixed to the page of a book.

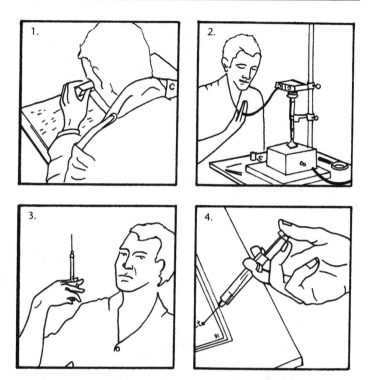

lens of a short focal length. After preparing the dots, he placed them in the loosened bindings of American magazines and then sent these innocent-looking issues to a general delivery address in Paris.

Linguistic Concealments

SEMAGRAMS

Semagram, based upon the Greek *sema*, "sign," and *gramma*, "written" or "drawn," is one of the two main forms of linguistic concealment. The replacement elements of the ciphertext or codetext are not letters or digits; rather, they can be the dots on dominoes, objects in a photograph positioned so as to convey a prearranged meaning, or a painting in which two forms such as long and short tree branches represent the dashes and dots of Morse code. To an observer, there is no obvious encryption present.

In one recorded case during World War II, German agents in England sent a message made with a knit sweater into Germany. Supposedly meant for a prisoner, it went instead to intelligence authorities. When the sweater was unraveled, its wool yarn was found to be full of knots. The yarn was straightened and the knots were compared to an alphabet printed vertically on a wall. Decipherment involved establishing a baseline such as the floor with the alphabet perpendicular to it. The end of the yarn was held to the floor beneath the alphabet, so that the first knot aligned with a letter. This knot was then placed on the floor and the second knot was checked for its letter matchup. In one instance, the knots revealed information about Allied naval vessels that were under construction and about to be launched.

Other semagraphic forms have included ways of concealing a map or missive within a seemingly innocuous drawing. Shown here is an apparently innocent drawing of a mottled leaf and its vein pattern. It actually conceals types of artillery emplacements and their locations along the "veins" that could represent canyons or river valleys.

During World War I, German intelligence developed some clever semagraphic drawings and designs to try to elude Allied postal censors and to communicate with their distant agents and supporters. In one instance of this method, a well-to-do Englishwoman whose brother, an airman, was a prisoner of war in Holland, was approached by a woman purportedly representing a war relief charity. While the visitor was soliciting a donation on the terrace of the country house, she made a sketch on the blank page of a book, depicting birds in a nearby meadow. After discussing the captured airman, the Englishwoman gave the visitor some money for her charity. The fund-raiser in turn suggested that the sketch be sent to the brother to give him a scene from his homeland. On this amicable note, the women parted.

Showing artillery
emplacements

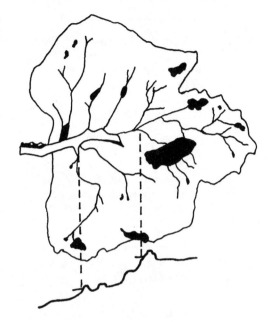

Later, when the charity worker's papers were found to be forgeries, the incident appeared to have been simply an elaborate ruse for a modest monetary theft. Nor did the donor realize the truth even after it was learned that her brother never received the sketch she sent. However, British authorities discovered that the charity worker had been drawing on paper preruled from a reduced version of a particular survey. The sketch was far from innocent: the position of the birds actually showed the location of a mine field providing a defensive barrier for a strategic naval site. When the drawing was sent to Holland, it was intercepted by a Dutch agent cooperating with Germany.

Though semagrams succeeded during World War I, the need for rapid, multidirectional orders in warfare led to far greater dependence on cryptographic methods adaptable to telephone and radio. The semagram eventually faded before the onslaught of electronic encryption.

Semagrams in Writing

Semagrams also involve attempts to conceal messages within openly visible writing. These forms can include the shapes or positions of letters, sometimes by slightly lowering characters or by creating patterns of spaces between letters and words.

One of the earliest proposals for such a system was made by Sir Francis Bacon, also known as Baron Verulam, Viscount St. Albans and lord chancellor to Queen Elizabeth I. (Some also believe he was the true author of Shakespeare's major works.) In addition to many other achievements, Bacon developed a clever form of steganography. Using 24 letters (*u* and *v* were interchangeable, as were *i* and *j*), he created the following alphabet using the first two letters of his name:

a	AAAAA	i	ABAAA	r	BAAAA
b	AAAAB	k	ABAAB	s	BAAAB
c	AAABA	l	ABABA	t	BAABA
d	AAABB	m	ABABB	v	BAABB
e	AABAA	n	ABBAA	w	BABAA
f	AABAB	o	ABBAB	x	BABAB
g	AABBA	p	ABBBA	y	BABBA
h	AABBB	q	ABBBB	z	BABBB

With this method, "the" becomes BAABA AABBB AABAA. Bacon called this process "bi-literal," and he proposed that it could be used to convey secret writing as well as audible signals with bells, horns and muskets (for example, one bell ring for *A* and two for *B*.)

The steganographic aspect was built by using different print typefaces and a cover text. In modern typeface we can illustrate this by having the *A*'s of the concealment in roman type in the covertext and the *B*'s of the masked words in italics. In order to hide the actual message "go" (AABBA ABBAB), the type of the covertext would be set as:

Be *here* soon.

The letters b and e are roman for the first two A's of "go." Then the *h* and *e* are italicized, the *r* and *e* are roman, and so on, according to the typeface variation designated for *A* or *B*.

HIDDEN MESSAGE: G   O

                     AA   BBAA BBAB

COVERTEXT: Be *here* *soon*

Bacon's method combines various aspects of steganography with concealment in apparently harmless words and semagrams with the slightly altered shapes of the letters themselves.

It should be noted that this method would be an even better concealment by having typefaces of very similar styles. Bacon's "bi-literal" fulfills the requirement of a semagram as its letters do not act as cipher replacements. Rather, the varied typeface shapes of its letters act as the substitution medium.

Based upon a idea similar to Bacon's bi-literal, the "secret adjunct design" can also be considered a semagram. It was developed in the United States in 1805 by Irish immigrant Robert Patterson, who had served during the Revolutionary War as a brigade major with a local militia group. He was concerned about the poor protection given to the communiqués of American ambassadors and set himself the task of developing a cipher for diplomatic use.

After his first cipher, an alphanumeric transposition, proved too difficult for general usage among American diplomats, Patterson developed a new "scheme of secret writing" and presented it to his friend and fellow scholar President Thomas Jefferson in December 1805. Using only two letters, *i* and *t*, he developed a "dot-and-dash" style using their secondary marks or adjuncts, the dot of the *i* and the cross or "dash" of the *t*.

Patterson proposed that two sets of 26 adjunct varieties be equated with letters of the standard English alphabet, with three or four indicating double letters while other combinations would represent often-used words. In a detailed description of the adjuncts, Patterson mentioned the possibilities of the lowercase *j* being an adjunct of *i*. Some of the variations for *i* were: dots (strong and weak); obliques descending left (/) and right (\); and commas (regular and reversed). Different positions of these marks were near to or slightly farther from the body of the letter, above the letter and to its left or right.

The adjuncts of the *t* consisted in variations of the horizontal line such as length, distance from the body of the *t*, crossing the stem, and so forth. Additionally, Patterson intended that spaces between words, at the conclusions of sentences and the end of the message could be indicated by preplanned characteristics (for example, the omission of an adjunct might signify a space between words).

The secret adjunct design

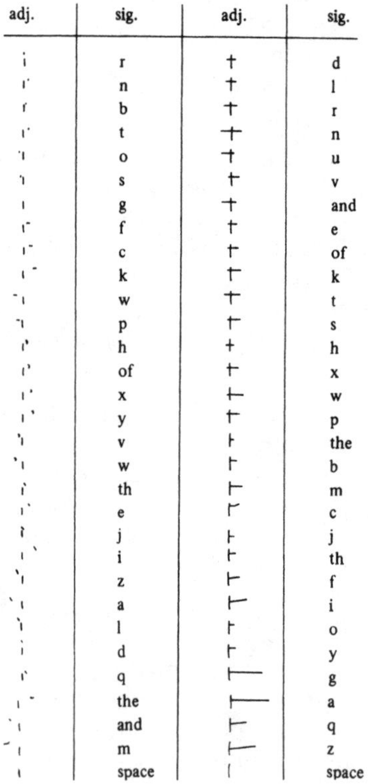

adj.	sig.	adj.	sig.
i	r	†	d
i˙	n	†	l
í	b	†	r
i˙˙	t	┼	n
˙i	o	ᛨ	u
ˀi	s	†	v
i	g	ᛨ	and
i˜	f	ᛏ	e
i˜	c	ᛏ	of
i˙	k	ᛏ	k
ˉi	w	┼	t
ˀi	p	ᛏ	s
i˙	h	✛	h
i˙	of	ᛏ	x
i˙	x	┝	w
i˙	y	ᛏ	p
ˀi	v	ᛐ	the
ˋi	w	ᛐ	b
i˙	th	ᛐ	m
i˙	e	ᛑ	c
�	j	ᛐ	j
i˙	i	ᛐ	th
ˋi	z	ᛐ	f
ˋi	a	ᛐ	i
ˋi	l	ᛐ	o
i	d	ᛐ	y
i˙	q	├──	g
i˜	the	├──	a
ˋi	and	├─	q
ˉi	m	├─	z
i	space	⌐	space

Unusual as this style was, no current record exists of its ever having been used by Jefferson or his diplomats. The varied clerical abilities and frequent writing miscues of State Department and embassy staffs alike may have worked against adopting a cipher that was dependent upon careful orthography.

Steganography often combines styles of codes and ciphers with physical concealment, and the Russian Nihilist cryptography is one such method. This version of visibly concealed writing involves patterns of spaces and shapes in some word endings and was associated with the anarchists in Russia known as the Nihilists. They were active from the latter 1850s until 1917 when the Romanov dynasty was overthrown.

The Russian Nihilist method uses the checkerboard method described earlier (see Ciphers section), but adds a steganographic level of concealment by disguising the missive as an innocent letter. Nihilist cryptographers depended on the fact that idiosyncratic styles of handwriting would be overlooked by a censor at a mail interception site, and their cryptographic method took advantage of some writers' habit of not connecting all parts of their longhand words. The encryptors added the pattern of downward or upward curves to indicate the end of a specific group of characters for the addressee. Following is an example of a letter written in Nihilist style:

Letter written in
Nihilist style

Arnold dear, it was good news to hear that you have found a job in Paris. Anna hopes you will soon be able to send for her. She's very eager to join you now the children are both well. Sonia

For example, the trio *arn* (with its upward *n*) was considered to be one character group. Each of these groups counted from left to right. Then two such groups at a time were paired as follows:

```
arn - 3, old - 3 = 33
deari - 5, t - 1 = 51
wasgo - 5, 0-1 - = 51
```

The full letter-derived numerical series reads: 33, 51, 51, 41, 23, 43, 33, 51, 45, 12, 43, 24, 11, 34, 34, 11, 34, 34, 42, 33, 11, 44, 42, 43, 33.

This number series was then compared with an alphabet square which had horizontal and vertical digits arranged to set up alphabetical coordinates. By moving across and down from the numerals, the addressee finds the letters at their intersecting points.

	1	2	3	4	5
1	a	f	l	q	v
2	b	g	m	r	w
3	c	h	n	s	x
4	d	ij	o	t	y
5	e	k	p	u	z

33 = n, 51 = e, 51 = e, 41 = d

23 = m   43 = o   33 = n   51 = e   45 = y

12 = f   43 = o   24 = r

11 = a   34 = s   34 = s   11 = a   34 = s   34 = s

42 = i   33 = n   11 = a   44 = t   42 = i   43 = o

33 = n

The dangerous communiqué reads: *need money for assassi-nation.*

Although inventive, semagramic writing is too involved for easy transmission and does not provide sufficient protection for high-level communications. It was largely abandoned as radio and electronic steganography came to the fore.

GRILLES

The grille originated with Girolamo Cardano, a 16th-century physician and mathematician who initiated the idea of hiding a message within innocuous phrases. His grille was made of a piece of stiffened material or metal into which rectangular openings of different lengths were cut to the same height as the chosen style of writing. The message was written in the holes by letter, syllable or word onto a piece of paper placed underneath the grille. Sometimes syllables and parts of words made wider cuts necessary. Once the intended correspondence was complete, the difficult task of building reasonable, innocent-looking sentences around it began.

The planned recipient used a matching series of cutouts to read the hidden meaning. He or she placed the grid on the paper, aligning the corners with each other and read the words made visible through the cutouts.

```
And there was mounting in hot haste the steed
The mustering squadron and the clattering car,
And swiftly forming in the ranks of war;
And deep the thunder peal on peal afar;
And near, the beat of the alarming drum
Roused up the soldier ere the morning star
While thronged the citizens with terror dumb
Or whispering, with white lips,—`the
foe! they come, they come!'
```

While this system was awkward to create and often revealed itself because of contrived cover phrases, it was used at

Secret message revealed by a grille

high levels of European government into the seventeenth century.

In 1777, during the American Revolution, British general Sir Henry Clinton applied a crude form of grille in correspondence with his fellow officer General John Burgoyne. From his headquarters in New York, Clinton sent a secret message to Burgoyne far up the Hudson River. The words were rather poorly concealed by what was called a "dumbbell cipher" because of its hourglass-shaped opening. Clinton's message informed Burgoyne that he was unable to join Burgoyne in a plan to divide the colonies along the Hudson River. Historians are not sure whether this correspondence reached General Burgoyne or, if it did, whether it actually affected his decisions. However, Clinton did not participate in the conflict, and Burgoyne was defeated by the colonists under Major General Horatio Gates in the Battle of Saratoga, the turning point of the war.

The illustration below illustrates the contents of one of these letters, dated August 10, 1777. The real message was to be found within the hourglass likeness representing a cutout in the covering paper:

Clinton's grille

You will have heard, Dʳ Sir I doubt not long before this
can have reached you that Sir W. Howe is gone from hence. The
Rebels imagine that he/is gone to the Southward. By this time
however he has filled Chesapeak bay with surprize and terror.
    Washington marched the greatest part of the Rebels to Philadelphia
in order to oppose Sir Wms army. I hear he is now returned upon
finding none of our troops landed but am not sure of this.great part
of his troops are returned for certain I am sure this (illegible)
must be in vain to them. I am left to command here, half my force may
I am sure defend every thing here with as much safety I shall therefore
send Sir W. 4 or 5 batⁿ I have too small a force to invade the New England
provinces, they are too weak to make any effectual efforts against me and
you do not want any diversion in your favour I can therefore very well
spare him 1500 men./I shall try something certainly towards the close
of the year not till then at any rate. It may be of use to inform you that
report says all yields to you. I own to you that the business will
quickly be over now. Sʳ W's move just at this time has been Capital
Washingtons have been the worst he could take in every respect I
sincerely give you much joy on your success and am with great
sincerity. . . .

The basic grille did not really shuffle or transpose letter positions, and in the fifteenth century this form evolved into what came to be called a turning grille. In this version the grille was turned regularly at planned intervals, for example a turn 90° left after each complete sentence, to change the apertures' positions and thus the letters' locations.

Also called "trellis grilles," from the openings in crossed wooden latticework garden structures, the turning types were known to have been applied by Dutch leaders for secret missives in the 1740s. They even saw a brief time of duty in the hands of Kaiser Wilhelm's forces during World War I. In late 1916 French cryptanalysts began to notice changes in what had become complex German substitutions. The methods had been too complicated for the Germans as well and they switched to transposition with grilles. The grilles varied in size, and each was given a codename: Anna (25 letters), Berta (36 letters), Dora (64 letters) and Emil (81 letters). However, the grilles' use was short-lived and curtailed after four months—to the dismay of the French, who had just begun to solve them.

In general, the grille was either square or rectangular in shape and was usually made of paper or cardboard. The square version seemed to have led to better spacing, and thus easier operation for the turning technique.

In the common six-by-six (36 letter) square, openings were made in one quarter of the cells (nine). The grille was placed on blank paper and the first nine letters of the message were inscribed in the holes. After a 90° turn of the grille, a second group of nine letters were written. This process was continued until each of the 36 openings was filled. A longer message required repeating the procedure with a new square. If the communiqué was shorter, the unneeded cells were blacked out to avoid confusions, such as forgotten or mistaken letter arrangements. The sender then removed the device and transcribed the transposed letters, usually in rows from left to right.

All of the facts about the exact direction of the terms, the sequence of the turns (for example, all right, all left, or alternating), and the transcription direction (for example, left to right or alternating rows) were sent to the recipient. He or she used a matching grille and the prearranged information to decipher the text.

NULL CIPHERS

Null ciphers are a type of open code in which only a few chosen words or letters are significant. The chosen characters may be the first letter after every comma, the fourth letter in every fifth word in a series of paragraphs, or the last word before each new paragraph. All of the other letters surrounding the arranged ones are nulls and thus meaningless. Their purpose is to make the openly seen text appear natural to enemy observers.

Null ciphers are often considered to be a code or a cipher, but really are a blending of these categories. This null form sometimes acts like a cipher as it is made up of a chosen series of letters and involves combinations of letters. The letters are not transposed nor are they substituted for other characters. In their final arrangement these letters are combined to form full words. The use of letters or syllables as equivalents for complete terms is more often a code than a cipher function. Null ciphers also fit into a steganographic category as their purpose is to hide a message within a seemingly innocent text.

Despite huge volumes of mail, wartime postal censors in the 20th century were expected to catch suspicious wording that might contain a null cipher. The presence of rather contrived or stilted phrasing was often a crucial clue, and the censors were sometimes aided by placing the suspicious words beneath each other.

```
Inspector number five
detaches the new
forms found with
this shipment. Stop.
Acknowledge earliest opportunity
the first new
contract you receive
from their courier
Howton. Stop.
```

The plaintext order, *strike now*, becomes obvious in a pattern of the third letter in every third word.

A very simple null cipher can be made by creasing a piece of paper vertically into thirds. The message is written vertically down the creases, and nulls are filled in around it to make an innocent message. To reveal the message, the recipient simply reads down the creases.

Some common examples of null ciphers can be found in chronograms and acrostics. Chronograms are inscriptions in which certain letters express a date in Roman numerals when placed in their proper order such as: *mercy mixed with love in him* (MCMXLVII = 1947). Not all chronograms have perfectly ordered numerals and some are purposely painted, etched or chiseled to highlight the digits on a building to give its construction date in an artfully secretive way. Although they are not the most practical mediums for ciphers, the numerals of a seemingly obvious chronogram could also be the superencipherment for hidden codewords or meanings. Continuing with our example, $1 + 9 + 4 + 7 = 21$ and $19 \times 47 = 893$. The numbers 21 and 893 could signify entirely different meanings ranging from other numerical values to encryptions representing people's names.

Acrostics are verses or arrangements of words in which specific letters, syllables or words of successive lines, verses and even chapters are aligned to spell a special word, motto or slogan when read in order. During the First World War, the Germans tried to disguise a secret dispatch in a press cable as follows:

```
President's embargo ruling should have immediate
notice. Grave situation affecting international
law. Statement foreshadows ruin of many
neutrals. Yellow journals unifying national
excitement immensely.
```

The first letters of each word spell, *Pershing sails from N.Y. June 1.*

Fortunately for the cause of the Triple Entente, General Pershing had set sail on an earlier date and returned to Europe to help lead the Allies to victory against Germany and Austria-Hungary.

Some null ciphers function by indicating a certain number of letters after each punctuation mark. One such example was used in England's Civil War. Sir John Trevanion was a Royalist who had been captured and imprisoned by the Puritans and was being held in a castle in Colchester, a city northeast of London. Because his fellow Cavalier, Charles Lucas and George Lisle had been executed before him, Trevanion was no doubt contemplating his own execution as well.

While awaiting his fate, he received the following message from a friend:

Worthie Sir John:

Hope, that is ye beste comfort of ye afflicted, cannot much, I fear me, help you now. That I would say to you, is this only: if ever I may be able to requite that I do owe you, stand not upon asking me. 'Tis not much that I can do: but what I can do, bee ye verie sure I wille. I knowe that, if dethe comes, if ordinary men fear it, it frights not you, accounting it for a high honour, to have such a rewarde of your loyalty. Pray yet that you may be spared this soe bitter, cup. I fear not that you will grudge any sufferings; only if bie submissions you can turn them away, 'tis the part of a wise man. Tell me, an if you can, to do for you anythinge that you wolde have done. The general goes back on Wednesday. Restinge your servant to command.

R.T.

The identity of Trevanion's friend R.T. remains a mystery, but there was apparently nothing suspicious about the missive for Trevanion's jailer. Even the reference to the unnamed general did not draw any unexpected scrutiny or keep the letter from being delivered to the prisoner's cell. But Trevanion found a simple yet cleverly hidden cipher pattern within these words.

The third letter located after each mark of punctuation spelled the message: *panel at east end of chapel slides*.

Trevanion seemed to become very interested in prayer after finding this message. But what seemed to the Roundheads to be an hour of piety was actually 60 minutes that the clever Cavalier used to make good his escape.

In the 19th century, with the rise in popular interest in cryptography, Arthur Conan Doyle incorporated a null cipher similar to that of Sir John Trevanion's in his story "Gloria Scott." This tale is described by Holmes as being the first case in which he was ever involved.

The father of Sherlock's college friend Victor Trevor had suffered a fatal stroke after receiving a strange note. Holmes recognized the name *Hudson* among the words, and because he had visited Trevor's country home earlier, he knew that this Hudson had seriously upset the elder Trevor before. Holmes applied his mind to the task of unraveling the following message:

```
The supply of game for London is going steadily
up. Head keep Hudson, we believe, has been now
told to receive all orders for fly paper and for
preservations of your hen-pheasant's life.
```

Holmes tried switching the locations of words and reading them backward, but this did not reveal a message. The key was that only every third word was significant, disclosing the message:

```
The game is up. Hudson has told all. Fly for
your life.
```

Once this warning was understood, other facts about the case fell into place and Holmes achieved his first success.

The frequent use of nulls for covert purposes was impractical because it often resulted in stilted language that was obvious and vulnerable to analysis. However, servicemen from many countries and in many wars used similar methods to try to evade military censors and communicate with distant family and friends. One amusing instance from World War II involved a G.I. who tried to inform his parents that he was stationed in Tunis, the capital of Tunisia. Prior to his departure, he had arranged to spell out the name of the city by changing his father's middle initial on the envelope. On the envelopes of the first letter, he addressed it to his father, using the fake middle initial *T*. In the second letter he used the middle initial *U* and so forth. However, he neglected to date the letters, and when they arrived out of order, his frantic parents scoured their atlas for "Nutsi."

A much more serious example involved an American prisoner of war in Japan who succeeded in getting a postcard with a null cipher past the camp censor:

FRONT:

Frank G. Jonelis, 1st Lt. U.S.A.
Zentsuji war prisoners camp
Nippon
Mr. F.B. Iers
c/o Federal Bldg. Company Room 1619, 100 Main St.
Los Angeles, California U.S.A.

BACK:

Dear Iers:
After surrender, health improved
fifty percent. Better food etc.
Americans lost confidence
in Philippines. Am comfortable
in Nippon. Mother: invest
30 percent salary, in business. Love
Frank G. Jonelis

An FBI agent discovered that only the first two words of each line were the message: *After surrender, 50 percent Americans lost in Philippines; in Nippon, 30 percent.*

JARGON CODES

Jargon codes have been practiced since cryptology's earliest days. Although also classified as a type of code, they are also considered a form of linguistic steganography because they are designed specifically to seem harmless to both casual hearers and military censors.

Jargon codes use innocent sounding codewords to transmit information. A request for an unusual song to be played on the radio could in fact be a signal for a sabotage unit to begin their work. While grilles have long since faded as an open code method and null ciphers are only used occasionally by individuals, jargon codes are still active among espionage agents and armed forces units. Among civilian business associates and rivals, phone conversations, faxes and e-mail still have their share of jargon amid their open commercial exchanges.

Jargon developed with languages and often became more specifically connected with types of speech which were also linked with social classes and occupations. A French code of the 1600s had a jargon-based theme including: Rome—*Jardin* (garden), The Pope—*La Roze* (the rose), and the Cardinal de Retz—*La Prunier* (the plum tree). These words could be clearly stated or written and did not sound or look like cryptic terms. As the practice became more standardized, seemingly innocent groups of words replaced others in entire sentences. For example, the message *She met him that afternoon* could actually mean "The enemy agent retrieved the package on Wednesday." Or the entire group of words could be a signal to initiate an action with no direct link to any single word.

An inventive example of a jargon code comes from World War I. An alert English censor became wary of large orders for cigars wired each day from British coastal cities by two "Dutch businessmen." The censor initiated an investigation that led to the arrest of Wilhelm Roos and Heicke Janssen on charges of spying for the Kaiser. It was discovered that an order such as "6,000 Coronas for Dover" was the code for *six heavy cruisers in port*. In July 1915 the pair faced a firing squad at the Tower of London, their cigar scheme going up in rifle smoke.

Soldiers also resorted to creating some of their own jargon codes amid the confusion of battle with telegraph and telephone wires broken, radio waves static or garbled, and human and other couriers like pigeons and dogs lost. One such "diamond in the rough" was a baseball code. It was credited to the 52d Infantry Brigade and was announced on April 1918 for use from its own headquarters through its companies and battalions. While not sanctioned by the American Expeditionary Force, it certainly must have puzzled eavesdroppers.

CASUALTIES

Strike out = Killed

Base on balls = Seriously wounded

Hit by pitched ball = Slightly wounded

Balk = Accidentally wounded

Put outs = Missing

Major = Commissioned officer

Minors = Enlisted men

ARTILLERY/TRENCH WEAPONS

Leonard using slow ball = We were bombarded by trench mortars

Cobb at bat = We bombarded

Sent in a pinch hitter = Our artillery laid down a barrage

INFANTRY

They tried hit and run game = Enemy raids
We tried hit and run game = Our raid

OUR ATTACK

Baker drives to the outfield = Strong attack
Baker drives to the infield = Small attack

LOCATIONS

Home plate = Regimental Headquarters

EIGHTH BATTALION

Right field = Right company in the first line
Short stop = Right company in support
Second to third base line = Communicating
trenches on left

Another example occurred on November 19, 1941, when a U.S. Navy interception outpost, Station S at Bainbridge Island (in the Puget Sound near Seattle), intercepted a radio message sent from the Japanese foreign office in Tokyo to its embassy in Washington, D.C.

The staff at Bainbridge sent the encrypted contents by teletype to the Navy's top-secret analysts at OP-20-G in the nation's capitol. The cryptosystem solvers determined that the Japanese code was one called J-19 and they were able to break it soon thereafter. Cryptanalysts revealed that the message, circular 2353, was a special plan for emergency notifications. If diplomatic relations were about to be severed along with international communications, a warning phrase about weather conditions was to be placed in the daily Japanese Language news broadcast. Known informally as the Winds code, it was a jargon code arrangement with phrases linking winds to compass points and countries according to their respective directions from Japan. Following is the decrypted message:

Regarding the broadcast of a special message in an emergency:

In case of emergency (danger of cutting off our diplomatic relations) and the cutting off of international communications, the following warning will be added in the middle of the daily Japanese language short-wave news broadcast:

1. In case of Japan-American relations in danger: *Higashi No Kaze Ame* [east wind rain]

2. Japan-Soviet Union relations: *Kita No Kaze Kumori* [north wind cloudy]

3. Japan-British relations: *Nishi No Kaze Hare* [west wind clear]

This signal will be given in the middle and at the end as a weather forecast, and each sentence will be repeated twice. When this is heard, please destroy all code papers, etc. This is as yet to be a completely secret arrangement.

Forward as urgent intelligence.

Some U.S. naval officers and intelligence staff persons believed that the broadcast of *Higashi No Kaze Ame* (east wind rain) would signal an imminent attack on the United States by Japan. On the Roof Gang radio monitors were instructed to listen to every Japanese weather broadcast. But in the swirl of other intercepts relating to several possible points of conflict in the Pacific, it still is not clear whether the radio alert was ever broadcast. When Japanese planes attacked Pearl Harbor on the morning of December 7, 1941, the assault came as a total surprise.

In the very clever orchestration of their espionage and diplomatic concealments prior to World War II, the Japanese used other series of jargon codes. On November 26, 1941, Tokyo sent Japanese ambassador Kichisaburo Nomura and special envoy Saburo Kurusu, who were based in Washington, D.C., an open code for telephone calls to facilitate their reports on the difficult negotiations with the United States. The code contained such terms as *Miss Kimiko* (President Roosevelt), *Miss Fumeko* (Secretary of State Cordell Hull), *The Marriage Proposal* (negotiations), and *The Birth of a Child* (crisis imminent).

As the talks in Washington were reaching a critical stage, Tokyo sent other open codes to its embassies. One, the *Ingo Denpo* ("hidden word") code, was set up to conceal plain-language messages in the event that obviously coded telegrams were banned. Some of its terms were *Arimura* (code communications prohibited); *Hattori* (relations between Japan and [name of country] are on the verge of catastrophe); *Kodama* (Japan) and *Minami* (America). These cables were indicated by completing them with the English *stop* instead of the Japanese *owari* (end).

During the First World War as noted above, cable censors closely monitored flower orders. Words regarding the delivery of specific flowers to certain sites on particular days were rife with cryptographic possibilities. First the flowers' names and delivery dates were banned. Later during the conflict, all international floral-linked communiqués were forbidden. This type of caution was not so overzealous as it might seem when compared with a flower-linked code that preceded World War II's most infamous attack.

On December 6, 1941, an FBI phone tap on a landline link with a transpacific radiotelephone call exposed a curious open code involving flowers. The call was between the editor of Japan's militaristic newspaper *Yomiuri Shimbun* and its

Honolulu correspondent Mrs. Motokazu Mori. The call was quite curious because the two openly discussed military activities around Pearl Harbor among other comments ranging from liquor, to social conditions and the weather. When the editor asked, "What kind of flowers are in bloom in Hawaii at present?" Mrs. Mori's cryptic answer was, "Presently, the flowers in bloom are fewest out of the whole year. However the hibiscus and the poinsettia are in bloom now."

The FBI and Hawaiian-area military intelligence personnel pondered the meaning of the flower references, but could make no sense of them. The real meaning of the Mori flower code has remained a mystery, with conjecture including references to observation planes, battleships at anchor and aircraft carriers returning to the harbor.

The Doll code was a jargon code used by a World War II spy for the Japanese, Velvalee Dickinson, a Madison Avenue doll-shop owner. Beginning in January 1942, Dickinson sent letters in jargon code to addresses in South America. The contents apparently contained facts about dolls, their clothing and repair needs. One letter spoke eloquently about a wonderful doll hospital and broken English dolls. Another referred to a Siamese temple dancer while a third mentioned seven Chinese dolls. They bore postmarks from different locations in the United States and appeared to be news exchanged by hobbyists. In fact, they were intended for Japanese agents and contained facts about American warships.

Questions first arose when a letter was returned from Buenos Aires, Argentina. Marked "unknown at this address," the correspondence was taken back to the apparent sender, a resident of Portland, Oregon. The woman had no knowledge of the communication and took it to the FBI. Four other letters were addressed to the same South American location and were returned to senders who had been Dickinson's customers. Her former clients were horrified to realize that she had used their addresses and imitated their signatures on these letters.

The FBI's Technical Operations Division cryptanalysts discovered that the condition of the dolls actually referred to repairs being made on U.S. ships in certain naval yards. The numbers of miniatures mentioned matched the actual totals of war vessels far too closely to be mere coincidence.

Bureau agents arrested the recently widowed Dickinson in January 1944. Threatened with the death penalty for espionage, she pleaded guilty to censorship violations by using illegal codes. In August 1944, she received a sentence of ten years in prison and a 10,000-dollar fine.

As the Second World War continued, a plethora of real and false jargon codes filled the airwaves. The British Broadcasting Corporation (BBC) sent innocent-sounding "personal messages" all over Europe to resistance forces opposing the Nazis. Pivotal BBC jargon codes for the French resistance before D-Day initiated widespread sabotage acts against German-held sites. The BBC's French-language transmission of "It is hot in Suez" started the Green Plan, the destruction of railroad equipment and facilities. "The dice are on the table" set the Red Plan rolling; its results were severed phone, cable and other communications links. The most important of these messages, phrases from the poem "Chanson d'Automne" by Paul Verlaine, announced the D-Day invasion and heralded the eventual end of the war.

Modern Transmissions Security

Not long after Alexander Graham Bell patented the telephone, civilians, militaries and businesses were all trying to find ways to make speech more secretive. *Ciphony* is a term used to describe a primary form of oral secrecy involving speech modification with scrambler devices. Because these techniques affect parts of a communication, they have an effect similar to the one ciphers have on segments of words. The word *ciphony* therefore has its roots in *cipher* and *telephony*, the science of telephonic transmission.

The human voice is made up of various frequencies, which can be measured in cycles per second. To scramble a message, these frequencies can be altered—a type of substitution. One method of scrambling is voice frequency band shifts wherein the frequencies are moved higher or lower than their original. Another method, band splitting, segments the frequency into smaller bands. Once the bands are segmented into subbands, a scrambling mechanism rearranges and alternates them in fractions of seconds.

Other scrambling methods include inversion, covering messages with "noise," amplitude modification and time variation. Inversion involves turning human speech "upside down." Low frequency vocal tones pass through the inverter and come out much higher, while the opposite happens to high frequency tones. Noise is achieved by electrically placing music or other sounds over the intended message, much like a null in a cipher that mixes false letters or symbols with the real ones. The recipient requires a descrambler identically synchronized with the originating scrambler in order to separate out the unwanted noise.

The third type of scrambling method, amplitude change, is also called "wave-form modification." This technique has a changeable electric current that creates alterations in the voice's volume level. Large differences in volume effectively disguise the message from eavesdroppers. The person receiving the message unscrambles it with a current arranged to reverse the fluctuations.

Time variation ciphony is accomplished with magnetized tape. In time-division scramble, or TDS, a recording mechanism copies parts of the speech in timed intervals and shuffles them before transmission. The recipient's descrambler is equipped with an equal number of recording heads that return the sound segments to another magnetic tape in their proper order.

Scrambling devices began to achieve popularity in the late 1920s and 1930s and saw widespread use during World War II. Expanded by Cold War and private industry concerns, ciphony now encompasses such transmissions security factors as: citizen's band sets that are made secure by pseudorandom changes in the frequency spectrum between caller and receiver; the NSA's Electronic Secure Voice Network of phone scramblers; and encrypted and digitized voice signals such as pulse code modulation.

Since World War II, the modern communications expansion with television, cable, satellites and the Internet has made close mail scrutiny and media limitations extremely difficult to maintain. However, the NSA, the successor to previous U.S. military intelligence groups, does keep a number of communications mediums under surveillance. The FBI also conducts court-approved wiretaps of domestic transmissions wherein secret messages may be concealed.

The variety and advanced levels of these modern technologies has instigated another steganographic category, called transmissions security. Transmissions security is an electronic form of concealment derived from steganography. Like this method of covering written text, transmissions security tries to hide the existence of secret messages in electrical exchanges, whether or not they are encrypted. Present day forms of transmissions security include laser technology, quick burst/spurt broadcasts, frequency hopping, and spread spectrum techniques of diminished signals covered with specially produced noise.

LASER AND INFRARED TECHNOLOGY

Laser technology has given secret inks a new place in cryptology by increasing their levels of security. When the already very specialized liquids are treated by a reagent, they are only fully recovered when viewed under a particular type of ultraviolet light calibrated to certain wavelengths. The secrecy is now not only in the ink's components, but also in the ultraviolet mechanism's settings.

Infrared light has also been used as a type of transmissions security during the past decade. Reportedly the Stasi, the former East Germany's secret police, had developed an infrared system through which their agents conversed in once-divided Berlin during the late 1980s. Unlike the mobile sending capabilities of the burst/spurt systems (see below), the infrared method was not very flexible. Its transceivers, or send-and-receive transmitters, were required to be in a line-of-sight mode in order to work. The direct beam carried the speaker's voice frequency. It was restored by the receiving mechanism and returned to recognizable sounds in the recipient's earphones. Its great advantage lay in the fact that it can transmit completely concealed messages that did not need to be scrambled by changed frequencies or mixed with background noise.

Laser beams also became a steganographic implement in the late 1980s when the CIA prepared a high-tech system for their agents. A microlaser was used to engrave encryptions in the borders of advertisements on the pages of magazines. The magazines were popular, easily obtainable issues that could be kept in plain sight without drawing undue attention. With a powerful magnifier, the agents could see the cipher and use the necessary keys to decrypt the dispatches.

PULSE CODE MODULATION

In pulse code modulation (PCM) soundwaves are converted into a sequence of binary electrical impulses. In a typical example, the speech is sampled several thousand times a second. Each sample is identified with one of a series of amplitude levels, represented by a multielement binary pattern. The result is a sequence of binary digits. With advances in technology, streams of these binary signals have reached the level of 28,000-plus bits per second.

The binary impulses can be encrypted by a process similar to that developed for the teletypewriter by Gilbert Vernam, which involved the addition of key pulses to plaintext message letters to produce the ciphertext. Binary digits are combined with a binary key stream produced by a key-generating computer. The result is the encryption, which offers strong speech protection. Computers are also used to store key pulses on magnetic tape.

BURST/SPURT TRANSMISSIONS

During the Second World War, Nazi U-boats sent early versions of burst transmissions. They reportedly recorded their messages with magnetic wire equipment and radioed them by high speed to Germany. After the fall of the Third Reich, this audio advance and others were studied by the Western Allies and the Soviet Union. When these once-united titans became Cold War enemies, burst transmissions became another tool in that confrontation.

Modern burst/spurt transmissions are very highspeed broadcasts. Information about mechanisms used by the United States during the late 1980s has only recently become publicly available. One such mechanism, the GRA 71 burst system, used a small tape recorder. With dot-and-dash keys, an agent typed encrypted Morse code that was recorded on the tape, while another mechanism ran the tape at 40 times its normal speed. This high-speed supercompressed message was then broadcast by radio. At this rate a 1-minute encryption took only 1½ seconds to send. At the recipient site, radio monitors listened constantly with recorders rolling constantly for the specific communiqués. The extreme speeds at which these messages were sent meant that they were difficult to detect by signals intelligence monitors. This technique had many military and espionage applications and was used to convey dispatches to submarines and undercover agents.

SPREAD SPECTRUM

Early radio transmissions used only a narrow band of the frequencies of the electromagnetic spectrum, but a much broader bandwidth began to be applied in the 1950s, and the term *spread spectrum* was coined. Today semiconductor chips, containing thousands of transistors, have made it possible to transmit clusters of digitized data in apparently random sequences over numerous channels. A series of inventors have contributed to the development of these processes, which include time jumping or hopping, frequency hopping, direct sequence and a hybrid of the direct and frequency types.

In the time-jumping version, data is generated by the sending radio, which transmits parts of the information at different times. The recipient's radio is equipped to pick up the transmissions at the preplanned times.

Frequency hopping is similar, in that it also segments the data. The sending set emits its information groups in a planned order and on a number of varied frequencies. The recipient's equipment recognizes the pattern of the emissions, gathers the sequence of data groups and reassembles them as the full communiqué. Frequency hops move messages from frequency to frequency at speeds greater than a thousand times a second.

Direct sequence involves the transmission of very brief parts of the dispatch known as chips. These tiny packs of data are generated as fractions of the full message. The recipient's equipment recognizes the correct chip order, recombines and builds them up to the original again. Additional variety and security have been added to the direct sequence with an encrypted signal and a frequency-hopping variation. In this high-security transmission, the message bits are sent at the same time, but on various frequencies, while the extra security cover sounds like a jumble of noise to an outsider. The receiving set is tuned to the correct frequencies needed to recover the signals.

This combination of the direct and frequency styles is fast becoming a fourth variant of spread spectrum transmissions. Aided by developing high-speed data processors, this hybrid is also reportedly adaptable for better digital encryption. These improvements will make it easier to check flawed exchanges and correct errors for first-time transmissions.

According to some analysts, these processes will enable consumers to communicate without middle person billing charges. Spread spectrum advances will permit millions of radio communicants to exchange digitized packets at low power and on multiple frequencies without interfering with others in the Ethernet, the Internet's counterpart in the airwaves. Such possibilities may very well change the entire scope of communications, including licensing for television, radio, CBs and cell phones as well as standard home phone bills. Ethernet versions of the security-versus-privacy debate regarding the Internet have already begun.

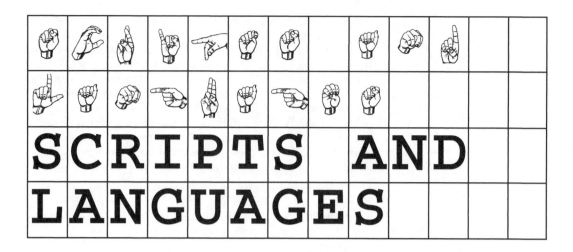

SCRIPTS AND LANGUAGES

The languages in this chapter are different from codes and ciphers, as the vast majority were not intended to be secretive. Both oral language and written scripts in fact developed to enhance communication between individuals and allow the conveyance of more complex types of information.

Writing has been defined as a system of graphic images or symbols meant to convey thoughts. The vocalization of thoughts in the form of rudimentary utterances preceded writing, yet in this chapter a choice has been made to emphasize the graphic factors. Script and language however, are inextricably bound together and factors from phonetics to linguistic varieties are interwoven in the contents of this chapter.

Protowriting

"Protowriting" refers to means of written communication before the development of scripts, and includes ice-age images, pictograms, tallies and representational drawings. The terms *graphic*, *pictograph*, *ideogram* and *logogram* are often used to describe protowriting. The word graphic is broadly applied when referring to any man-made marks on a surface.

Pictograph, from the Latin, *pictus*, meaning "painted" and the Greek, *graphein*, "to write," refers to such marks when they are formed into a recognizable image. The term *ideogram*, from the Greek *idea*, "idea," and *gramma*, "written or drawn," describes a mark that represents an idea. The words *ideograph* and *pictogram* can also be used interchangeably with, respectively, *ideogram* and *pictograph*. Some scholars believe proto-writing should be thought of as standing for words rather than ideas. They argue that the signs represent syllables or phonetic values and prefer to use the term *word sign* or *logogram*, from the Greek *logos*, "word," and *gramma*.

Man's first graphic marks were probably the carved and painted images on cave walls such as those discovered in southern France. Found throughout the 1900s, these renderings of bison, horses and staglike heads are an estimated 20,000 years old.

Several of the figures also have symbols with them. A painting of a bison found in Marsoulas, France, contains a series of connected vertical lines which resemble the letters *u* and *v* and even a candelabra. Despite this discovery, however, historians do not believe that a true writing system was in use at this time.

Various ideograms include descriptive-representational devices and identifying-mnemonic devices. Descriptive-representational devices are drawings of familiar objects, such as animals, objects or people, arranged in a meaningful configuration. For example, the passing of a week may be illustrated by seven moonlike images. Descriptive-representational devices were often used to convey instructions or to record an event.

Identifying-mnemonic devices connote ownership. Markings such as diagonal wedges chipped into a piece of stone identified a specific person's possession or an object's creator.

Some of the most well-documented examples of pictographic writing are Native American pictograms. Drawing from studies in the 1800s by the Bureau of Indian Affairs, William Tomkins collected a number of pictographs used by the Sioux and Ojibwa tribes.

The Ojibwa and Sioux symbol drawings are basic representations of nature, people, animals and objects. For example, the sun and moon are depicted as follows:

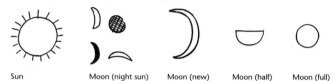

Sun   Moon (night sun)   Moon (new)   Moon (half)   Moon (full)

Sunrise and sunset are similarly depicted, with a directional mark added:

Sunrise   Sunset

Time is given more detail, as follows:

Three nights   Three years

The following chart indicates one year's progress:

January (snow moon)	February (hunger moon)	March (crow moon)	April (grass moon)	May (planting moon)	June (rose moon)

July (heat moon)	August (thunder moon)	September (hunting moon)	October (falling leaf moon)	November (beaver moon)	December (long night moon)

Animals also appeared frequently in Native American pictographs. Some examples are shown below:

| Antelope | Eagle | Duck | Horse | White buffalo |

Both nomadic and agricultural tribes had to be constantly aware of the climatic conditions. The following are pictographic interpretations of some weather conditions.

| Cloud | Cold and snow | Lightning | Flood | Storm and wind |

The characters in Sioux pictographic stories are arranged in a spiral formation, beginning in the center of the spiral and reading counter-clockwise to the end. This form is used in Lone Dog's Winter Count and certain other famous Sioux documents.

The pictographic account below is a replica from a Sioux document.

Sioux pictographic story

The story may be interpreted as follows:

"An Indian trader by the name of Little Crow went on a journey. He traveled for three nights until he came to a river. He traveled at night was because he was in enemy country. At the river he secured a canoe, camped there that evening, and at sunrise the next morning started down the river and traveled two suns [days]. He now traveled in daytime, because he was in friendly territory. He was an Indian trader in shells, which were used for wampum and ornamentation. At the end of the fifth day's travel he reached the village where the shells were obtainable. He rested there for three days in conference with the chief. As a result he traded for a large amount of shells, and at sunrise on the fourth day he loaded his canoe and started down the river and traveled for two days. On the second day a storm came up, with rain and lightning. He saw the lightning strike a tree and set it afire. As a result of the storm he became sick, so he searched and found some medicinal plants and waited there a couple of days until he felt better. He then traveled at night and hid away in the daytime. He knew that the country abounded in game because he heard foxes and wolves. He finally reached home, though some days late. Twenty braves of the tribe came out to meet him, including their chief, Standing Bear. They were happy that he had had a safe and successful trip, and the tribe celebrated his return."

Tallies developed around the same time as these descriptive-representational and identifying-mnemonic devices. These were physical objects formed, arranged, or marked in some way that had significance both for the giver and for the recipient. For example, the recipient of a cord wrapped around a stone could unwrap the cord once every sunrise for fourteen days or until the tally giver returned. If the tally giver did not come back, the recipient could take an agreed-upon action.

As well as being markers of time, tallies were also used to count objects. This practice existed from the 12th century through the early 19th century in England, where wood tally sticks with specially cut notches were recognized as receipts by the British Exchequer until the 1830s. For example, the cut for 1 pound (£) was the width of a ripe barleycorn.

The development of counting was instrumental in the progression from protowriting to more elaborate ideogram marks

Tally stick

and pictograph images. Many scholars have concluded that a primary impetus for writing was man's need to count to keep track of transactions. As recording the exchange of foodstuffs, tools or weapons grew more complex, the simpler memory aids and tallies were no longer sufficient.

One very early form of counting was done with clay tokens. These rather common-looking objects are dated as far back as 8000 B.C.E. through 1500 B.C.E. and have been found at various sites across what are present-day Iran and Iraq. As well as being a means of counting, they may have also served as a more permanent record of a series of transactions. Their varied shapes and designs could have been used to designate separate amounts of a particular item and may have been used for bartering as precursors to money. This idea that objects could represent things other than themselves was a crucial step in the development of true writing.

The earliest known clay tablets were found in Mesopotamia, the plain between the Euphrates and Tigris Rivers, and are believed to be linked to increased trade activity among the Mesopotamian city-states. Inscribed by the Sumerians around 3300 B.C.E., they contain numerals and symbols believed to be used for calculating purposes. The numerals were impressed in soft clay in the style known as cuneiform (Latin: *cuneus*, "wedge" and *form*, "shape") in the third millennium B.C.E.

While not considered a complete writing system, these symbols are considered to be quasi-pictographic. Around this time a critical development in the history of writing occurred. Ancient Sumerian scribes realized that sequences of pictographic symbols could not be fully accurate or complete representations of ideas; symbols had to be linked with the sounds of everyday speech. They began to add signs that had

Clay tokens

phonetic values to the primary values. For example, the illustration below shows that the symbol for "reed" was altered to become "reimburse" because the Sumerian sound *gi* for reed (*A*) was also used in "reimburse" (*B*).

Cuneiform symbol "reed" (A) and "reimburse" (B)

A. B.

This was the basis of the rebus principle, wherein a pictographic sign represents a sound rather than an idea. Although pictograms, by their very nature, also have phonetic values, phonetic representations differ in that the sounds are often not connected to the sign's content.

The word *rebus* is believed to have originated with the Latin word *res*, or "thing." The literal meaning of the word was "by things," referring to the fact that the sounds of the symbols were combined to create new meanings. The first known use of rebuses began with the Sumerians, but rebus forms also appeared in other ancient scripts and again in Europe during the Middle Ages.

In England, the rebus was first used as a form of signature and as decorations on important papers like deeds and birth records. They were also emblazoned on coats of arms to depict family names.

A rebus by Giovan Battista Palantino c. 1540

The uses for rebuses were expanded in France during the 1500s. In the provinces, carnivals and pageants often included performances that incorporated symbolic puzzles. They were called *de rebus quae geruntur*, "concerning things that are taking place," and satirized everything from local life to the national government. The shows in one particular region, Picardy, were so popular that they came to be titled *rebus de Picardie*.

The evolution of scripts has moved along a similar continuum all over the world: ideas developed into a word-picture form and then to simple phonetics, while rebus styles predominated until a syllable-based script emerged. Signs became increasingly language-specific, as symbols now represented sounds rather than ideas. In order to understand them, each symbol had to be translated into the sounds which formed a word, and then retraced to the original idea. This is the same process upon which epigraphers embark on the task of deciphering extinct scripts.

Extinct Scripts

Although there have always been fewer scripts than spoken languages, many ancient scripts have also been lost due to poor preservation or wartime destruction. However, the loss of a script is not always associated with the fall of a civilization. In many cases, the language and script may continue to evolve until they are so far removed from the original script that they are indecipherable. When a script is no longer used and understood by its people, the script is considered to be extinct. The fragility of the scripts material, often papyrus or pottery, means that few examples of extinct scripts survive to the present day.

Some ancient scripts, however, have survived the test of time. In some cases, a ruler's achievements were carved in stone monuments, painted on pottery or written on scrolls. The sheer number of the copies and the varied forms containing them assured the survival of some for posterity. In other instances, benevolent leaders may have ordered the restoration and preservation of older scripts. Decrees to have scribes copy and store such texts, even if not incising them anew on temples, did ensure that lesser-used scripts were transferred to new generations.

Probably the most important factor in the survival of a script was a long period of peace. The sacking of an enemy capital often meant the unplanned or purposeful destruction of artifacts considered valuable to the loser. Such losses included torching scroll repositories and defacing or toppling temples. At times the entire cultural life of a city was wiped out by an invading army.

Such was the case of the Romans when they defeated Carthage in 146 B.C.E. at the end of the Third Punic War (149–146 B.C.E.) Carthage was a large Mediterranean trading center in North Africa near present-day Tunis, Tunisia. When the Romans invaded, they destroyed every vestige of Carthegenian culture and very little of Carthage, from structures to scripts, survived.

Other scripts lasted because they provided efficient and easy sight-to-sound equivalencies. If a script's images could be easily interpreted, it was more likely that a significant portion of the populace would remember it and teach it to their children. Conversely, the more complex the script, the less likely it was to be understood or passed on to future generations. A large population and a national leadership that inculcated interest in literature also helped maintain the foundation of writing and reading that kept some scripts alive.

Scripts and the languages they conveyed often did become extinct when their civilizations were ended by famine, natural disaster, plague or conquest. However, epigraphers have succeeded in deciphering some of these scripts because examples were written or carved on objects that were kept in sacred or hidden places. They may also have been inscribed on monuments that survived time and weather, or were copied enough times to stay extant in spite of weather, neglect or intentional destruction.

DECIPHERING AN EXTINCT SCRIPT

The Egyptian hieroglyphs, especially those discovered at Tutankhamen's grave are undoubtedly the best-known example of ancient writing that has been solved by the scholars known as epigraphers. While archaeologists are well known for their explorations into past treasures, epigraphers work to solve sandstorm-blurred carvings or faint pottery art inscriptions.

Epigraphers have much in common with cryptanalysts, but their task is both easier and more difficult. On the one hand, the ancient scribes were not trying to conceal their words, and actually worked on behalf of pharaohs and potentates who wanted to record their achievements for history. Decryption is also more difficult, however, because at times the complete language must be reconstructed.

The challenge of deciphering writings of extinct languages is twofold, as either the writing or the language may be unknown. In the first case, the spoken language may be known but the script is not. In the second, the writing may be understood but the language is unknown. There are even cases in which neither the language nor the script is known. With these puzzles and seemingly impossible obstacles, epigraphers require the same curiosity, analytic ability and knack for hard work that typifies cryptanalysts. Thanks to such people, several extinct scripts have been solved using cryptanalytic techniques.

Epigraphers are sometimes aided by the discovery that supposedly "unknown" languages are in fact branches of recognized linguistic families and can be studied by similar techniques. If a script is known but the language is not, there are two deciphering possibilities. If the language can be proved to be part of a recognized family, it can be solved by searches for the known family's characteristics. If the linguistics are of an unknown origin, an entire lexicon must be built. Decipherment is considered virtually impossible unless a bilingual text, which contains parts of an understood language equated with the language in question, is found.

By the same token, if neither the script nor the language is understood, decipherment is also considered impossible. A discovery of a similar language or script link would have to yield sizable contributions in order for epigraphers to even begin analyzing the script.

Bilingual texts and language links have been instrumental in solving extinct scripts. In the early 19th century James Prinsep, a Sanskrit scholar at Oxford, revealed the secrets of two ancient scripts by analyzing bilingual text imprinted on Persian coins. One of the coins' inscriptions was in Greek, brought to Persia by Alexander the Great. Using known Greek names as phonetic clues, Prinsep solved a portion of Pahlavi, a language spoken and written in Persia from about the third to the tenth century C.E.

Buoyed by this success, Prinsep began analyzing a text regarding Asoka, a king of India of the Maurya dynasty, who ruled in the third century B.C.E. The inscriptions were in the unsolved Brahmi, an alphabet in a known language of the Pakrit people. Prinsep came across writing on temple items and surmised that they were a gift to the temple. Working with the Pakrit word for *gift*, he compared it with the unknown Brahmi script. This breakthrough was the foundation for later Brahmi solutions, and epigraphers discovered that Brahm was the root language for Indic scripts.

Bilinguals have also aided in the solution of arcane writings. An obscure tongue known as Sidetic, from a coastal city called Side in southern Asia Minor, had its roots in Greek, but for a long time the available text did not contain sufficient details for extensive analysis. After the discovery of a longer bilingual in 1949 analyst Helmuth T. Bossert was able to make a partial solution of Sidetic.

In 1929 in Ugarit, Syria, Hans Bauer excavated a number of tablets containing a script known as Ugaritic cuneiform. Bauer had had an extensive background in decryption, and had worked as a cryptologist during World War I. Within a week of his discovery, Bauer found answers to the phonetic values of half of the symbols. He then applied his cryptanalytic skills to segments of signs that seemed to represent basic forms such as prefixes and suffixes.

Bauer compared the Ugaritic cuneiform with a language called West Semetic, which he presumed—correctly—had been used in the Ugarit area. Aided by frequency and distribution lists, he compared the West Semetic sound values with the still unknown syllables. He was thereby able to isolate and identify introductory prepositions and seek typical words. After finding the Ugaritic words for *king* and *son*, *mlk* and *bn*, Bauer found related terms about royalty and kinship. With these techniques, similar to those used to find related military terms in World War I dispatches, Ugaritic cuneiform was well on its way to solution.

CUNEIFORM

Predating Egypt's hieroglyphics, the intricate script called cuneiform was developed in the ancient empire of Mesopotamia, about 3400 B.C.E. From 1928 on, excavations were conducted by the German Oriental Society and the German Archeological Institute. In 1928 the first clay tablets illustrating early Mesopotamian pictographic symbols were found at Uruk, on the Euphrates River south of Babylon. They dated from about 3300 B.C.E., making them the earliest example of a true script yet discovered.

It was not until about 2500 B.C.E. that the distinctly wedge-shaped cuneiform began to predominate over the older pictography. In order to save time and materials, Sumerian scribes began to simplify the more elaborate pictographic style.

Symbols became abstracted from the objects which had inspired the original pictograph. Historians believe that this development reflected a general tendency toward utilitarianism rather than visual beauty.

Cuneiform was written on slightly damp clay tablets, using styluses made from reeds. These could be cut to create tips

Cuneiform's wedge shapes

The evolution of cuneiform

Original pictograph	Pictograph in position of later cuneiform	Early cuneiform	Classic Assyrian	Meaning
				Heaven, God
				Earth
				Man
				Pudenda, Woman
				Mountain
				Mountain woman, slave girl
				Head
				Mouth, to speak
				Food
				To eat
				Water, in
				To drink
				To go, to stand
				Bird
				Fish
				Ox
				Cow
				Barley, grain
				Sun, day
				To plow, to till

that were circular, flat, diagonal or pointed to facilitate different styles of writing. The scribe began at the top left-hand edge of a tablet and wrote downward to the base edge. He or she (women scribes did exist) resumed marking at the top of a second column and repeated the same procedure until a tablet's side was filled. When one side was complete, the scribe turned the tablet over on its base edge and began forming stylus marks beginning in the top right-hand corner. The columns' direction was reversed as the scribe moved from right to left. Upon the completion of a tablet, it was allowed to harden, or was fired in a furnace, thus preserving the record for posterity.

Surviving tablets from the second millennium B.C.E. show that some standard signs were rotated 90° east. A horizontal inscription pattern replaced the vertical lines, although the signs remained aligned in columns, and a left-to-right script was standardized. These changes were seen most often in vernacular documents, and the older directional style was used for many years on monuments and ceremonial artifacts.

Cuneiform is made up of a basic trio of elements: pictograms, phonograms and determinatives. Pictograms signify an object or an idea linked with an object. In English, phonograms, or phonetic symbols are the alphabet letters. In cuneiform, phonograms include rebuses and syllables that represent a word. Determinatives are signs added to the ends of words that serve to clarify potentially ambiguous meanings.

A type of marsh grass, for example, was graphically indicated by its blades. The names of two such plants were /naga/ and /te/, but were also known as /nidaba/ and /eref/ (virgules / indicate pronunciation). The latter names happened to be those of a goddess and a town, each linked with marshlands.

When the scribe wanted to indicate the plant /te/, he or she placed the determinative graph /u/ (plant) before the graphic image. To signify /nidaba/ (marsh goddess), the scribe prefixed the determinative graph for /dingir/ (god) to the plant sign, /te/.

Some of the ancient Sumerian counting practices have survived to the modern era. To quantify time or to measure angles, the Sumerians used a sexagesimal system founded on multiples of 60 (which became the modern day 60 seconds, 60 minutes and 360°). Their numbers included the following images to count people, animals or textile products:

Cuneiform number symbols					
	5	4	3	2	1
	50	40	30	20	10

The sexagesimal system				
	$60^2 \times 10$ (36,000)	60^2 (3600)	60×10 (600)	60

Despite wide time gaps and cultural differences, cuneiform continued to be used by scribes in the empires of Babylonia (2700–538 B.C.E.), Assyria (seventh century B.C.E..) and Persia (sixth century–331 B.C.E.). Frequent warfare as well as disastrous river floods caused great social upheavals and the weak cultural cohesion in the region collapsed during the invasion of Alexander the Great in 330 B.C.E. The unique identity of the Sumerians and their language were irreparably shattered. Still, a uniformity of style and function had been solidified in the cuneiform writing, and text had been inscribed in enough places for some of it to survive to the 20th century.

Decryption

Cuneiform was all but forgotten until 1618, when Garcia Silva Figueroa, Spain's ambassador to Persia, made an important discovery. While studying ruins near a site called Shiraz, he found similarities between the many crumbled edifices and descriptions he had read in the writings of ancient Greek and Roman writers. The words in those texts had described the formerly great city of Persepolis, once the capital of Persian kings. He correctly believed that the Shiraz location was Persepolis. It was later determined that a location he had found named Takhtī Jamshid had in fact been the palace of King Darius.

Despite Figueroa's findings, no explorer or linguist followed his footsteps for over 150 years. The first cuneiform inscription was not published until 1657, and curiously, the signs were viewed as decorative or repeated attempts at artistic variations rather than distinct writing. In the 1700s some scholars noticed that there were varying scripts amid the signs and some speculated that they might be forms of Egyptian, Greek, Latin or even Ogham inscriptions from Britain. In spite of some growing curiosity, no detailed study of the signs was carried out.

It was not until the 1770s that a German adventurer, Carsten Niebuhr, began serious study of cuneiform. He noticed two important factors—that a number of the characters were repeated and that the ends of lines were uneven. These discoveries enabled him to deduce the left-to-right flow of the inscriptions and to define the trio of writing forms. Niebuhr is also credited with beginning the process of separating and categorizing specific signs.

Soon afterward Frederik Münter, a language scholar, suggested that the three scripts were alphabetic, syllabic and ideographic, and that each expressed the same facts in three developing stages of the Persian language, namely, Zend, Pehlevi and Parsi. The three scripts were later found to be the Old Persian, Neo-Elamite and Neo-Babylonian languages.

Solution of cuneiform was aided by partial translations of Old Persian from documents kept by Parsi descendants in India by A. H. Anquetil Duperron in 1771. In 1793, using some of the inscriptions found at Persepolis, Silvestre de Sacy compared Greek versions of the kings' texts to solve inscriptions of Persian kings dated from the third century C.E.

As these breakthroughs began to coalesce, a German schoolteacher named Georg Grotefend began his own study of cuneiform. He suffered an early setback when his findings were rejected by the prestigious Göttingen Academy because he was not an Orientalist. In spite of this rebuff, Grotefend continued to make contributions to the field. He realized that the script was alphabetic because of the length of the "words" and the limited number of different signs in each. He also noticed that a frequently seen oblique mark served as a type of word divider and that the three most often appearing signs were vowels.

These presumptions aided him in his search for name identities. Grotefend believed that the names of kings would be present in inscriptions, and continued Münter's search for a typical Persian title, "king of kings." Grotefend realized that two words were needed for this, a short one and a longer version of the same word.

Grotefend began to seek a royal lineage of kings and sons with the same or very similar titles and the name change following the lineage. He knew from reading de Sacy's work that the inscriptions always included a formal declaration of paternity in the formulaic style: *A*—great king, king of kings, son of *B*. As he knew of the Persian royal dynasty known as the Achaemenids from Greek histories, Grotefend related them to cuneiform genealogies and concentrated on certain figures on the entranceways at Persepolis.

Grotefend made a critical step when he determined the name of Xerxes, son of Darius, by its position in the genealogy. He had been aware that the names he was searching for were Greek spellings, not Persian. To identify the sound values of the signs, he consulted the partial Old Persian translations of Duperron. This material indicated that the Persian spelling of Xerxes would be "Khshhershe." When he found the cuneiform word *Khscheï(tm)*, Grotefend also realized that the cuneiform and Persian words began with the same sign.

From comparisons of partial translations of related writings and educated guesswork, Grotefend slowly built up an alphabet of Old Persian. Though his interpretations have been criticized by recent scholars, he made the important discovery that signs represented sound values.

In 1823 Rasmus Rask, a founder of comparative linguistics, discovered a relationship between some sound values, their signs and Sanskrit, the classical Indic literary language. This permitted an even better reconstruction of Old Persian.

The next advancement was made by Henry C. Rawlinson, who had been an officer in the East India Company's army from 1826 to 1833 and a military adviser to the brother of the Shah of Persia until 1839. Through this service he learned Arabic, Hindustani and modern Persian and it was he who solved cuneiform's famous text.

Rawlinson saw the huge trilingual inscription of Darius the Great at Behistun (now Bisitun in Iran) amid many stones on the side of a mountain. The carvings were a monumental effort by scribes and other stoneworkers who had cut ten columns of cuneiform into a cliff in the Zagros Mountains of western Iran. Inscribed some 490 feet above ground level, the text was written in Old Persian, Elamite and Babylonian.

At great personal risk, Rawlinson copied the Old Persian by hand while standing on a cliff ledge. Aided by a Kurdish assistant, they suspended themselves from the cliff in order to make papier-mâché impressions of the scripts. After some ten years of this arduous, risky enterprise, the entire trilingual had been copied.

By this time a number of scholars were working on extensive translations, including Rawlinson himself, Edward Hincks, an Irish parson, and a German professor named Christian Lassen. Rawlinson is credited with finding the names of Darius's subjects who were also described in Greek histories of Persia. His linguistic skills also helped in comparisons of Old Persian and other known precursors of European languages. Lassen had previously noticed that some of the cuneiform symbols only appeared in front of certain vowels, and Rawlinson and Hincks realized that the spelling of cuneiform symbols was affected by their pronunciations. This finding enabled Rawlinson to put together many more syllables and word combinations, and by 1846 he had a full translation of the Old Persian part of the Behistun rock carving.

Eventually the Babylonian and Elamite sections also revealed their meanings. The Babylonian signs' relationships with

The Behistun
inscriptions

Hebrew and other Semitic languages proved important, as did
the realization that sign meanings changed depending on
their use in sentences. By the 1850s a series of cuneiform
translations had been accomplished.

In spite of Rawlinson's success, some historians believe
that he borrowed some of his important findings from the
unobtrusive, publicity-shy Hincks without giving him due
credit.

■ **527**

EGYPTIAN HIEROGLYPHS

Egypt's hieroglyphs (Greek: *hieroglyphika grammata*, "sacred carved letters") date from 3100–3000 B.C.E., and are slightly "younger" than cuneiform. The development of hieroglyphics coincided with the start of Egyptian dynastic history when the great leader Narmer (Menes) united Upper and Lower Egypt around 3100 B.C.E. Although a number of pots with painted pictograph-like designs that predate this period have been found, this artistry was not necessarily a precursor of hieroglyphics.

Hieroglyphics approximately 700 signs are a mixture of logographic and phonographic representations. In other words, they are both semantic symbols that represent ideas and phonograms for sound writing. The images are pictographs in the sense that they do signify a recognizable object, but they could also represent a sound completely different from the recognized image. Signs could also indicate groups of consonants with varying sound values.

Because the ancient Egyptian language was largely based on consonants and their vowels were unmarked, the many different signs used to express consonants were often confusing, even to the scribes. To distinguish between similar phonetic signs, they used determinants, such as plant or water, to signify the proper realm of the word.

In addition to the hieroglyphic script, the Egyptians also used two scripts which were descended from the hieroglyphs. They were the hieratic (Greek: *hieratikos*, priest) and demotic (Greek: *demotikos*, popular, common use). The former was reserved for religious texts, and the latter was used for descriptions of more worldly events.

Hieroglyphs were written in alternating left-to-right and right-to-left lines, a style of writing known as boustrophedon (from a Greek word describing the way an ox-drawn plow moves through a field). Reading direction was indicated by the position of the signs. If the images were facing to the left, the reading direction was from left to right. Inscriptions also flowed from right to left if appearances were enhanced by doing so. For example, the images at an entrance were placed

A cartouche of Tutankhamen

in the most pleasing direction of sight when opening a door and entering a room.

The following comparison with Greek Letters exemplifies the signs and phonetic sounds of ancient Egyptian:

A reconstruction of hieroglyphs pronunciation

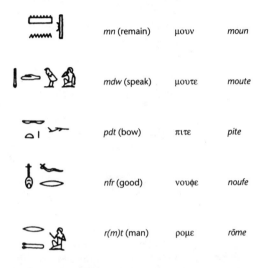

mn (remain)	μουν	moun
mdw (speak)	μουτε	moute
pdt (bow)	πιτε	pite
nfr (good)	νουϕε	noufe
r(m)t (man)	ρομε	rōme

Although exact pronunciation is uncertain, the approximations are based on clues from Coptic, which is written in Greek letters. Other clues come from the Assyrian and Babylonian languages, which did distinguish the vowels in their writings and which contain vocalized forms of Egyptian words.

Decryption

Hieroglyphs had been the subject of much conjecture and many mistaken beliefs. In July 1799, the tragedy of war brought good fortune to the solution seekers when French soldiers serving Napoleon made a key discovery in Egypt. This was a slab of black basalt called the Rosetta Stone, although it was actually found at the town of Rashid in the Nile's western delta.

The Rosetta Stone weighed about three quarters of a ton and measured 3 feet 9 inches high, 2 feet 4½ inches wide and 11 inches thick. Its contents included the inscriptions of two languages, Egyptian and Greek, and three scripts. The Egyptian scripts were in hieroglyphs (top) and demotics (middle), while the Greek (bottom) was in the capital letters of the Greek alphabet. Dated about 196 B.C.E., the inscription was a copy of a religious decree commemorating the first anniversary of the coronation of Ptolemy V.

Copies of the inscription were made and sent to scholars throughout Europe. Silvestre de Sacy and a Swedish diplomat named Johan D. Akerblad solved some proper names by comparing the Greek segment with the Egyptian demotic, but the first real success was not until 1814. An English physician named Thomas Young noticed that the demotic script had both phonetic and nonphonetic signs and also found connections between the demotic and the hieroglyphic inscriptions.

His identification of the name "Ptolemy" exemplifies his deciphering strategy. He knew that Ptolemy was a non-Egyptian name as the Ptolemies were descendants of one of Alexander the Great's generals. From previous studies, he also knew that Egyptians spelled foreign names alphabetically, and he agreed with other analysts that royal names were contained in elongated ovals, or cartouches.

Young compared the Greek letters for Ptolemy with the hieroglyph and finally identified the correct phonetic values of six of the hieroglyphs. Emboldened with this success, he continued his efforts from 1814 to 1818. Like code and cipher analysts, he counted signs, made lists and kept track of others' progress. During this time, he identified all or part of 80 demotic terms and their hieroglyphic counterparts. Using the known Greek words, he was also able to translate the majority of his matchups. His notebooks became a valuable body of facts for others, and he generously shared his findings with the man actually credited with making the largest breakthrough, Jean-François Champollion.

Champollion's cartouches:

the cartouche of Ptolemy (top left) and his royal title (bottom left) both found on the Rosetta Stone;

and the cartouches of Ptolemy (center) and Cleopatra (right) found on the Philae Obelisk

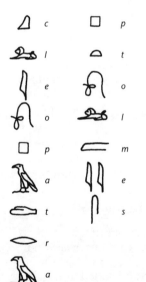

△ c	□ p
🦎 l	◠ t
∤ e	𓂋 o
𓂋 o	🦎 l
□ p	⌒ m
🦅 a	∥∥ e
◠ t	∩ s
⌣ r	
🦅 a	

Phonetic representations of Cleopatra and Ptolemy found on the Philae Obelisk

Champollion's early education in Greek and Latin came to his aid when he began detailed analysis of the hieroglyphics, and he was soon internationally recognized for his work. When Champollion received a letter from Thomas Young, he could not believe that the hieroglyphs could be phonetic. In 1822, however, after a detailed count of the images, he realized that the glyphs outnumbered the Greek words by a margin of three to one. This discovery finally disproved the lingering belief that hieroglyphs represented whole words. Champollion was then able to write the Greek names in hieroglyphics and give phonetic values to the symbols.

A second generous act further aided Champollion. In 1815 William J. Bankes had found and excavated an obelisk. He had taken it to Britain and sent a copy of its inscription to Champollion in early 1822. The Philae obelisk was a bilingual with Greek text on the base and hieroglyphs on the column. After careful analysis, Champollion discovered that one of the two royal cartouches on the obelisk was that of Ptolemy (the other was Cleopatra's). He checked it against Ptolemy's cartouche from the Rosetta Stone and found a near perfect match. From this point onward, he was able to do more comparative translations with the Greek and their corresponding demotic, hieratic and hieroglyphic counterparts.

Champollion's understanding of Coptic, an Egyptian literary speech extant since the fourth century C.E. and still used in Coptic church rituals, was essential to his research. Thanks to

Cartouche of Ptolemy
found on the Rosetta Stone

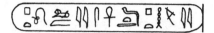

the efforts of foresighted archivists in the 17th and 18th centuries, previous Coptic manuscripts had been saved. Later study showed that Coptic was built upon a modified Greek alphabet and included signs from Egypt's demotic script. Coptic therefore provided a bilingual for Egyptian hieroglyphs. Champollion translated Greek terms into Coptic and tried to match them with the glyphs' images. Coptic provided a key guide to pronunciation because the ancient Egyptian scripts only spelled out consonant values.

In September 1822 Champollion wrote a letter to the Paris Academy which assured his fame. Titled "Lettre à M. Dacier relative à l'alphabet des hiéroglyphes phonétiques," it revealed a substantial increase in the number of phonetic hieroglyphs developed by Thomas Young. It also contained accurate decipherments of the names of the majority of Egypt's Roman emperors.

Although Champollion continued to add to his list of findings, and wrote a grammar of hieroglyphics, he apparently never formally recognized Young's important contribution.

HITTITE CUNEIFORM AND HIEROGLYPHS

Recorded in the Bible and in the records of other ancient civilizations, the Hittites flourished in the region of Asia Minor and what is now present-day Syria, from around 2000 to 700 B.C.E. The Hittites also borrowed from and added to the Sumerians' script. They also had a picture script similar to the Egyptian hieroglyphs.

Specifics about the Hittites' uses of cuneiform only became known after a discovery in 1906. An archaeological project around the city of Boghazköy, which had been the Hittite capital Hattusas, found some 10,000 cuneiform tablets which contained a large amount of script written in an unknown Hittite style.

Decryption

The Hittites' cuneiform was solved largely due to the efforts of one man. When Czech Assyriologist Bedřich Hrozný found a logogram for bread, *nu*, he began a translation of the text *Nu-an-ezzateni, wadar-ma ekuteni*. He then realized that *wadar* was very similar to the German *wassar*, and English water and that *ezzateni* resembled the German *essen* and English *eat*. Hrozný had discovered Hittite's Indo-European base, which meant that it was linked with the family of languages spoken in Europe, southwestern Asia and India. Other scholars took up the cause and built upon Hrozný's findings. The decryption of Hittite cuneiform was concluded by 1933.

Epigraphers are still working with the Hittites' form of hieroglyphics. Experts believe that it was primarily applied for display, as it is generally found on seals and rock inscriptions. One major clue is the Tarkendemos seal which was found in the late 1800s. The circular silver object contains two scripts, cuneiform and hieroglyphic. From Hrozný's and others' earlier Hittite cuneiform solutions, comparisons can be made in the positions and forms of the hieroglyphic symbols. Decryption of the hieroglyphs is still proceeding.

LINEAR B

The earliest scripts of the Aegean Sea region are believed to have been influenced by the Egyptian hieroglyphs. The first symbols have been traced to Crete from about 1900 B.C.E. From this undeciphered form grew an unsolved script that has been designated Linear A. Linear A was a predecessor to Linear B, whose decipherment is one of the great stories of archeological history.

To tell that story, it is first necessary to take a side trip to the island of Cyprus, where one of Linear A's primary descendents, a Cypriot script, was active from the seventh to the second centuries B.C.E. In 1871, epigrapher George Smith partially solved this script. He studied its 55 characters and from their arrangements in the script and relationships with each other, he realized that the symbols stood for syllables. He then analyzed a bilingual inscription that contained Cypriot and known characters of Phoenician. Smith saw two points where the Phoenician word for "king" was located and found the Cypriot words in those same positions in the Cypriot text. The Cypriot signs had variations that Smith attributed to influences of declension. He correctly guessed that one was the nominative case (designating the subject of a verb) for "king," and the other was the genitive case (a relational case chiefly showing possession) for "of the king."

Smith then studied languages wherein the next-to-last syllables of "king" had a nominative to genitive change. He found such an example in the Greek *basileús* (nominative) and *basiléos* (genitive). Using this pattern and aided by the comparison of traceable proper names in Greek, he began piecing together the phonetics of 18 Cypriot characters on some medallions. After his research on this topic stalled, Smith was lured away to join a search for some Babylonian tablets, but his works would provide an important clue to solving Linear B.

In 1900 an archaeologist named Arthur Evans located the famous ancient city of Knossos on Crete and found the palace of King Minos along with numerous clay tablets. The tablets were inscribed with unknown signs and Evans called them Minoan (after Minos) for the language he believed they represented. The script was designated Linear Script of Class B.

Between 1901 and 1904 Evans discovered over 3,000 clay tablets. These appeared to be inventory lists as they listed numbers of people or commodities. Upon close perusal, Evans discerned logograms and other pictograms that seemed to stand for male and female animals.

Evans then compared the recognized phonetic values of Smith's partially deciphered Cypriot script with the unknown sounds of Linear B. The most similar symbols and sounds are shown below (1):

The Cypriot
sound clues

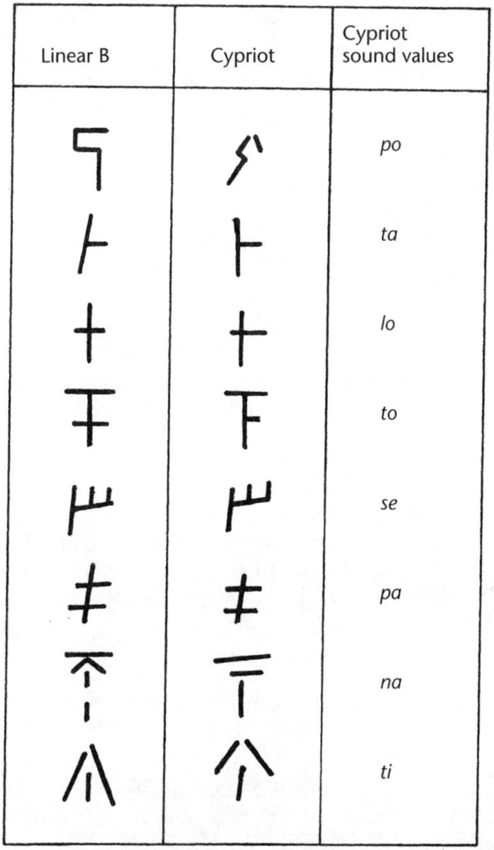

Linear B	Cypriot	Cypriot sound values
⅂	ƒ'	po
⊢	⊢	ta
+	+	lo
⊤	Ⅎ	to
⊔	⊔	se
≢	≢	pa
⊼	⊤	na
∧	∱	ti

He eventually noticed a link between certain signs and the logograms of horses' heads. Two of the horses were without manes and he presumed that they were young horses. Based on the sound values shown above, the signs were pronounced "pōlo." This was the Greek word *pōlos* meaning "foal." Evans's guess had been correct. Amazingly, he could not accept his own findings. He had formed so many preconceived notions about Crete's people and culture as separate entities from Greece that he could not believe that the Minoans spoke and wrote a form of Greek.

Due in part to this belief, Evans's dream of decipherment was not fulfilled before his passing. Unfortunately, he also hindered research by withholding his tablets from other scholars for many years.

Decryption

In the 1940s Dr. Alice J. Kober took up the challenge of deciphering Linear B. She published her first article as an assistant professor in the classics department at New York City's Brooklyn College. In "The 'Adze' Tablets from Knossos" (*American Journal of Archaeology* January–March, 1944), she developed her first ideas about words that contained symbols that looked like an adze.

In the same journal during the spring of 1945 she demonstrated evidence of inflection in the final signs of certain words. This was important because these changes in words and forms, especially in endings, signified changes in grammatical relationships like tense, case, gender, number or person. If the script was indeed an inflectional language, this would give the epigrapher clues about similar patterns in other words.

An example of an inflective language with changing word endings is Latin. Following is an example of its cases:

DECLENSION OF LATIN WORD, *AQUA* (WATER)

nominative	aqua	aquae
genitive	aquae	aquárum
dative	aquae	aquás
accusative	aquam	aquás
ablative	aquá	aquás

In 1946, Kober started identifying some of the signs that made up the inflections.

She located a symbol that often ended a word and referred to it as "7," designating it the ending for Case I. She then looked in other listings of words whose stem contained the same basic symbol groups. She found a second ending which looked like a number 5. She called it "40" and made it her Case II ending.

	A.	B.
Case I	ᴪᵞᏁᙚ	᠖ᛞᚼᙚ
Case II	ᴪᵞᏁ5	᠖ᛞᚼ5
Case III	ᴪᵞꓕ	᠖ᛞᛨ

Kober's inflection pattern

Kober developed models to indicate the stem-signs of some nouns that showed declension aspects. The letters *P* and *Q* will stand for the symbols of one noun stem, and the letters *R* and *S* will stand for the symbols of the other:

CASE I   PQ 7   RS 7

CASE II PQ 40   RS 40

CASE III PQ   RS

Kober argued that the Linear B signs represented pure vowels or consonant-vowel groups. Syllables like *ba* (consonant-vowel) were permitted, syllables like *ab* (vowel-consonant) or *abd* and single consonants like *g* were excluded. She then suggested that the stems of the example nouns ended in consonants. The stem was followed by an inflection which began with a vowel. A single sign, called a joining or bridging symbol, joined the stem's consonant and the inflection's vowel. There were also expected variations within the joining signs. She labeled joining sign variations for the PQ noun "2" and "59," and variations for the RS noun 36 and 20.

CASE I   PQ 2 7   RS 36 7

CASE II PQ 2 40   RS 36 40

CASE III PQ 59   RS 20

To illustrate her proposals with a real noun, Kober used the noun *sadanu* from Akkadia, an ancient country north of Babylon. The stem was *sad-* and the case endings were *-anu*, *-ani* and *-u*. She blended this with her example of the PQ noun to illustrate her proposal.

CASE I	PQ	2	7
	sa	da	nu

CASE II	PQ	2	40
	sa	da	ni

CASE III	PQ	59	
	sa	du	

Kober demonstrated that the signs identified as 2 and 59 both showed the fixed consonant of the stem (*d*). The signs were made up of a consonant-vowel pair (*d* and *a*) and one part of this pair linked the stem and became part of it, the *d* of the stem *sad*. While she could not yet put names and pronunciations to the actual Linear B words in 1946, Kober did demonstrate the consonant-sharing characteristics of the language.

Kober then turned her attention to the nouns in her Case I groups PQ and RS. These had the same case endings and provided an identical vowel to their linking signs, 2 for PQ and 36 for RS. The linking/bridge signs differed because PQ and RS had varied stems which gave different consonants to the bridge signs, but the vowel stayed the same in both.

This can be illustrated with a created word, *gihanu*, using the same declension as Kober's real word *sadanu* and the same endings *-anu*, *-ani* and *-u*.

CASE I	PQ	2	7	RS	36	7
	sa	da	nu	gi	ha	nu

CASE II	PQ	2	40	RS	36	40
	sa	da	ni	gi	ha	ni

CASE III	PQ	59		RS	20	
	sa	du		gi	hu	

While the stem consonants of signs 2 and 36 vary (*d* and *h*), the case-ending vowel (*a*) is the same. This is also true for signs 59 and 20 with the vowel *u*. Kober had found that some of the Linear B symbols had vowels in common, others had like consonants and some signs had both in common. She represented these in the pattern shown below:

Consonant	Vowel 1	Vowel 2
1.		
2.		
3.		
4.		
5.		

Kober's syllabic grid

This pattern aided solution of as yet untranslated symbols. The signs had similarities that could be compared by aligning all similar vowels in a column and all similar consonants in a row. When the vowel and consonant characteristics were shared, the letters could be tried in sections of different words with these similarities to see if they made sense.

While Kober had not yet assigned phonetic values to the symbols, she had discovered gender differences in some of the signs containing lists of humans and animals. For example, to designate a female, extra marks were added to the respective male sign.

By the latter 1940s, some historians believed that she was well on her way to a complete solution. Although her life was tragically ended by cancer in 1950, her system is the basis upon which the final solution of Linear B became possible.

The full decryption was made by Michael Ventris a few years later. Ventris showed an aptitude for languages from childhood and was fluent in French, German and Greek as well as Polish and Swedish. He had first seen Linear B at the age of 14 in 1936 when he went to see an exhibit set up by Sir Arthur Evans. His curiosity stayed with him throughout the upheavals of World War II.

In 1950 he consulted scholars who were analyzing Linear B and received help from most of them. One influential contribution was made by Emmett L. Bennett, Jr. A cryptanalyst during World War II, Bennett's doctoral thesis had studied the Minoan Linear script from Pylos. In 1950 he helped clarify the numbering and measuring systems of Linear B and Linear A and made important arrangements of the signs and symbol variations. He also gave an order to ease confusion when referring to the signs in conversations and in print.

Ventris began circulating what he called "Work Notes" to other scholars in January 1951. Included in them were Alice Kober's primary findings. He called her analytical principle, with the connected consonants and vowels, a "grid," using a variation with the vowels at the top of the columns and the consonants at the sides in his own work. He continued to distribute 20 of his Work Notes until June 1952.

Ventris practiced cryptanalytic techniques when he prepared statistical analyses of sign position in different words, whether initial (beginning), medial (middle, near middle), or final. From such listings he was often able to locate letters that were nearly exclusive for the beginning positions of words. He also tabulated the high frequency of a trio of symbols at the initial positions of words. These were symbols that appeared to be a double axe, a throne and a capital A with an extra bar. He believed that these were pure vowels since pure vowels most frequently begin words in syllabic languages.

Using Arthur Evans's "Cypriot clue," Ventris found some consonant and vowel similarities in some Linear B symbols that looked like the Cypriot forms. He also combined some ideas from Kober's work and the decryption of other scripts. He guessed that particular signs, which were repeated after the symbols hypothetically representing personal names, represented towns or occupations.

From comparisons of the Linear B and Cypriot symbols, he found similarities in the Cypriot signs that were equivalent to *na* and *ti*. Using his expanded Kober grid, he found specific vowel and consonant locations for *i* and *n*. He then compared this finding to a symbol in a four-sign place-name. The word combined the consonant *n* and the vowel *i* to form *ni*. Ventris also solved an initial vowel (*a*) for this place-name. With "a_ni_" and a knowledge of Cretan history, the town of Amnisos came to mind. The symbols did indeed stand for Amnisos, the port for Knossos. Soon Ventris was able to extract the names of other Cretan towns. Placing these findings in the grid, he determined the phonetic values of other symbols through the linked vowel and consonant listings.

Although Ventris was reading words that were archaic Greek, he, like Evans, had not been able to believe that the Minoans spoke Greek rather than an unsolved tongue such as that of the Etruscans (which he had earlier believed was the basis of Linear B). In July 1952, he began a solution partnership with John Chadwick, a Cambridge University Greek scholar. In a 1953 article, they cautiously concluded that there was clear evidence of Greek in Linear B.

Ventris's and Chadwick's work was validated when U.S. archaeologist Carl Blegen compared Ventris's findings with a three-lined clay tablet that he had found at Pylos on the Greek mainland in 1952. In 1953 Blegen reported that the discoveries of Ventris and his predecessors had enabled him to read what has come to be known as the "tripod tablet," from its description of three-legged pots. The language of Linear B was a thousand years older than the words of Plato, but it was Greek nonetheless.

At the height of his success, Ventris's life was abruptly ended in 1956 by a car accident in England. Yet his legacy lives in his decryption of Linear B.

MAYAN GLYPHS

The ancient Mayans lived in southeastern Mexico and Central America, and their descendants still live in the Yucatán Peninsula of Mexico, Belize, Guatemala and Honduras. About 1500 B.C.E. proto-Mayan speakers had begun to settle the lowlands of Central America, but it was not until about 250 B.C.E. that villages became larger cities with scribes and a standardized writing system. The Mayans' script, a form of hieroglyphics developed around 700 B.C.E. is believed to have come from a people to their west who the Mayans believed were their mythical ancestors. Other historians credit the Mayans' language and writing development to a group of people known as the Olmecs, who inhabited Mexico before the Mayans.

The manuscripts and buildings which contained examples of Mayan glyphs were considered heathen and destroyed when Spain conquered Mexico in the 16th century. Only four bark-paper manuscripts (codices) survived the vicious onslaught, although many more carvings on monuments and temples remained.

In the early 1860s a remarkable book was discovered in Madrid. Charles-Etienne Brasseur de Bourbourg translated it into French and published it in 1864. Originally titled *Relacion de las cosas de Yucatán* (An account of the things of Yucatán), it had been written in 1566 by Fray Diego de

Landa. Landa was a Catholic priest who by all accurate accounts had tortured the Maya, but who had been fascinated by their history. In the late 1560s he had been recalled to Spain to face accusations about his cruelty and he wrote this text to answer his accusers. He was cleared and instated as bishop to the Maya in 1572 until his passing in 1579.

In his text Landa included his version of the Maya's alphabet and calendar system. His alphabet was flawed as his comprehension of the Mayan language was rudimentary and he did not understand the Maya's syllabic writing. Still, some consider his rudimentary alphabet the "Rosetta Stone" of Mayan glyphs, since more accurate interpretations can be traced to studies of his work.

Landa's translator Brasseur de Bourbourg also aided Maya research. He served as an abbot in Guatemala and in 1869 published an important Maya codex which he had also discovered in Madrid. The Madrid Codex was found in two parts, known as the Troano and the Cortesianus, and it is believed to date from around 1400 B.C.E. Its 56 pages were illustrated on both sides of a doubled fig leaf and depicted activities such as pottery and weaving and hunting. Its figures are not as well formed as those of the Paris and Dresden Codices, which were also named for the cities where they were found after being brought to Europe.

The Paris Codex was found in a little-used corner of the Bibliotéque Nationale in 1859. It has been dated between C.E. 800 and 1500 and deals primarily with Maya rituals and ceremonies. Made of paper from tree bark, its 11 separate leaves fold out in what is known as a screenfold, or accordion fold, to make 22 pages containing columns of images.

Also screenfolded is the Dresden Codex was probably written between C.E. 1100 and 1500, before the conquest of Mexico by the Spanish. While Hernando Cortés may have taken this codex back to Europe with his other spoils, it effectively disappeared until 1739 when it was sold from a private collection in Vienna to the Royal Library at the Court of Saxony in Dresden.

The Dresden Codex's cover was made from jaguar skin and unfolded to a length of nearly 12 feet. Its 74 pages were made of bark paper coated with gesso, a type of plaster of Paris, which formed a good surface for paint. Its images of Maya gods, symbols, animals and astronomical events were especially important to advances in decryption of the Maya's script.

Decryption

Solutions to the codices' glyphs came together from different, often conflicting sources over many years. These included Fray Diego de Landa's flawed alphabet and calendar interpretations, in which he linked the names of days and months with hieroglyphic images. Another significant step was made by Turkish-born Constantine Samuel Rafinesque-Smaltz. During a much traveled career, he published his own periodical, the *Atlantic Journal and Friend of Knowledge,* in the 1830s. He was the first to understand the true values of the dots and bars that represented the Maya number system.

Rafinesque-Smaltz also proposed connections between the codex manuscripts, the monument inscriptions and the language of the Maya's descendents. Although in the 1800s no systemized analysis had been started, this proved important in the 20th century when studies of 30 Maya-related languages yielded terms that fit some ancient inscriptions.

Another significant contribution was made in the 1880s by Dresden librarian and linguist Ernst Förstemann, who studied Landa's calendar and Rafinesque-Smaltz's numerical system and discovered that the fantastic symbols of the Maya's astrology had actual astronomical meanings. From the Dresden Codex, Förstemann found that the Maya used a vigesimal (base-20) number system, and that the images of the Dresden Codex involved almanacs that covered 260 day periods. The Dresden also contained charts for calculating the cycles of the planet Venus, tables concerning lunar eclipses (an inauspicious event to the Maya) and a calendar system covering many centuries, all of which had been used through the Maya's empire. Forstemann's solutions provided insights into previously unexpected aspects of Mayan lives, history and language.

Researchers were still basing their study on Landa's flawed alphabet into the latter 1800s, and their research was colored by preconceptions about the Mayan script. Sir Eric Thompson, a leading Mayanist, asserted for years that the writing was neither syllabic nor alphabetic; rather, it was used in cult rituals.

In 1952 new insight came from an unexpected source, from behind the Iron Curtain. A Russian language scholar named Yuri V. Knorosov argued that the glyphs could be read phonetically. Although he had not yet personally visited the Central American building sites, he presumed that Landa's "alphabet" had a syllabic nature because the phonetic values of the Maya's signs equated at least in part with the spelling of Spanish names.

Knorosov followed a proposal made 76 years earlier by a researcher named Leon de Rosny. In 1876 Rosny had used the Landa "alphabet" when studying the symbol for "turkey."

cu   *tz(u)*
Turkey glyph

Rosny had read the first part of the glyph as *cu* after Landa's similar alphabet drawing. He then guessed that the full glyph was read as *cutz(u)* because the regional Mayan word for turkey was *cutz*. Knorosov then compared the second sign of the turkey glyph with the first one in the symbol for dog. It was identical.

As the Mayan word for dog is *tzul*, the second dog sign was given the phonetic value *l* like Landa's symbol for the letter "el."

tzu   *l*
Dog glyph

Using this process, Knorosov deciphered similar phonetic sounds for other symbols. However, as he was published only in Russian for a long time, his work went unheeded by many contemporaries and received mixed reviews when it was finally published in English. Still, Knorosov is credited for initiating a new direction in research into Mayan glyphs.

Over the succeeding decades, the glyphs' many combinations of main signs and affixed symbols have made exact meanings difficult to pinpoint as terms had both ideogram and syllable equivalents. As words can be written in many ways, decryption of the glyphs naturally is slow. Mayan art is used to try to understand the pictographic meanings. If an image's form is recognized, epigraphers try to match it with a known Mayan term. From this link, they try to spell out a full phonetic reading.

Mayan scholars now generally agree that some 85 percent of the images can be understood, although the Mayan pronunciation is not exact.

Unsolved Scripts

With a large body of knowledge accumulated from past investigations and successes, it may seem surprising that any group of carved symbols or written inscriptions have remained unsolved. Yet a number of scripts still hold their secrets. Some of the most fascinating are Proto-Elamite, Linear A and Etruscan, and the scripts found on the Indus seals, the Phaistos Disk and the Voynich Manuscript.

PROTO-ELAMITE

Elamite was the semipictorial script of the language of Elam, which was once spoken in what is present-day western Iran. It is the precursor of the Elam form inscribed in Mesopotanian cuneiform. Proto-Elamite symbols have been found on clay tablets across both Iran and Iraq. The earliest known

A bilingual text: linear Elamite (left) and Akkadian cuneiform (right)

record of Proto-Elamite dates from about 3000 B.C.E. As of this time, both the script and language of Elam remain a mystery to epigraphers.

INDUS SEALS

In 1921, after excavations at Harappa in northern India uncovered the ruins of a brick city, archeologists realized that the Indus River Valley of northwest India and Pakistan had once held a thriving civilization between 2800 and 1600 B.C.E. More digs revealed that this city, Harappa, and another major city, Mohenjodaro, had been the equals of the better-known contemporaneous cities of Mesopotamia and Egypt.

Developed in Harappa, the script of the Indus Valley was used to record trade, remember important events and celebrate leaders.

The script is linear and primarily consists of stylized picture signs. As examples are most often seen as brief inscriptions on seal stones and less frequently on pottery, painted amulets or bronze tools, some historians believe that the script was used to mark property or record transactions. Others suggest certain god and goddess images were the predecessors of Hindu deities.

An Indus seal

Indus fish signs

The glyphs that hold much interest for epigraphers are the fish signs, found frequently in various forms.

These signs have also appeared with star or astral symbols in arrangements that may have represented constellations. As the Tamil word for fish also means "star," some scholars have traced these fish signs to the Tamil language, a branch of the Dravidian languages that existed before Sanskrit. The Dravidian speech is non-Indo-European and is still spoken by the peoples of southern India and Ceylon.

The leading expert on the Indus script, Asko Parpola, has also discussed the fish-star connection using the examples of a "roofed fish" and a black star. Both signs were interpreted as representations of the planet Saturn. Parpola linked these images to later signs which depicted Saturn as a godlike figure riding a turtle or another aquatic creature with a covering, a "fish" covered by a "roof." Although the script itself had become extinct, many cultural ideas had survived and been passed on to other peoples. The analysis of such interconnected symbols are one of the few tests that can be used in decryption efforts.

Current findings suggest that the Indus script is similar to other pictographic writings made before 2500 B.C.E. Researchers also believe that the people of the Indus Valley were of the Dravidian family and that their religion was related to those of ancient western Asia and later India. Experts believe that partial decryption of the Indus script is still possible, as many traces of Indus Valley culture and religion survive in modern-day India and because languages like Tamil give valuable information about Dravidian speech.

A Cretan seal hieroglyph

LINEAR A

When Arthur Evans discovered Linear B, he also found two other unintelligible scripts. The older form, a hieroglyphic script, developed around 1900 B.C.E. and appeared primarily on seals. The other, Linear A, was mainly inscribed on clay tablets.

The oldest example of Linear A has been traced to 1800 B.C.E. Linear A was used in Crete and nearby islands between 1850 and 1450 B.C.E. The precursor of Linear B, Linear A is believed to have been the script of Crete's classic Minoan civilization centered at Knossos.

Debates arose as to the origin of the Minoan people. Some scholars believed that they had migrated from Greece or some other part of the Aegean Sea. Others thought the Minoans may have been descendants of the Hittites who once dominated Asia Minor.

Between 1994 and 1995, at Miletus, Turkey, archeologist Wolf-Dietrich Niemeier found markings on clay pottery that were subsequently identified as Linear A. This discovery has sparked renewed interest as to whether the Minoans began in Asia Minor and had a maritime trade empire stretching to Crete, or whether the Cretans colonized parts of Asia Minor. Another possibility is simply that the pottery had been brought to Asia Minor by a sea trader.

A Linear A
fragment

Linear A remains an enigma with a partially known script, no known language and tantalizing evidence of its existence over a wide region.

THE PHAISTOS DISK

In 1908 an Italian archaeologist named Luigi Pernier discovered a chalk-encrusted terra-cotta disk, and soon this seemingly insignificant artifact provided epigraphers with yet another mystery.

Named for the Minoan palace in southern Crete where it was discovered, the Phaistos Disk sparked immediate interest. Excavations in the same ruins revealed fragments of a Linear A tablet and many believed that the newly found disk was the long-sought key to Linear A. Ninety years later, the disk's script is still indecipherable.

The Phaistos Disk

Dated about 1700 B.C.E., the Phaistos disk is made of clay, measuring 6½ inches across and ½ inch thick. Its 45 symbols occur around 241 times on both sides of the disk and are arranged in 61 lines placed from the outside spiraling clockwise inwards. The figures are stamped rather than carved into its surface, and portray human forms, plants, animals and weapons.

Since creating a series of hand punches or stamps to impress the images on the wet clay is considerably more time consuming than merely inscribing the glyphs, it seems likely that the stamps were reusable. If the disks were mass produced, however, scholars question why no other copies or similar objects have been found. Adding to the mystery is the fact that the symbols do not resemble the early Minoan hieroglyphs, Linear A or the known Linear B scripts.

As of 1998, two solution seekers have made progress in decrypting the disk. Twin brothers, Reverend Kevin Massey-Gillespie and Keith A. J. Massey believe that a different ancient script can provide clues to the Phaistos disk. A large civilization had once flourished at a site called Byblos in modern day Lebanon, and cultural and historical ties indicated numerous contacts between Byblos and Cretan cities. A Semitic-based writing known as Proto-Byblic script existed contemporaneously with the disk and contained similar figures like human forms and weapons.

The Masseys are not the first to see similarities between the Phaistos and Byblos scripts. Maurice Dunand, in the 1940s, and Victor Kenna in 1970 also compared aspects of the symbols, and in 1946 Edouard Dhorme published the first listing of consonantal values for the Proto-Byblic script.

In epigraphy, as in cryptanalysis, possible solution angles often develop from unconventional approaches. As the figures were stamped from right to left, many scholars insisted that the glyphs should be read the same way. The Masseys argue that the direction of stamping did not necessarily influence reading direction, nor does the Egyptian hieroglyphs' model (figures' facing direction indicating the reader's starting place) necessarily apply to the disk. Basing their analysis on a left-to-right reading, they believe an intelligible Greek language-based text has begun to appear.

The Masseys are carrying out a detailed, systematic study, applying consonantal values to every sign in the disk. Should their progress continue, the Phaistos disk may finally be solved in the near future.

ETRUSCAN

The Etruscans lived in a central Italy bordered by the Arno and Tiber Rivers between 700 and 100 B.C.E. They were eventually absorbed by the growing Roman empire and their language and culture ceased to be distinctive. Although the Etruscans were once believed to have originated from Asia Minor or points to the north of Italy, historians now agree that their civilization might have been indigenous to Italy.

Unlike many earlier and contemporaneous people, the Etruscans never built an empire. Rather, their influence spread through peaceful colonization and trade. The cultural bridge between early Greek settlements in southern Italy and the people to the west and north, the Etruscans' historical presence has been established through archaeology, Greek and Roman writings and the remnants of their script.

The alphabet shown is a model only, as some of its letters are not in Etruscan. The Etruscans had no voiced stops (*b*, *d* or *g*), but scholars include these signs in the alphabet to reproduce the Greek model on which the Etruscan examples were based. This is a standard practice as the Etruscan language was written with Greek alphabet letters.

Very little is known about the Etruscan language, although researchers have compiled an alphabet, and archeologists

A model of the Etruscan alphabet

Transcriptions and phonetic values

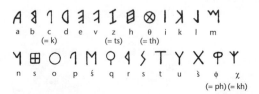

have found some 13,000 inscriptions, including engravings on mirrors and gems, across central Italy. While epigraphers can read the phonetics of the script, they do not know the meanings of the words.

The Etruscan language is unusual as it has no relationship with any of the languages of Europe. Neither does it have any connection to other currently existing languages that arose from the 4000 B.C.E. migration from central Europe that eventually extended from India to Ireland.

However, researchers have not given up hope of a solution. In 1885 a script was found whose symbols and arrangement appear to be connected to Etruscan. Found at Kaminia on the island of Lemnos in the northern Aegean, the script was engraved on a stele, an upright stone slab or pillar engraved with an inscription or design and used as a monument.

Dated around the sixth century B.C.E., the stele contained 98 symbols that made up 33 words, that were once thought to be Etruscan. However, Italian archaeologists working at Lemnos in 1928 discovered parts of similar inscriptions on pieces of pottery; this find proved that the stele's language was actively used in Lemnos, indicating that the stele was probably not transported from another Mediterranean site.

In 1964, bilingual plaques were discovered at Pyrgi, a seaport some 25 miles west of Rome. Dated around 500 B.C.E., one section of the text is written in a recognized Phoenician language and the other in an Etruscan/Greek script. The Pyrgi inscriptions have been translated as an Etruscan ruler's thankful offering to a goddess.

However, such discoveries often only provide shards of clues. As of 1995 the Pyrgi plaques had only added one new word to the Etruscan solution, *ci*, the number three.

The Lemnos stele

RONGORONGO

Rongorongo is a hieroglyphic script inscribed on wooden tablets developed around 400 C.E. on Easter Island (Rapanui), the South Pacific, famous for its huge carved wooden statues. The script, named for an Easter Island word meaning

Rongorongo script

"chants" or "recitations," has been found inscribed on wooden tablets, only 21 of the tablets have survived for analysis.

Rongorongo consists of tiny, amazingly regularly formed glyphs about 1 centimeter high and 10 to millimeters wide. The primary signs are stylized images of creatures or objects. The symbols were carved along the lines of shallow grooves that were cut lengthwise with a tool made of shell, obsidian flakes, sharks' teeth or birds' bones.

The vast majority of the glyphs are anthropomorphic. The figures face forward or sideways with their limbs in varied positions, holding objects ranging from staffs to barbed string. Large popping eyes in some images may represent coils of hair in others. Zoomorphic symbols—fish, lizards and especially birds—were often used.

The direction of the Rongorongo writing and its reading is boustrophedonic (left to right, right to left) with a 180 degree turn. Beginning at the lower left-hand corner, a reader would proceed from left to right. At the end of the line, the reader would turn the tablet around before beginning the next line. This style of inscription meant that every other line appeared to be carved back-to-front and upside down.

The tablets were first seen by a European in 1868 when newly converted islanders sent tokens of respect to Tepano Jaussen, Tahiti's Catholic bishop. One gift was a long twine of human hair wrapped around an ancient piece of wood. When he examined it more carefully, Jaussen realized that the wood was covered with symbols.

When Jaussen tried to investigate the tablets on Easter Island, he discovered that a large number of the wooden objects had either been burned, being thought of as pagan relics, or had been hidden to prevent destruction. After much disappointment, Jaussen finally found an Easter Island native named Metoro who was working in Tahiti. Metoro managed to chant or read some of the few remaining tablets and Jaussen's arduous efforts to transcribe the words led to about 300 pages of notes. From this material, a compilation of a few hundred hieroglyphs called the Jaussen list was published, and became the subject of many failed decipherment attempts.

In 1958 German epigrapher Thomas Barthel gathered the known Rongorongo information in his text titled *Grundlagen zur Entzifferung du Osterinselschrift* (Basis for the decipherment of the Easter Island script). He did succeed in identifying two and a half lines on one tablet as symbols for a lunar calendar, but his claims to a broader decryption collapsed when it was learned that he was merely repeating Jaussen's record of Metoro's chants.

Claims to solutions have continued into the 1990s, but many of these proved to be hoaxes. Despite the pictographic nature of the symbols, Rongorongo is clearly a true script. There is an order to the Rongorongo carvings that follow rules of arrangement. Certain symbols repeat and connect, but not at random. Other signs are never associated or joined. This consistency shows intention and planning rather than haphazard carving. Without new material or a proven and solvable bilingual for comparison, however, the chances of a solution for Rongorongo are not promising.

VOYNICH MANUSCRIPT

Named for U.S. rare book dealer Wilfred M. Voynich, the Voynich Manuscript has puzzled decryptors for centuries. Many analysts believe it was written sometime between the 13th century C.E. and 1608, when it was bought by Jacobus de Tepenecz, the director of the botanical gardens of Holy Roman Emperor Rudolph II. The text then was transferred from Tepenecz to an unidentified person and then to Johannes Marci, rector of the University of Prague. Marci in turn sent it to one of his former teachers, Athanasius Kircher, who had published a cryptology text, but Kircher was unable to solve the text. After Kircher, the manuscript was filed and

A page from the
Voynich Manuscript

forgotten in a personal or church library until it was acquired
by the Jesuit College of Mondragone in Frascati, Italy. In 1912
Voynich bought the text, intending to solve the mysterious
script.

Voynich made photostats of the manuscript and distributed
them to persons with similar interests. In spite of professional
and amateur attempts, the book remained unreadable and
even its language base is a mystery. In 1962 the manuscript
made the headlines when New York City book antiquarian
Hans Kraus bought it for a reported $24,500. He tried to resell
the manuscript for $159,250 but found no takers. In 1969 he

donated the manuscript to Yale University's Beinecke Rare Book and Manuscript Library.

The book measures about 6 by 9 inches, and according to various sources, contains between 204 and 235 pages with either 28 or 31 pages lost. A number of unusual, fanciful illustrations accompany its complex text, and seem to indicate the content of each chapter. Based on these drawings, scholars assume the writing is about astronomy (zodiac signs), biology (images of human figures), cosmology (celestial spheres and stars), herbs (fantasy flora), pharmaceutics (vases and parts of plants) and recipes (many brief paragraphs).

The script does not appear to be cryptic (though that is not out of the question), and seems to be late-medieval writing. Many of the same letters repeat, as do groups of letters and words, with occasional variations in endings. Recent analysis has indicated that the Voynich Manuscript contains two distinct styles.

The dual aspect was first identified by Captain Prescott Currier, a U.S. military cryptology expert who was an associate of William Friedman. The variations have been dubbed "A" and "B," and indicate that the text was written by at least two different people, or by a single individual at different times.

Analysts have presumed that the spaces in the text were intended to be separators, and so consider the character clusters "words." These have been studied for frequency, and the number of appearances tallied. Curiously, words either appear in both A and B or they appear in B only. Very few words appear solely in A. Such frequency counts have as yet been unable to establish whether the manuscript was encoded or enciphered.

As with the elegant Etruscan plaques, the spiraling Phaistos disk and the Indus seals, epigraphers are unsure what purpose the fanciful Voynich Manuscript served. For dedicated analysts, the quest for answers leads onward.

Cryptic Scripts

RUNES

Two scripts, runes and ogham, have been called cryptic because their origins are uncertain and because each was used as a concealment system. The word *rune* has origins in the Old Saxon, *rūna*, "magician," Old Irish *rūn*, "secret," and Middle High German *rūne*, meaning "secret" or "whispering." Found inscribed on clasps, weapons, memorials and rings, the accounts of runes' uses indicate connections with cult and secretive styles of writing. Runes have also been linked with cult leaders who used the signs as symbols of special knowledge.

Runic inscriptions date to the third century B.C.E. and are found in locations as far apart as England and Italy, as well as the Scandinavian countries and areas around the Black Sea. Some scholars believe that runes were created in eastern Europe by Goths, the Germanic people who conquered areas from the Danube River to Italy in the third through the fifth centuries C.E., while others link them to carvings in the Alpine valleys of southern Swiss Alps. Most runic scholars, however, agree that the symbols' forms show a descent from the Etruscan script.

A clearly recognizable characteristic of runes is their angular form with almost no curved lines, and some argue that these forms indicate the first writing implements that created them, namely, wooden sticks or bones. The script itself is

The Runic alphabet

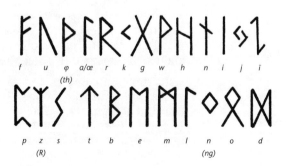

f u φ a/α r k g w h n i j $\ddot{\imath}$
(th)

p z s t b e m l n o d
(R)   (ng)

called "futhark" after the first six alphabet signs (*th* was originally one letter), and an alphabet of 24 signs and sound values was standard. The sounds are approximations as the languages most closely linked with runes do not have equivalents in modern English.

The earliest runic forms appear to be written both left to right and right to left, with no distinction between uppercase and lowercase letters. The runic alphabet developed from an old Germanic alphabet with 24 symbols made up of three groups of eight signs. Letters were grouped as 1–8, 9–16 and 17–24.

This alphabet was used from C.E. 200 to 750. In the fifth and sixth centuries, the Germanic Anglo-Saxons took this script to the British Isles where the total of signs increased from 28 to 33. This growth indicates that the alphabet was expanding to be able to describe life in a new land, incorporate new words and phonetic values.

The increase in symbols seems to imply that the signs were used openly, rather than cryptically, and some historians claim that they were only considered cryptic because very little was known about the early Germanic tongues. Most scholars, however, describe runes as being cryptographically applied.

For concealment purposes, a series of marks were substituted for the runes. They identified the symbols' group (1 to 3) and the numeric order of its place in that group (1 to 8). Two cryptic versions were isruna and hahalruna.

In isruna, a short vertical stroke indicated the group numeral and a longer mark gave its group place number. Thus *f*, group 1, letter 1, was replaced by one short and one longer vertical mark. The rune for *m* was the third letter in the third group and was represented by three short and three long marks.

Hahalruna was based on a single vertical line. Diagonal strokes indicating the group were placed to the left of the vertical and group place was represented by strokes to the right. The letter *h*, group 2, first letter, was cryptically represented by a vertical shaft with two diagonal marks to the left and one diagonal stroke to the right.

The encryptions may have been applied to dispatches during warfare, for trade secrets or to protect the activities of cults. A famous example of runic inscriptions is the Rök stone in Sweden, a 13-foot high granite stone containing more than 700 runic characters as well as cryptographic forms. Runes appear to have faded from active use around C.E. 1200.

OGHAM

Created by the Celts, Ogham has been categorized as Primitive Irish, the oldest form of Irish. As unusual as this script was in its standard forms, some versions of it were also encrypted. Ogham dates from about C.E. 350 to 600 and has been found on stones ranging from Cork and Kerry in Ireland to Wales and the Isle of Man.

The reasons for ogham's development are uncertain, but a number of historians have agreed that it was used for concealment. In Gaelic, the word *ogham* referred to cryptic speech in which letter names could be substituted for the letters themselves.

In his 1937 book *The Secret Languages of Ireland*, R. A. Stewart Macalister proposed that ogham was founded upon phonetic rearrangements of earlier letter pronunciation practices. He argued that certain aspects of alphabets and languages used by diverse groups such as the Greeks, Romans, Druids and Celts all have links to the ogham script and the Old Irish language. Macalister believed that ogham was actually a system of notating hand gestures. The script's sequence of marks were adapted from the positions of the hands (the base line) and fingers (the projecting lines). Some historians argue that ogham was a private signaling code used by the Druids, the often secretive Celtic religious order found in Britain, Ireland and France.

Examples of ogham

Ogham can be arranged vertically or horizontally, but the basic pattern of the writing style is the same. Like runes, the ogham marks were divided into four "families," which were four groups having five letters each.

The letters are indicated by groups of one to five strokes or notches that are arranged along a center stemline. These are considered to be in five configurations: above the line, below it, perpendicular above and below, diagonal above and below. The fifth group has varied designs that have been interpreted either as diphthongs or as signs with consonant-like values.

A 15th-century collection of history, genealogy and folklore known as the *Book of Ballymote* lists a number of methods of enciphering ogham. Following are descriptions of various forms:

OGHAM NAME	DESCRIPTION
head of quarreling	additional lines that intersect the ogham marks
interwoven	additional straight or looping extensions to the ends of ogham marks
well-footed	a sequence of dots at the ends of lines above and below the stem line
host	each letter tripled
point against eye	reversed alphabet
serpent through the heather	an irregular line placed atop and beneath consecutive characters
vexation of a poet's heart	lines as brief strokes projecting just beyond a rectangle
twinned	each sign doubled

`fraudulent`	`signs are substituted by the next symbol in the group`
`sanctuary`	`a mark is placed between every regular pair`
`outburst of rage`	`a mixed substitution`

While ogham faded from active use in the late 600s C.E., ogham, like runes, continued to be considered pagan symbols. Some Catholic missionaries and church leaders actively discredited them and had inscriptions removed or destroyed.

Modern Pictographic Scripts

Pictograms continue to be the basis for several modern scripts such as those of China and Japan. In China, a formal script began around 1300 B.C.E. and has remained active to the present day. The signs have remained similar in form from early scratches on bones, and stone carvings, but more modern forms were naturally affected by writing with brushes on paper, as well as by later mass-produced printing.

Each character can be reduced to as few as nine strokes, but some strokes have variations and signs can consist of as many as 20 strokes. The script is aligned in vertical columns from the top of the right side of a page downwards and the columns are read from the right to left.

Chinese characters are believed to have started as a pictographic form, like cuneiform. These word-signs were clarified or enhanced by determinatives and over time these combinations evolved into groups of syllable signs.

The development of Chinese script differed from that of cuneiform and hieroglyphics as it did not evolve into a phonetic writing system. Instead, the number of syllabic symbols remained constant and more word signs were developed to accommodate new needs.

Early Chinese bone scratches

Chinese pictographs

Some 2,500 signs were once used, but as new vocabulary was developed the number increased to around 50,000 characters. Many of these are actually combinations of signs, which together represent another word.

There are two types of grouped signs, compound and complex. In compound signs two or more signs are combined to form the sign for a third word. For example, the character for *tree* is written twice to make the word *forest*. The characters for the words *woman* and *son* make the sign for the word *love* or *goodness*.

Compound signs are not necessarily aligned side by side. Different meanings are indicated by placement above, behind, or below each other. Sometimes one part may indicate the

The characters for *nü,* "woman," and *tsu,* "son," create the character *hao,* "love," "goodness"

woman

son

love, goodness

meaning of the entire sign, while another part gives the pronunciation.

Complex characters consist of a determinative combined with other symbols. These give more specificity or detail to the initial sign. For example, the word *wei*, "leather belt," combined with a determinative for "cloth" creates the character for "curtain." *Wei* with the determinative for *man* means "greatness."

Chinese characters are grouped in six categories:

1. Pictographs *(xiong xing)*, which include about 600 signs which are the foundations of Chinese script.

2. Representative or symbolic pictures *(zhi shi)*, which stand for abstract terms and are borrowed from words similar in meaning.

3. Symbolic compounds *(hui yi)*, which are created by repeating the same character, or joining two characters to form a new meaning. For example, the character for "child" repeated twice means "twins." The characters for "sun" and "noon" when placed together form the sign for "bright."

Symbolic compound characters

sun   noon   bright

4. Inverted or rotated characters *(zhuan zhu)*—for example, the sign for "child" turned upside becomes "childbirth."

5. Sound indicators *(xie sheng)*, which are often called semantic-phonetic. They consist of two parts, a determinative symbolizing the main concept and a rebuslike character indicating pronunciation. For example, the sign for "shining" is a combined "sun" and "moon" sign to which the phonetic for "fire" is joined.

6. Borrowed forms *(jia jie)*, which are similar to sound indicators. These are often ambiguous characters such as *zu*, which can represent both "foot" or "to suffice."

The shapes of signs are very important in Chinese culture. Since each character has a meaning, Chinese believe that

■ 563

characters often resemble their meanings and have their own personality.

JAPANESE

Scholars have long debated whether a writing existed in Japan before the introduction of Chinese script, but archaeological evidence indicates that Chinese characters were the main influence in Japanese script. In the third century B.C.E. the Japanese came into contact with the more advanced Chinese led by the Han dynasty. Although there is no evidence of direct sea commerce between the two, cultural ideas did mingle through Chinese and Japanese dealings in Korea. Religious and cultural ideas were slowly exchanged by each nation's elite, and the Japanese eventually adopted Chinese writing and papermaking after C.E. 600.

The Japanese refer to the Chinese characters as *kanji*, which is derived from the Mandarin word for Chinese characters, *hanzi*. There are two main ways of pronunciation, known as the *kun* reading and the *on* (or Sino-Japanese) reading. When a Chinese graph (character) is applied to represent a Japanese term with the same or similar meaning as a Chinese term, the Japanese call it kun. The kun form is native to Japan while the on is more closely linked with the Chinese.

The kun and the on variations are exemplified by the words for "ocean" and "sea water." The Japanese word for ocean is *umi*. This character is also the Chinese sign for the Mandarin *hai*. For "sea water," however, the Japanese on word is *kai sui*. This reading is a derivation of a previous Chinese pronunciation for sea water, and is therefore considered an on, or Sino-Japanese, reading.

On readings are required when groups of kanji have been borrowed by the Japanese. Following are some kanji read as kun (when single words) and read as on (when more words are used):

In addition to graphic images, phonetics are also used. In early Japanese writing characters with kun readings began to be used with characters applied for phonetic values. When

	海	面	星
kun reading	umi	omote	hashi
meaning	ocean	face	star

	海水	水面下	水星
on reading	kai sui	sui men ka	sui sei
meaning	sea water	underwater	Mercury

Japanese *kun* and *on* readings

this was first done a thousand years ago, Chinese and Japanese words had more phonetic similarities. The Japanese called this early combining of writings and phonetic graphs *man'yogana*, which literally meant "phonographic script of the Myriad Leaf type." This name was based upon a poetry anthology called the *Manyosahu* (Myriad leaf collection), which combined Chinese graphs representing a symbol with Japanese words of alternating consonants and vowels. A Chinese graph represented a consonant-vowel pair and so longer Japanese terms were represented by more than one Chinese graph.

The numerous graphic and phonetic problems resulting from these combinations took several centuries to clarify. The resulting script is simpler and more efficient than the primarily pictographic script from which it evolved. Eventually, the Japanese used simplified Chinese signs applied in a syllabic manner to spell out their own polysyllabic words. They also developed special notations to give the order for reading characters, added certain Japanese phonetic elements, altered Chinese phonetic values to represent Japanese syllables and

Two Japanese syllabaries: hiragana (upper rows) and kata kana (lower rows)

a あ ka か sa さ ta た na な ha は ma ま ya や ra ら wa わ
  ア　カ　サ　タ　ナ　ハ　マ　ヤ　ラ　ワ

i い ki き shi し chi ち ni に hi ひ mi み ri り
  イ　キ　ツ　チ　ニ　ヒ　ミ　リ

u う ku く su す tsu す nu つ hu ぬ mu ふ yu む ru ゆ る
  ウ　ク　ス　ツ　ヌ　フ　ム　ユ　ル

e え ke け se せ te て ne ね he へ mu め re れ
  エ　ケ　セ　テ　ネ　ヘ　メ　レ

o お ko こ so そ to と no の ho ほ mo も yo よ ro ろ o を n ん
  オ　コ　ソ　ト　ノ　ホ　モ　ヨ　ロ　ヲ　ン

simplified the phonetic characters so that each had a fixed phonetic value.

During the eighth through the tenth centuries, two "syllable alphabets" or syllabaries developed. They are called *kata kana* (side kana or partial kana) and *hiragana* (easy or plain kana). Kata kana developed as a squarish straight-lined script style made from parts of Chinese characters and was used in official texts, dictionaries and histories. It is also used for words borrowed from foreign tongues and functions like italic type in alphabetic scripts.

Hiragana is a more flowing script derived from the cursive Chinese signs. It is less formal and is more often used than kata kana.

In the early 20th century, Japanese began spelling their words with roman letters known as *romaji*, as symbol scripts were difficult to use with Morse code. Romaji later became accepted by the Japanese media and was used to describe news of world events and international business dealings.

■ 566

Scripts and Signs for Special Interest Groups

Unlike the organic development of "true" languages, certain scripts were developed for a particular purpose. Invented forms were created for many reasons, among which were a leader's wish to inculcate his rules or goals through the script, a people's desire to differentiate themselves from or surpass a rival populace by having a uniformly understood script, the conversion interests of a religious group or the need for a more efficient writing system. More recently, they have also included scripts and signs for special-needs groups.

Many ancient legends attribute the invention of a script to either a deity or a hero. Meanwhile, an actual historical figure, Darius the Great, who ruled Persia from 521 to 486 B.C.E., has been credited with inventing the Old Persian cuneiform. More modern-day scripts have been created all over the world. In West Africa a syllabic script, the Mende, was created by a Muslim tailor named Kisimi Kamala around 1930 to allow people of humble origin, like himself, to teach themselves to read and write.

In southwestern China during the 1800s, Samuel Pollard created a script to teach a group of non-Chinese called the Miao. A syllabic style that read from left to right, the signs contained a number of geometric forms. Pollard was aware of the Miao's desire to distinguish themselves from the predominant Chinese culture and emphasized words, phrases and references that were particularly applicable to the Miao. As a Christian in an era of missionary zeal, Pollard also included religious themes in his teaching of the script. According to linguists, this script is still used in Miao region.

In North America, two invented scripts include Cree and Cherokee. English Methodist missionary John Evans developed a script called Cree, a syllabic script read from left to right, between 1840 and 1846 when he lived in the Hudson Bay area of Canada. Teaching with his newly developed Cree allowed him to preach to people who had been speaking in varied tongues and using pictographic images. Using Cree, Evan pursued his missionary efforts and converted many of

The Cherokee script

D_a	R_e	T_i	ᴔ_o	O_u	i_v
S_ga ᴆ_ka	Ᵽ_ge	y_gi	A_go	J_gu	E_gv
ᵮ_ha	?_he	ᴅ_hi	ᴋ_ho	Γ_hu	ᴀᴙ_hv
W_la	ᴕ_le	ᴘ_li	G_lo	M_lu	ꭾ_lv
ꝺ_ma	ᴕ_me	H_mi	Ꝫ_mo	y_mu	
Θ_na t_hna G_nah ᴧ_ne		h_ni	Z_no	ꝯ_nu	O_nv
ꞁ_qua	ꝏ_que	ᴾ_qui	ᵛ_quo	ꞷ_quu	Ɛ_quv
ᵾ_sa ᴕ_s	4_se	b_si	ꝸ_so	8_su	R_sv
ᴌ_da W_ta	S_de Ꞇ_te	ᴊ_di Ꞃ_ti	ꞈ_do	S_du	Ʇ_dv
ꞵ_dla ᴧ_tla	ᴌ_tle	C_tli	Ꝫ_tlo	ꝴ_tlu	P_tlv
G_tsa	ᵥ_tse	ꞣ_tsi	K_tso	d_tsu	C_tsv
G_wa	ᴕ_we	Θ_wi	ꝏ_wo	ꝸ_wu	6_wv
ꝏ_ya	ᴮ_ye	ꞡ_yi	ꞑ_yo	G_yu	B_yv

the area's inhabitants to Christianity. Inuit on Baffin Island still practice a form of Cree.

The Cherokee script was invented in 1821 by a man known variously as Sequoya or Sikwayi. This alphabet gave syllabic values to the letters of the Latin alphabet combined with created signs, so that the 85 symbols signified 6 vowels, 22 consonants and about 200 sound groups. For example, *D* represents a long or short *a*; *T* signifies *i*, *h* stands for the syllable *ni*, and *P* for the consonants *t 1 v*.

In the 1820s, most members of the Cherokee nation of North Carolina could use the script and continued to do so when the Cherokees were forced to move to Oklahoma in the 1830s. Its use eventually declined, although there have been some attempts by the Cherokees to revive it.

SHORTHAND

A good example of a system invented to serve a particular interest is the script called shorthand. Shorthand is a system of speed writing whereby signs or symbols replace conventional letters and words. Such methods have been variously called brachygraphy (short writing), phonography (voice writing), and stenography (narrow writing).

Until the stenotype machine allowed machine-aided stenography, shorthand had served news reporters, staffs of legislative bodies, and had been used in court trials and most prominently in business. Shorthand was not a modern creation and in fact preceded business machines by many centuries.

The three primary shorthand methods are hieroglyphic, orthographic and phonetic. Hieroglyphs, such as those of the Egyptians, are the oldest form. The orthographic form is an abbreviation of normal spelling wherein shortened words are combined with special signs. In the phonetic version, a series of symbols represent sounds, and words are written based on their phonetic values.

The ancient Egyptians, Persians, Hebrews and Greeks had some types of quick writing. Many historians link the beginning of rudimentary shorthand with the Greek Xenophon, who wrote the memoirs of Socrates in the fourth century B.C.E. He apparently used a rudimentary form of abbreviated writing in the hieroglyphic style to record his notes. In the Roman empire, a freedman named Marcus Tullius Tiro created his *notae Tironianae* (Tironian notes) around 58–63 B.C.E. The private secretary to statesman and Stoic philosopher Cicero, Tiro used his education and influence to write and distribute a form of shorthand dictionary. Tiro's method was taught in Roman schools and used by leaders such as Julius Caesar. It lasted until the Middle Ages, when it began to be linked with occult practices and met with religious opposition. Nevertheless, when papal scribes began to take an interest in abbreviated writing and discovered older shorthand texts, the method was accepted once more.

In 1588, British physician Timothy Bright published *Characterie: An Arte of Shorte, Swifte, and Secrete Writing by Character*, which described a system using combinations of lines, half-circles and circles to represent the letters of the alphabet.

Bright's system demanded a great deal of study from its users, however, and did not prove to be rapid in practice.

Some 30 years later, English translator Thomas Shelton invented a system called tachygraphy (from the Green *tachys*, "swift," and *graphein*, "to write"). Described in his book *Short Writing* (1620 or 1626), Shelton's method consisted of nearly a dozen forms that differed only slightly from longhand. Digits were also assigned to some 260 typical words. As well as facilitating speed-writing, Shelton's invention also had aspects of secrecy. Words could be concealed by using dots for some vowels and having consonants replace other vowels in the middle of words.

The most famous applications of Shelton's system was in the diary of Samuel Pepys. Beginning in 1660 this British civil servant wrote for over nine years in tachygraphy, which he augmented with passages in Greek, Latin and other Romance languages as well as with nulls of his own invention. While this shorthand/language combination was not all that unusual, the diary itself is noted as a literary classic as it contained a fascinating account of the world in which he lived. His documentations of daily life in Great Britain were finally revealed between 1819 and 1822 by a Cambridge University scholar named John Smith. The diary made Pepys, and Shelton, famous.

In Colonial Virginia, William Byrd kept a similar journal on a part-time basis between 1709 and 1741. Byrd's method was later identified as one created by William Mason, a London artist and stenographer. Mason's 1672 shorthand textbook, *A Pen Pluck't from an Eagle's Wing*, was revised and appeared in 1707 as *La Plume Volante* (The flying pen).

All these systems were orthographically based; for years, inventors had been trying to create a system of speed-writing based on phonetics, but without much success. The little-known William Tiffin was credited in 1750 with an early phonetic system, but it was two others, Sir Isaac Pitman and John R. Gregg, who made the most influential changes in shorthand. In 1837 Pitman published *Stenographic Sound-Hand*, whose basis was fully phonetic. It contained special sound

A page from Samuel Pepy's diary
using Shelton's shorthand

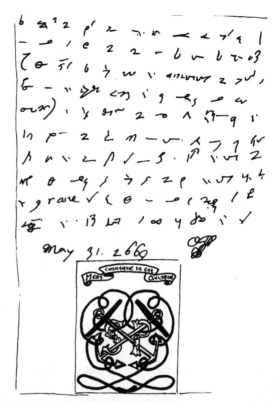

indicators for consonants and vowels; groups of pairs called
"breath" and "voice" letters, respectively indicated by light-
or dark-shaded symbols; dots and dashes; straight, sloped and
shaded lines; and abbreviations of common phrases. In 1852,
Pitman's brother Benn brought this system to the United
States where, with some modifications, it became the most
widely used method on the highly commercialized East Coast
by the 1890s.

In Ireland, Gregg was the Pitman brothers' primary rival. His
pamphlet *Light-Line Phonography* (1888) was also a phonetics-
based system of characters with circles, hooks and loops. It
kept the well-known cursive movement of longhand while

Three styles of shorthand: Gregg (top), Pitman (middle) and modern (bottom), representing the phrase "Friendship is a treasured possession."

Gregg

Pitman

Modern

eliminating Pitman's shading techniques and his odd placement of characters above or below the writing lines.

In 1893 Gregg came to the United States and taught his method throughout the Midwest and South, where shorthand was virtually unknown. Gregg's efforts were rewarded when his style finally superseded that of Pitman.

During World War I, shorthand systems were modified to add a further element of concealment to encrypted messages. Allied intelligence contained special departments that studied German systems. Postal censors scanned the mails for suspicious examples, which were then turned over to translators and encryption analysts. America's Military Intelligence Division had a section that tested for invisible ink and could read several shorthand styles, including the German Schrey, Stolze-Schrey and Gabelsberger.

In World War II, the written message was superseded by the much faster radio transmission. Even when the mails were used to pass intelligence, the microdot was preferred over any type of conventional written communications, including shorthand. Today, with the invention of machines that greatly facilitate dictation and note-taking, shorthand is rarely practiced.

SCRIPTS FOR THE SIGHT-IMPAIRED

Braille

Sight-impaired persons have been greatly aided with symbols that they can read by touch. Two such systems were developed by Louis Braille and the lesser-known William Moon. Braille, visually disabled himself, developed a system known as Braille to introduce literacy to the blind. Born in Coupvray, France, he was blinded as a baby when he picked up a sharp tool and accidentally cut one eye while playing in his father's harness-making shop. Poor hygiene led to infection and soon Braille was totally blind. His disability made him the object of ridicule and cruelty, yet his perseverance allowed him to continue his education with sighted children.

He continued his schooling at the National Institute for the Blind in Paris and trained there to become an instructor. The best known means of touch reading at the time were unwieldy texts with raised letters and, continually frustrated by the lack of better learning tools, Braille applied himself to finding a better method. In this search, he discovered an unusual system that had been originated by a French army officer.

Captain Charles Barbier had devised what he called "night writing" or sonography, a system composed of embossed dots and dashes, in order to provide the military with a means of reading messages in the dark. Ingenious as the method was, its detailed arrangement of dots and dashes made it too complicated for the time restraints of the battlefield.

Convinced that the principles of sonography could be used to benefit the bind, Braille's breakthrough came when he began using raised dots alone. He realized that simple arrangements would facilitate tactile learning, and dots would allow better spacing on paper. By reducing the Barbier designs by half and removing the dashes, Braille created the foundation of what is now known as the Braille cell.

When Braille's system was officially published in 1829, it included formations for music, mathematics and science, opening up hitherto unavailable avenues for the vision-impaired. Braille continued his work at the National Institute and continued to make improvements as he taught. Using simple instruments, a metal rule with open spaces, a stylus, a board and paper, his students became skilled at taking notes and composing paragraphs. Heavy paper was placed between the rule and the board and as the rule was passed over the paper, the stylus was applied to poke through the openings and create the dot groupings.

The foundation of Braille is a cell composed of six possible dot points, from which a total of 63 dot variations can be made. For orderly arrangement the six possible dots per cell are numbered 1 through 6.

The Braille cell

1	● ●	4
2	● ●	5
3	● ●	6

Each of the 26 letters of the English alphabet is indicated by a differing arrangement of dots. The pattern for *a* to *j* is made from dots 1, 2, 4 and 5:

Letters A to J

A ●   C ● ●   E ● ●   G ● ●   I ● ●

B ●   D ● ●   F ● ●   H ● ●   J ● ●

Letters *K* to *T* are formed with the addition of dot 3; *U* to *Z* are formed by using combinations of all six dots:

Letters K to Z

K   O   S   W

L   P   T   X

M   Q   U   Y

N   R   V   Z

Capitalization, numbers and punctuation are among the many variations made possible by this system. Punctuation marks are indicated by dots in the lower part of the cell. Capital letters are indicated by dot 6 being placed just of the left of a letter formation.

Dots 3, 4, 5 and 6 are used to form a "reverse L." When this appears in a text, it indicates that the dot groups that follow are numbers. The dots for *a* to *j* are then used as the numbers 0 to 9:

Numbers 0 to 9

1	6
2	7
3	8
4	9
5	0

Despite the advances Braille had made in helping the vision-impaired, he was not officially acknowledged for his contribution until two years after his death on January 6, 1852.

Moon type

William Moon was born in Horsemonden, Kent, in England in 1818, and unlike Braille, he lived to see his efforts receive official recognition. Moon had suffered from vision problems since the age of four, and had prepared for his eventual blindness by studying different training styles. When he completely lost his sight at the age of 22, he devoted himself to developing another kind of raised sign. While he admired the Braille system, he realized that it was most beneficial to the young with their swifter learning capabilities. Older people, not trained in Braille from childhood, tended to have difficulty adapting to its use.

Moon found a solution in an adaptation of the roman alphabet, which he varied so that signs would be more clearly discernible. Moon type is based on roman capital letters and their variations. By the 1880s the Lunar/Moon type had been

Moon type alphabet

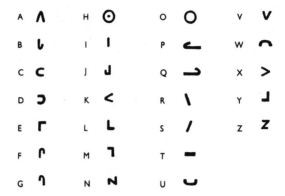

standardized, and had been incorporated into the education of the blind.

SIGN LANGUAGE

Signs have also been invented to help the hearing-impaired, who would otherwise be separated from general communications. One of the early successful forms was developed by Charles Michel de l'Epée.

Born in Versailles, France, in 1712, Epée was an abbot who frequently traveled between Paris and the town of Troyes. On one such return trip to Paris he stopped at the home of a parishioner whose two daughters were both deaf. As he observed the girls' awkward attempts to communicate by waving, facial expressions, nods and touching, Epée first conceived of sign language.

He began to watch the deaf in the Paris markets and on quiet country roads and noted a wide variety of personal, nonuniform motions. Needs and interests were also vocalized by making noises to get people's attention. Having watched all kinds of pointing, crossing of fingers and clapping, Epée realized that five fingers, if held and moved in various planned ways, could express a wider range of meanings. If these positions were equated with letters of the alphabet, whole words could be signed.

Epée's sign language

Letters A–I

Letters J–R

Letters S–Z, "and"

Numbers 1–10

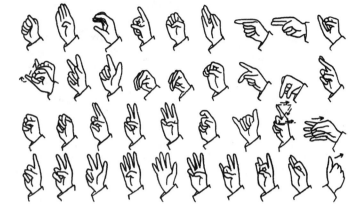

International sign language

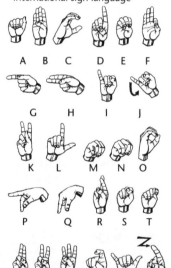

Epée taught his one-handed alphabet to his parishioner's daughters, who were able to converse in a systematic way for the first time in their lives. In 1755 Epée began a school for the hearing- and speech-impaired. From among priests and public instructors, he began a struggle to recruit volunteer teachers and won the favor of Louis XVI. The tiny school was funded from the royal coffers and grew to become a major institution in Paris.

In 1776 Epée published a book about his sign language and the proper instruction of the deaf and the mute. He was writing a more elaborate dictionary of hand signs when he died in December 1789.

The international sign language shown at the left uses movements similar to Epée's invention.

Languages: Invented, International and Artificial

Spoken languages have also evolved or have been created for special purposes. They have sometimes arisen from language variations or have been constructed by linguists in the search for an international language.

Argot (French: *argot*, "claw," "spur") is believed to have started as a special language among the thieves' dens of Paris around 1100. The term argot, or jargon (Old French: *jargon*, *gergon*, "a chattering"), has since been used to describe criminals' slang as well as the special vocabularies and idioms of people in similar occupations or lifestyles.

From the late 1700s through the 1800s, pickpockets and thieves created and nurtured different kinds of rhyming slang particular to their wandering and illegal activities. In many of the larger cities in the British Isles, this jargon served to conceal comments about everything from passersby to criminal plans. These statements were often simply words that rhymed with the ones they replaced. This pastime spread in popularity until it even found a niche among the country gentry for a time.

Following are some examples of phrases:

1. Go'or vest mar fer some blotch.

2. Gi'up pears fer yer tup o' fisherman's daughter.

Although there were no set rules of rhyming or grammar, the following are conceivable translations:

1. Go over to West Bar for some scotch

2. Get up the stairs for your sup (sip) of water.

This slang was intentionally made difficult to comprehend, with no discernible pattern of sentence structure or pronunciation, and an individual could not be certain that he had conveyed his message correctly. Many word fanciers began to lose interest when these slang practitioners started to go to extremes in trying to vary their creations and still make them rhyme. As their popularity waned, the predominant users remained the "dips" (pickpockets), forgers and swindlers.

Other types of secret languages that had alternating periods of popularity in Europe during the 1800s were Turkey Irish and Opish. Traced to school games in Europe and England, both were quite simple methods involving the addition of an extra syllable or letter pair to a word. Neither was capable (on paper) of masking any written intentions from a skilled codebreaker for very long. When verbalized rapidly by an experienced speaker, however, both tended to sound like foreign tongues and provided amusing verbal entertainment.

Turkey Irish added the letter pairs *ac* before every vowel in a word. A word like *wary* became *wacaracy* (with *y* as a vowel). When a word had back-to-back vowels, like *guard*, the letter pair was placed before the first vowel to create *gacuard*. Modern variations place the syllables *illig* or *ag* before each vowel. Using the sentence "Call me at ten" the message becomes *Cilligall millige illigat tilligen* or *Cagall mage agat tagen*.

Opish operated on a similar principle: the letters *o* and *p* (hence the name) were placed after each consonant. For example, the Opish word for *patrol* would be *popatopropolop*. Whether or not it served a real concealment purpose, it was certainly enjoyable to say. A sentence containing even a few words in Opish would be quite difficult for an amateur eavesdropper to discern.

Altered word usages and curious phrases are found in two other invented "languages," Pig Latin and Macaronics. Pig Latin is not an adaptation of Latin, rather, it is a purposeful alteration of English words. In Pig Latin, the first consonant, or cluster of consonants, is combined with the syllable *ay*. This new syllable is placed at the end of the word. The first vowel and any additional letters remain in their set position. In the example *Wednesday*, *ednesday* combines with *W* and *ay* to form the word *ednesday-Way*. The pattern is used for every word to build a Pig Latin sentence.

The verbal burlesque known as Macaronics mixes vernacular words in a Latin context. The term was created in 1517 by the poet Teofilo Folengo, who in that year wrote a poem entitled "Liber Macaronices." In it he described "macaronic art," mixtures of forms and meanings. Macaronics had no set rules and could be used in a medley of styles. For example, *verbum* (Latin for "word") could have an *s* appended to it, be combined with the word *they*, and be turned into the nonsensical phrase, *they verbums*. With a further stretch of the imagination this became *they were bums*. Because of such loosely connected links, macaronic word play did not lend itself to verbal codes as easily as Opish or Turkey Irish, and quickly lost its popularity.

Another form, called Agro, formed its own vocabulary rather than simply manipulating English words. Agro used the English alphabet with *c* and *q* omitted or regarded as the letter *k*. The letter *g* was always pronounced as a hard *g*. Opposites were expressed by reversing their syllables and plurals were formed by repetitions of single words. Possession was indicated by placing the letters *ro* before or after the given word. A change is made in the first vowel of a verb to indicate tense: *a* for present, *e* for future, *o* for past.

Some sample words and their various forms follow:

VOCABULARY

house	gomo
hour	sak
list	zuk
paper	folo
week	som
with	hok
need	rajo

OPPOSITES

bring	far
send	raf
come	kad
go	dak
here	rik
there	kir

NUMERALS

1	alta
2	boda
3	koda
4	doda

6 foda
7 goda
8 hoda
9 ita
0 joda

PRONOUNS

I/me   alk
she/her   ke
we/us   alk alk
it   ko
you   bol
they   kaka
them   keke
he/him   ka

The sample Agro sentence *alk rof folo folo bol rojo* means "I brought the papers you needed."

On a more serious note, from time to time throughout history, certain nations' predominance has led to their languages becoming widely known and used outside their borders. Languages that were constructed in attempts to create a common international language have been categorized as international or universal languages.

Latin is arguably the most prominent international tongue. Although it is no longer actively spoken, its influence has continued from the days of the Roman empire. Latin remains the basis of the Romance languages, with hundreds of word roots in active use.

Other languages have become international for other reasons. In the 1600s French was accepted as the special tongue of diplomats. In the Mediterranean a form of hybrid speech was developed from a combination of French, Arabic, Turkish, Italian, Spanish and Greek. In modern times, the number of international languages may expand as immigrant populations take up new residences and bring their own language to their adopted countries.

Where international languages have not naturally evolved, people have tried to artificially create them. This is particularly true in fields of business where English has become predominant during the 20th century. From the 1500s onward scholars have been considering what are termed a priori languages. Philosophers like René Descartes and Wilhelm Leibnitz in the 17th century theorized that a generalized tongue based on logic could be constructed. These two philosopher-mathematicians gave credibility to the search for an artificial international language.

SOLRESOL

Attempts to develop a universal language have covered the gamut of the misguided, the bizarre and the unusual. In 1817 Jean-François Sudre invented a type of artificial language called Solresol, cryptic to all but its practitioners. Its seven pitches or syllables based on the word *solhfa* (Sudre's created name) could be united in groups of five at once, with variations of order and stress. Its vocabulary was constructed by combining the syllables that identified the notes of the musical scale. This would express words in the artificial language. For example, a term like *sollasi* meant "ascend," and *silasol*, "descend." Sudre was able to "converse" with his students using violin notes. Though clever, this method was limited because it was linked to music, which is generally regarded as a source of enjoyment and not as part of a word system and Solresol soon faded into obscurity.

VOLAPÜK

The failure of Solresol demonstrated that artificial language could not be connected to a specific nation or discipline. In addition, artificial languages needed a broad vocabulary that incorporated the most efficient elements of many well-known tongues. Volapük was the first attempt to gain a fairly wide measure of acceptance.

Volapük, literally "world speech," was created by Johann Martin Schleyer in 1879. A Catholic priest in Baden, Germany,

Schleyer was fascinated by world trade and believed in the progress of the brotherhood of mankind. Inspired to make a personal contribution to that progress, he applied himself to the search for a universal language.

His dedication led him along a path that others either had not taken or had not completed. Schleyer invented a universal alphabet that included the speech sounds of every major language. He also compiled a dictionary of vocabulary and grammar using the following basic principles:

1. Vowels—*a, e, i, o, u, ä, ö, ü* (pronounced as in German)

2. Consonants—most pronounced as in English.

3. Root words—mainly Latin, English and German.

4. Adverbs and adjectives—indicated by suffixes: *-ik* (adjective); *-o* (adverb)

5. Verbs—a prefix indicated tense, and personal pronouns were added as suffixes.

Although at first Volapük had many avid practioners, it soon fell out of use. Schleyer was protective of his creation and refused to change rules even if they proved to be inefficient. In addition, the divisiveness between the many academic factions who wished to standardize Volapük undercut its chances for expansion beyond a very limited range.

ESPERANTO

Probably the best-known and most stable artificial language is Esperanto. Created in 1887 by Polish oculist Ludwig L. Zamenhof —derived the name from the French *esperer* or the Spanish *esperar*, meaning "to hope"—Esperanto's success is due in part to Zamenhof's willingness to share its developmental changes with other scholars and artificial language devotees. When used verbally, Esperanto was also more fluid than its predecessor Volapük. The following rules are the foundation of Esperanto:

1. Its alphabet consists of 28 letters, each of which has only one sound.

2. There are six standard accented letters: *c*, *g*, *h*, *j*, *s* and *u*.

3. Stress in each word falls on the next-to-last syllable.

4. Verbs do not decline for number or person, which are indicated by an accompanying pronoun.

5. Verb endings express function or tense.

6. Nouns end in *-o*, adverbs in *-e* and adjectives in *-a*.

Following is a list of basic Esperanto words:

NOUNS

ENGLISH	ESPERANTO
family	familio
mother	patrino
father	patro
daughter	knabino
son	knabo
brother	frato
sister	fratino
parents	gepatroj
brothers/sisters	gefratroj
sons/daughters	gefiloj
boys/girls	geknaboj
child	infano
gentleman (Mr.)	sinjoro
lady (Mrs.)	sinjorino
home	domo
kitchen	kuirejo
dining room	mángocambro
bedroom	dormocambro
room	cambro
garden	gardeno
car	automobilo

VERBS (PRESENT TENSE)

ENGLISH	ESPERANTO
is	estas
says	diras
works	laboras
speak	parolas
responds	respondas
sits	sidas
stands	staras
eats	mangás
goes	iras
sends	sendas
uses	uzas
writes	skribas
visits	visitas
loves	amas
thinks	pensas

NUMBERS

ENGLISH	ESPERANTO
one	unu
two	du
three	tri
four	kvar
five	kvin
six	ses
seven	sep
eight	ok
nine	náu
ten	dek

FUNCTION WORDS

ENGLISH	ESPERANTO
and	kaj
the	la
who, which	kiu
of, from	de
in	en
that (one)	tiu
how	kiel
to	al
about	pri
where	kie

BIOGRAPHIES

Alberti, Leon Battista (1404–1472)

The illegitimate but favored son of a wealthy Italian, Leon Alberti was a model Renaissance man, adept in art, music, architecture and writing. Among his architectural achievements were the designs of the original Trevi Fountain and the Pitti Palace. He also wrote architectural treatises, poetry, and fables.

It seems only natural that a man of so many interests would be fascinated by the means to conceal missives, but unlike other knowledgeable men who saw secret writing as an amusement, Alberti applied himself to its mastery.

He earned his reputation by solving codes of increasing degrees of difficulty. Then his curiosity directed him toward the developing field of ciphers. As this means of transposing and substituting began to be used more frequently in the letters of other creative people, Alberti studied them and originated important concepts of his own. He wrote about letter frequencies in Latin and Italian sentences, and his study of how these aspects affect cipher solving is considered the first such presentation of cryptanalysis in the West. Applying his skills to cipher and code creation, he took cryptology a significant stride forward.

Alberti's primary contribution to code and cipher creation was through a polyalphabetic disk. The disk's larger and smaller plates shifted alphabets and a few numerals to change their alignments with each other. This created polyalphabetic replacements which was a real advancement for concealers. By also bringing his disk's numerals into the arrangements for masking letters, he gained credit for creating enciphered code, another first.

Alberti was a visionary whose work was not fully appreciated in his day. It was not until the late 1800s that nations began to encipher their codes regularly over the broad range of diplomatic and military communications.

Bazeries, Etienne (1846–1931)

Born in 1846 in a French fishing village on the Mediterranean, Etienne Bazeries entered the army at 17 and served valiantly in the Franco-Prussian War (1870–71) and during tours of duty in Algeria in the late 1870s.

Bazeries's interest in cryptology developed when he became fascinated with breaking the encryptions in the personal columns of newspapers. In 1890 his claim that he could solve the prominent French military cipher of that day was met with derision. But after he made good on his boast and also succeeded in breaking a replacement system, French military officials took notice.

While still involved with the military, Bazeries's assignments included judging others' cryptosystems, studying the nomenclators of former French rulers such as Louis XIV and Napoleon, and solving the encryption of an anarchists' cabal. During the latter case in 1892, Bazeries revealed that the anarchists had used an alphabet shift concealment called the Gronsfeld cipher.

After he retired from active duty in 1899, the French army employed him as a cryptanalyst. He solved numerous encryptions, but never succeeded in persuading the military's higher authorities that French official encryptions were also susceptible. As a security proposal, he developed his own "cylindrical cryptograph," which was a series of alphabet-bearing disks on a spindle. Bazeries's invention was rejected after a bitter rival broke the cover and revealed the message.

In 1901 Bazeries wrote a scornful but insightful text about his critics, enemies and the military hierarchy. Titled *Les Chiffres secrets dévoilés* ("Secret ciphers unveiled"), it was considered a significant text in cryptologic history.

The outbreak of the First World War and its patriotic fervor brought Bazeries back to help solve the encryptions of the Kaiser's forces. By this time the French administration did have a better appreciation of cryptography and cryptanalysis, but historians differ in crediting Bazeries directly for this change. Perhaps some officials could never admit that the man nicknamed "the Napoleon of ciphers" had been right all along. Bazeries lived to see the victory of France and her Triple Entente allies. After 1924, he spent the rest of his years in quiet retirement.

Cardano, Girolamo (1501–1576)

Born in Milan in 1501, Girolamo Cardano reportedly had an overriding desire to be remembered by future generations. A mathematician and physician, he wrote feverishly to establish a written legacy with some 240 texts (131 published and 111 in manuscript form).

Oddly, he did not pen a specific concealment-building or -solving book. Rather he put his ideas about these subjects into two texts about science, *De Subtilitate* and *De Rerum Varietate*. The former, printed in 1550, was a compilation of explanations and art regarding scientific phenomenon. The latter, published in 1556, contained similar themes and was a sequel by today's terminology.

Among the types of concealments he mentioned was an autokey wherein the plaintext letters enciphered themselves. This was an important new idea in cryptography. But the method was flawed because he had the key begin anew with each new word. This led to multiple possible decryptions making the specific intended letters difficult to understand.

His physical masking methods included invisible inks and the use of a grille, a piece of stiff material such as wood or metal with precut holes that permitted spaces for letters or words to be written in them. Though often contrived sentences had to be formed around the specific letters, the style became popular among emissaries of different European countries during the 16th and 17th centuries. Because of these applications and his popular science texts, Cardano did achieve the immortality he craved.

Champollion, Jean-François (1790–1832)

Born on December 23, 1790, Jean-François Champollion demonstrated a very early aptitude for foreign languages. He studied Middle Eastern languages in Paris from 1807 to 1809 under the skillful guidance of Silvestre de Sacy, himself a noted scholar of ancient writings. During these studies, Champollion devoted himself to his lifelong passion—the solution of the Egyptian hieroglyphs.

The same year he completed his studies with de Sacy he was accepted as a professor of history at Grenoble in southeastern France, where he worked until 1815. His support of Napoleon Bonaparte hurt Champollion after the dictator was finally exiled and Champollion was prevented from teaching from 1815 to 1817.

Building upon the findings of Swedish diplomat J. D. Åkerblad and English physician Thomas Young, Champollion compiled the research that would ensure his fame. In September 1822 he wrote a letter to the prestigious Paris Academy. His "Lettre à M. Dacier" described phonetic hieroglyphs and contained accurate decryptions of the names of Egypt's Roman emperors.

The letter caused a sensation and Champollion's situation quickly improved. He became recognized as the premier scholar of hieroglyphs. His many successes included discovering the differences between the hieroglyphs' pictographic images and phonetic complements, proving the linguistic link between Ancient Egyptian and Coptic, and formulating an early grammar and lexicon for the hieroglyphs.

As he added to his findings, he studied collections of other antiquities and visited centers of culture such as Florence, Venice and Rome. He also acquired art for the Louvre and organized an expedition that visited Egypt in 1829–30. After being awarded membership in the Académie Française in 1830, he was also given the first chair of egyptology at the Collège de France.

At the height of his career, he succumbed to a stroke at the age of 41.

Chappe, Claude (1763–1805)

Born in Brulon, France in 1763, one of five brothers, Claude Chappe was encouraged by an uncle to pursue scientific interests. Aware of message systems such as that of England's Robert Hooke, he became determined to develop a signaling method of his own.

During the 1790s he persistently promoted his "aerial telegraph" among businessmen and politicians alike. His eldest brother, Ignace Chappe, who had been elected to the Legislative Assembly of France, finally helped obtain the necessary funding for the project.

Chappe's perseverence was rewarded when in 1794 the first series of signaling towers was completed between Paris and Lille, some 140 miles to the north near the present-day border with Belgium.

Chappe's mechanisms each had a crosspiece and attached indicators that were moved into different positions to represent message elements. After some practice with the control ropes and pulleys, the system's operators were able to send signals down the route to Lille in about two minutes. When accurate transmissions were verified, Chappe's fortune seemed to be assured.

A system was constructed to connect Paris with the Mediterranean port of Toulon. When its 116 stations transferred a communiqué in 20 minutes, the new telegraph seemed to promise fame and fortune for Chappe. Over the succeeding years more stations were erected to connect other towns in France as well as in bordering countries.

As Chappe's success and fame grew, so did his detractors and competitors. Chappe became embroiled in a series of costly suits from petty arguments over message miscues to claims that his invention was a copy. Overwhelmed by these controversies, he took his own life in 1805. His family maintained the networks in his honor until they were supplanted by the electrical telegraph in the 1850s.

Driscoll, Agnes (1889–1971)

Born in Illinois in 1889, Agnes May Meyer was an Ohio State graduate who taught mathematics and music in Texas schools. She joined the U.S. Naval Reserve in 1918. Her interest and training in languages, mathematics and statistics gave her a solid foundation for a career in cryptology.

Beginning with secretarial duties, she quickly rose to position of Chief Yeoman (F), the highest designation available for women during World War I. After completing her active duty in July 1919, she found employment as a stenographer in the office of the director of Naval Intelligence in Washington. Still considered a clerk, her big career move came when she accepted an invitation to join the Department of Ciphers in Geneva, Illinois, near Chicago.

Also known as Riverbank, the site was the residence of William and Elizebeth Friedman, who were fast-rising cryptology experts. Under the influence of William Friedman's research and writings, Meyer's cryptographic skills were honed.

By 1921 she had left Illinois and returned to employment with the Navy, where she established herself as a premier cryptanalyst in the Code and Signal Section. A woman in a man's world and often jokingly called "Madame X," Meyer nonetheless persevered with her work. She surprised many who believed she was a confirmed spinster when she married Washington attorney Michael Driscoll in 1924.

Driscoll was responsible for training naval officers such as Joseph Rochefort and Laurence Safford, who would be influential figures in cryptology during the Second World War. In 1926 Driscoll made the opening breaks in the Japanese codebook designated Red by U.S. analysts. She proved her skills again in 1931 when she successfully attacked Red's replacement codebook, designated Blue. In 1935 she recognized cipher traffic that appeared to be created by a machine. It was labeled M1, and Driscoll developed a manual process to decipher the intercepts. Working with graph paper and sliding cipher sequences, she found the combination *to-mi-mu-ra*. Further effort revealed that this was the name "Thompson," which appeared frequently in Japanese foreign-office traffic.

In 1936 Naval Intelligence used the decryptions to break an espionage cabal operating on both coasts. Pacific Fleet radioman Harry Thompson and cashiered Navy officer J. S. Farnsworth (Agent K) were passing gunnery and engineering data to Japanese agents. They were later convicted and imprisoned.

In 1939 Driscoll made her most important contribution to U.S. security when she solved Japan's JN naval encryptions. The JN25 was a very secure five-numeral, two-part code superenciphered by a series of numbers added to the digits equated with the codewords. It was Driscoll who spearheaded the attack on the additives, as she and her fellow analysts drove their first wedges into the JN25 in the autumn of 1939. After painstaking analysis targeting the underlying codenumbers and the codewords, decryptions of new JN25 messages were possible by September of 1940.

These solutions aided U.S. victories in the Pacific, and provided the Navy with crucial information in the naval battles of the Coral Sea and Midway. The early JN25 solutions also provided important clues to solving later Imperial Navy systems.

After the Second World War, Driscoll worked for the top-secret National Security Agency until her retirement in 1959. The retirement years of this remarkable woman were not documented, and she passed away in undeserved obscurity in 1971.

Evans, Sir Arthur J. (1851–1941)

Born in July 1851 at Hertfordshire, England, the son of archeologist Sir John Evans, Arthur Evans was educated at Harrow, the private preparatory school for boys near London. His education continued at the universities of Oxford and Göttingen, Germany.

He followed his archaeological interests for many years while living in the Balkans. In 1882 he returned to Oxford as a fellow of Brasenose College. In 1884 he became curator of Oxford's Ashmolean Museum and remained in that position until 1908.

Evans is best remembered for his long-time studies of early life in Crete. He first became interested in the island's history during a visit to Athens where he studied etched gemstones similar to others found on Crete. He began to consider the possibility that the etchings' script was an unknown form of writing.

Evans again visited Crete in 1894 but did not begin his excavations until March 1900 at Knossos, where he discovered the ruins of the palace of King Minos. Finding numerous clay tablets and other artifacts, he named the unknown civilization Minoan after the legendary king.

For some 35 years he continued to excavate this Bronze Age culture. Evans separated his discoveries into three primary epochs, Early, Middle and Late which spanned a time frame from 3000 B.C.E. to 1200 B.C.E., and described a series of phases of pottery art that he believed were indexes of artistic and technical progress. His most influential publications were *Scripta Minoa* and the four-volume *The Palace of Minos*.

His inability to decrypt the pictographic script that he had found at Knossos, Linear B, was a major disappointment, although he continued his efforts to solve the script. He was later faulted by other archaeologists and historians for not publishing all of his discoveries sooner and refusing to give others access to his tablets.

Nevertheless Evans was lauded for his achievements. Knighted in 1911, he was elected president of the Society of

Antiquities and served from 1914 to 1919 while also presiding at the British Association from 1916 to 1919. He passed away near Oxford in July 1941.

Fabyan, George (1867–1936)

George Fabyan is primarily remembered for his influence on William and Elizebeth Friedman, who became the most famous couple in cryptology.

A wealthy textile dealer and philanthropist, Fabyan established an early "think tank" at an idyllic 500-acre estate called Riverbank near Geneva, Illinois, some years before the start of the First World War. His varied interests included acoustics, chemistry, cryptology and genetics, and he invited numerous experts to his intellectual community, where they pursued their work as paid employees.

Fabyan hired a Cornell geneticist named William Friedman to improve the quality of his farm's animals and grains. Friedman accepted the position and adapted well to the surroundings, but it was Fabyan's interest in potential secrets in Shakespeare's writings that piqued Friedman's own curiosity.

Fabyan had hired scholars to try to solve an ongoing debate over the authorship of William Shakespeare's works. Some detractors of Shakespeare's abilities claimed that Sir Francis Bacon was the real author of some, if not all, of the works attributed to Shakespeare. Many were convinced that the actual author's identity was hidden in cryptic form within the texts of the plays and sonnets.

William Friedman became familiar with the Shakespearean scholars when he helped them with photographic enlargements of texts containing Elizabethan print. While doing so, he met Elizebeth Smith. They married in May 1917 amid their mutually growing interest in cryptology.

With the U.S. entry into World War I, Fabyan's Department of Ciphers accepted work from the U.S. government in the autumn of 1917. They studied some interceptions of Axis messages and taught U.S. Army officers about aspects of cryptology. While William Friedman taught, Fabyan published Friedman's writings, which were quickly recognized as the pillars of modern cryptology. After the war, however, when the two were briefly united again, Fabyan inexplicably alienated Friedman by not permitting him to resolve the ongoing Shakespeare question.

The Friedmans left Riverbank permanently near the end of 1920. Fabyan lived for another 16 years, his greatest accomplishment having been his association with the Friedmans.

Friedman, Elizebeth (1892–1980)

Elizebeth Smith was a native of Huntington, Indiana, and the youngest of nine children. Through a hectic childhood with eight siblings, she emerged a purposeful young woman who earned a degree in English at Michigan's Hillsdale College in 1916. Employed at the Newberry Library in Chicago soon after graduation, she was recruited for other employment by a wealthy textile merchant named George Fabyan who had set up a think tank called Riverbank in Geneva, Illinois.

Riverbank was the residence of Fabyan's protégés, who researched topics such as acoustics, genetics and the authorship of Shakespeare's works. Smith became involved in the latter pursuit as a member of the Department of Ciphers of the Riverbank Laboratories. Although she had had no previous cryptology experience, her college and library training served her well as she compiled data from others' studies of Elizabethan texts.

A young genetics researcher named William Friedman helped her make enlargements of the print under examination. He and Smith courted in the idyllic Riverbank setting and married in May 1917.

During this time William had become involved with the Shakespeare project and discovered a talent for cryptology. When America entered World War I, he solved some enemy communiqués and taught cryptologic basics to U.S. Army officers.

Elizebeth continued her Riverbank activities and waited anxiously while William departed for his tour of duty with the U.S. Army's Radio Intelligence Unit in France in the spring of 1918. When he returned, the duo resumed their Riverbank research activities. Fabyan reportedly interfered with their work with the Shakespeare controversy, however, and they were not permitted to complete their efforts. The Friedmans chose to leave Riverbank by the end of 1920.

Both found employment with the U.S. Army, and Elizebeth also did some cryptanalysis for the Navy and the State Department. While William remained with the Army, Elizebeth's skills found a new application in the latter 1920s when she did cryptanalytic work for the Coast Guard. She solved a number of the radio-sent encryptions with which bootleggers coordinated their sea operations during the Prohibition. Through the early 1930s she showed great personal courage by testifying in court cases about her solutions which resulted in the imprisonment of rumrunners and members of dangerous drug smuggling rings.

After the controversial Prohibition era ended, she combined an active family life with cryptanalytic work for the Treasury Department. During World War II, Elizebeth obtained cryptology-related employment with the Army and Navy. Elizebeth also worked for the International Monetary Fund and helped the IMF set up its cryptographic security built upon the one-time tape system.

She and William united their talents to co-author a 1957 text titled *The Shakespearean Ciphers Examined.* Although the publisher's chosen title hinted at the presence of ciphers in Shakespeare's works, their detailed studies of others' findings showed that there were no encryptions in Shakespeare's classics that proved anyone else's authorship.

After William's passing in 1969, Elizebeth lived quietly until her own death in 1980. Because of their mutual accomplishments, she has been the most often mentioned of women cryptologists.

Friedman, William (1891–1969)

Born Wolfe Friedman in Russia in 1891, William Friedman was the son of Rumanians who came to America in 1892. After they settled in Pittsburgh, Pennsylvania, Wolfe changed his name to William. His early interest in science grew over the years, and Friedman eventually attended the genetics program at Cornell University. While attending graduate school, he was hired by a wealthy textile merchant from Illinois named George Fabyan.

Friedman went to Fabyan's estate called Riverbank in Geneva, Illinois. During his genetics research with Fabyan's crops and livestock, he proved himself to be an unusually creative and meticulous scholar.

Friedman became acquainted with the science of cryptology through Fabyan's investigation into the authorship of William Shakespeare's works. Friedman took photos of Elizabethan era printing to help the scholars who were searching his writings for cryptic clues. While doing so, he met researcher Elizebeth Smith.

Their marriage in May 1917 coincided with William's increased involvement with Fabyan's Department of Ciphers of the Riverbank Laboratories. With the U.S. entry into World War I, the group began analyzing U.S. government intercepts of enemy messages. Friedman quickly rose to prominence when he solved a number of the transmissions.

His cryptology career blossomed as he began to teach U.S. Army officers about cryptology in the autumn of 1917. He developed a series of seven monographs for the training sessions and wrote an eighth monograph upon his return from World War I duty. Together these were known as the Riverbank Publications. They included instructions for reconstructing primary alphabets, solving running-key ciphers, and breaking cipher-machine-based encryptions, and they became key cryptographic texts.

Riverbank Publication number 15 was published in 1917 and entitled *A Method of Reconstructing the Primary Alphabet from a Single One of the Series of Secondary Alphabets*. The primary alphabet was exemplified as a mixed alphabet such as that in a Vigenère tableau to make a polyalphabetic encipherment method.

Publication number 16 dealt with solutions of running-key ciphers, polyalphabetic ciphers set up with lengthy texts in attempts to provide more concealment. Friedman demonstrated how to break them with a table of high-frequency keyletters and plaintext characters equated with known cipher letters and alphabets. Using only these letters, he applied an anagram process to find sensible text in both the key and the plaintext.

Among Friedman's other writings were cipher solutions, a description of cryptology-related literature, and methods of discerning cipher machine encryptions. Within these important documents was the landmark Riverbank Publication number 22. Written in 1920, it is described by cryptology historians as the single most influential document of its kind on the science.

Number 22 was entitled *The Index of Coincidence and Its Applications in Cryptography*. The fundamental techniques of the science are found within it. One method enabled an analyst to rebuild a primary cipher alphabet with no need to speculate about plaintext letters. The second technique, monumental in scope, linked cryptology and mathematics by presenting the distribution of letters as a curve with characteristics that could be quantified with statistics. This was a pivotal link between often arcane cryptology and accepted mathematical principles.

During this same year, Friedman coined the word *cryptanalysis* to rectify a long-term confusion between words like decode and decipher. These terms had been applied to describe both the actions of intended recipients who had been given the correct codebooks or cipher keys, and those of a third party interceptor. Cryptanalysis gave the latter's solution efforts a separately defined identity for the first time in cryptographic history.

Disillusioned with George Fabyan, the Friedmans left Riverbank in the latter part of 1920. William resumed his productive relationship with the U.S. Army when he accepted a limited Signal Corps contract to develop encryption processes in January 1921. Afterward he was hired through civil service as a War Department employee.

Friedman continued his relationship with the U.S. Army throughout his highly distinguished career. While Herbert Yardley's staff (the Black Chamber) were solving other countries' cryptograms, Friedman continued to test and develop Army field ciphers such as the M-94. When the offices of Yardley and personnel were closed down, the Signal Corps set up the Signal Intelligence Service with Friedman as its first director.

During peacetime he slowly built an organization that, although underfunded, was prepared for the challenges of the Second World War. He led the cryptanalytic team which solved the Alphabetical Typewriter '97, the Japanese diplomats' cipher machine. Dubbed Purple by the U.S. cryptanalysts, the information gained from its decryptions became known as Magic. These solutions and others achieved by the SIS did much to ensure the Allies' victory.

After the Japanese surrendered in 1945, Friedman maintained his association with the military during reorganizations of Army intelligence agencies and the new National Security Agency in 1952. He worked as a consultant for some years after his retirement in 1955. After receiving many high honors for his long-secret and unheralded work, Friedman passed away in 1969.

Hagelin, Boris (1892–1983)

Boris Hagelin was born in czarist Russia, where his Swedish father worked in the Caucasus region managing the Nobel family's oil interests. Educated in Russia and Sweden, Boris grew up with a special aptitude for using and creating machinery.

After graduating from Stockholm's Royal Institute of Technology with a mechanical engineering degree in 1914, he obtained employment at ASEA, one of Sweden's premiere electrical companies. After six years with this firm and another at the Standard Oil Company in New Jersey, his career took a major turn.

In 1916 his father and Emanuel Nobel, nephew of the dynamite inventor, had invested in Aktiebolaget Cryptograph (Cryptograph Incorporated) in Stockholm. The younger Hagelin was first sent to the company in 1922 to look after their interests. Three years later he combined this special information and his ability to take advantage of it to create an important business in cipher machines.

Hagelin had heard that Swedish military officials were contemplating purchasing the German cipher machine called the Enigma. Taking action, he altered parts of a mechanism created by Arvid Damm, Aktiebolaget's co-founder. Hagelin's version, the B-21, had keywheels with active and inactive pins. The pins' varied arrangements created a polyalphabetic ciphertext. Swedish Army officials approved it for purchase in 1926, and the company's prospects seemed much brighter.

When Damm died in 1927, the still struggling business was bought by the Hagelins and renamed Aktiebolaget Cryptoteknik. Boris Hagelin now directed the business and began developing a series of cipher machines that printed letters on paper. One such invention, the C-36, was a very compact mechanism ordered in large numbers by the French military.

International interest in the C-36 and other secrecy systems was the result of growing militarism in the 1930s. The rise of fascism in Germany, Italy and Japan had finally alarmed leaders disillusioned by the First World War or obsessed with the Depression.

In 1936 Hagelin started contacting U.S. cryptologic officials about the C-36, and he made two visits to America in the late 1930s to discuss technological and contractual matters. After World War II began in Europe in September 1939, communications security concerns were of utmost importance for Allied nations. Hagelin himself was threatened by the Third Reich when the Germans invaded nearby Norway in April 1940. By that time he was developing a machine for America and knew his options were rapidly narrowing. He and his wife decided to leave Sweden, and began a harrowing journey across Europe.

Hagelin and his wife took blueprints and two disassembled cipher machines by train from Sweden through Berlin, the very center of the expanding Nazi force. They traveled through to Axis partner Italy and the city of Genoa, where they bought passage on a ship bound for New York City.

After extensive tests and contract discussions, Hagelin's design was approved for the U.S. Army as a middle-range cryptographic system from division level down to battalion. Its identifier was Converter M-209 and its widespread use made Boris Hagelin cryptology's first millionaire.

In 1944 he returned to Sweden and bought an estate thinking that the encryption machine business was over, but for one of the few times in his life he was completely mistaken. The Cold War and the emergence of new Third World countries required continued communications protection. Hagelin maintained his business in Stockholm until Swedish legislation about appropriating inventions of national defense importance in 1948 forced him to relocate to Zug, Switzerland. Switzerland's neutral position allowed him and his productive employees to develop new generations of security systems into the computer era.

Hebern, Edward (1869–1952)

Born in Streator, Illinois, Edward Hebern grew up in an orphanage. He obtained a high school education and at age 19 went west to California, where he worked at a sawmill and then in construction in the Fresno area.

Hebern became fascinated with cryptographic techniques around the time of his 40th birthday. Beginning in 1912, he filed patents for security methods ranging from checkwriting mechanisms to typewriters with cipher keyboards. His position in cryptologic history was assured when he began work with rotors, a significant advance in polyalphabetic encipherment.

Hebern had experimented with electric typewriters linked by wires connected randomly between them. A plaintext letter touched on one keyboard, sent a current through the wires and activated a key on the second keyboard. This new letter was then printed in ciphertext. This was still a monoalphabetic encipherment, however, as the wheels continued to be plugged into the typewriter in the same connections.

In 1917 Hebern made drawings of a wheel-like device that contained contacts on its outer edges that were randomly connected with a web of wires inside the disk. This was the basis of the rotor that varied single-alphabet replacements to make polyalphabetic ciphers. When a keyboard letter was pressed, activating an electric current, the electricity moved randomly through the rotor's contacts and wires. The current then activated the ciphertext character on a second keyboard or on a board illuminated by lights. The rotor then shifted a notch, changing the pattern and introducing another alphabet to the process.

In 1918 Hebern put his idea into practice. Pursuing this line of thought, he developed "interval" rotor wiring, in which some contacts were wired in planned connections while other contacts and wires remained random. He also experimented with systems containing multiple rotors that shifted their positions as the letters were encrypted. He was so confident with his progress that he advertised an "unbreakable" cipher in 1921.

The U.S. Navy cryptology expert Agnes Meyer, however, solved the sample cryptogram. Yet, still encouraged by the attention, Hebern went to Washington to discuss his invention. Believing he had the Navy's support, he began and incorporated Hebern Electric Code, the first U.S. cipher machine company in 1921. He sold stock and built a factory in Oakland, California.

His vision exceeded his finances and by latter 1923 he still had no major Navy contract. Nervous investors overreacted when they did not receive quick returns on their holdings. An investors' revolt, an investigation and a trial followed in 1926. Hebern was found guilty of violating California's corporate securities act. Although the verdict was later repealed, Hebern could not raise enough money to pay his investors and the company was driven into insolvency and foreclosure.

Hebern tried to revive his business by incorporating another small company in Reno, Nevada. The Navy again studied and used some of his five-rotor machines in the latter 1920s and early 1930s, but Navy officials continued to deny him a long-term contract.

Believing that IBM had infringed upon some of his ideas, he brought a patent interference suit against them but lost the case in 1941. During the Second World War, he became convinced that the U.S. military branches had used his machines and developed others from his rotor advancements without duly compensating him. He initiated a suit in 1947 against the Navy, the Army and the Air Force, but the process became enmeshed in court proceedings for years.

Hebern died in 1952 before a resolution had been reached. In 1958 the U.S. government finally agreed to pay his estate $30,000, too little too late for this poorly treated inventor.

Hitt, Parker (1877–1971)

A native of Indianapolis, Parker Hitt had shown an aptitude for engineering and was studying civil engineering at Purdue when the Spanish-American War began in 1898. After serving in Cuba, he continued his career with the U.S. Army.

In 1911 at the Army's Signal School in Fort Leavenworth, Kansas, Hitt's interests took an important turn. By this time an infantry captain, he was impressed by a series of special conferences that included discussions about encryptions and communications. This experience encouraged to pursue a self-taught education in cryptology, and he developed a special ability in cryptanalysis. After graduation, he remained to become a Signal School instructor.

In addition to his teaching, Hitt studied U.S. Army encryptions and found them to be riddled with holes. He suggested that the British Playfair cipher be adopted, but his idea was rejected. Undeterred, he presented a version of the Etienne Bazeries cylinder to his superiors in 1914.

Hitt "unrolled" the letters from the disks of the cylinder and arranged them as strips of paper. On 25 such strips he placed two sets of mixed alphabets. The strips were numbered and put in a 7 by 3¼-inch frame according to a keynumber made from the series of strip numerals. The keynumbers changed and thereby altered the letters' arrangements. The strips were moved up or down to spell the first 20 letters of the message on a horizontal line. Any other jumbled horizontal line could be chosen as the ciphertext. In the 1930s and early 1940s versions of this system, designated M-138 and M-138A, containing 100 strips with as many as 30 slides used at a time, served the U.S. Army, Navy and State Departments in different security situations.

Hitt also published an important text on cryptanalysis. In 1916 the Army Service School's presses printed 4,000 copies of his Manual for the *Solution of Military Ciphers*. At 101 pages, it was the first actual book of its type published in America, since only some small pamphlets had been printed before. Although Hitt was commended for his work, the text was obsolete from its inception. The world war-generated encryptions had become far more difficult than Hitt had realized.

After the U.S. joined the Triple Entente in 1917, Hebern went to Europe with the staff of General John Pershing as the chief signal officer of the American Expeditionary Force's First Army. While he did not have direct cryptologic duties, he was often sought for consultations on concealing and revealing methods.

After his war service, Hitt was employed in the 1920s by the International Communication Laboratories, the cryptographic branch of the giant International Telephone and Telegraph. Like so many others who dealt with secret matters, he remained in obscurity, his important national service unheralded in his lifetime.

Kasiski, Friedrich (1805–1881)

Friedrich Kasiski was born in November 1805 in a western Prussian town and enlisted in an East Prussian infantry regiment at the age of 17.

He moved up through the ranks to become a company commander and retired in 1852 as a major. Although he had become interested in cryptology during his military career, it was not until the 1860s that he put his ideas on paper. In 1863 his 95-page text *Die Geheimschriften und die Dechiffrirkunst* (Secret writing and the art of deciphering) was published. A large part of its contents addressed the solution of polyalphabetic ciphers with repeating keywords, a problem that had tormented cryptanalysts for centuries. Although the Berlin publication received scant attention, historians have recognized it as an important addition to cryptology.

Kasiski was the first person to recognize that when a repeating section of a key meets a repetition in the plaintext letters, repeating ciphertext characters result. This was an essential aid to solving such polyalphabetic ciphers.

Kasiski's process led to methods by which analysts could count repeating sequences, total the spaces between them, and thereby learn how often the keys were reused and how many letters were in them. The number of letters in the key revealed the total number of alphabets in the polyalphabetic encryption. The analyst could then group all the characters encrypted by the first keyletter, all those concealed by the second keyletter and so forth. Such collections of letters, each governed by a single keyletter, were arranged as a monoalphabetic substitution and could be solved by established cryptanalytic techniques.

Disappointed by the lack of interest in his findings, Kasiski turned his attention to other activities including anthropology. He took part in artifacts searches and excavations and wrote numerous archeological articles for scholarly journals. He died in May 1881 not realizing the significance of his cryptanalytic findings.

Kerkhoffs, Auguste (1835–1903)

Born in January 1835 at Nuth, Holland, Auguste Kerkhoffs's full name was Jean-Guillaume-Hubert-Victor-François-Alexandre-Auguste Kerkhoffs von Nieuwenhof. The son of a prominent family, he received degrees in science and letters from the University of Liège. He then began a series of activities reflecting his expanding interest in foreign languages. He taught school in Holland, became a member of Dutch literary societies and then traveled through England, France and Germany as a secretary for an American journalist. In 1863 he obtained a high school teaching position in Melun, some 25 miles southeast of Paris, where he married in 1864.

He spent ten years in Melun studying topics ranging from archaeology to history and mathematics. From 1873 to 1876 he earned a Ph.D. at the Universities of Bonn and Tübingen. He financed his study by tutoring, and he continued to write on various subjects. While he eventually completed texts such as a Flemish grammar and a German verb manual, his major work was his *La Cryptographie militaire*. He began it in 1881, while teaching German at two Parisian schools, and he finally published it in 1883.

The 64-page text was a significant advance for cryptology. Kerckhoffs described the accepted parameters of the science, and gave accounts of modern enciphered code, mechanical encryptions, telegraphy and military dispatch secrecy. He also advised that wartime cryptography should have procedures that were easy to understand and efficient to operate.

Kerckhoffs is recognized as the first cryptology scholar to distinguish between the general cryptographic system (an alphabet tableau or codeword list) and its key. He realized that although the general system might be guessed or stolen by an agent, a strong key would still make the concealment very difficult to break.

La Cryptographie militaire is primarily noted for Kerkhoffs's two important cryptanalytic tools, superimposition and symmetry of position. Superimposition is a central method for solving polyalphabetic substitutions, while symmetry of position is a method of determining groups of ciphertext letters, a slightly indirect but clever way to try to identify the plaintext characters they covered.

In addition to these important analytic contributions, Kerkhoffs added one more tool for cryptographers. He promoted a device called the Saint-Cyr slide, named after Saint-Cyr, the national military academy of France, where it was part of the instructional courses. Making a connection that others had overlooked, Kerkhoffs realized that the polyalphabetic functions of the slide were similar to the alphabet tableau of Blaise de Vigenère and Giovanni Porta's cipher disk. This finding gave a new crypotographic validity to the slide and it enjoyed considerable popularity for a few years.

A successful educator and cryptology pioneer, Kerkhoffs had long worked for and promoted the cause of a shared form of speech for the globe called Volapük. Its failure to attract lasting use was a great disappointment to him.

Kullback, Solomon (1907–1994)

A native of New York City, Solomon Kullback had earned an MA in mathematics from Columbia University by 1929. In 1930 he was hired by William Friedman to work for the Signal Intelligence Service (SIS) within the U.S. Army Signal Corps. Having a knowledge of Spanish, he began as a junior cryptanalyst.

During the 1930s, the SIS developed encryption systems for the Army, solved foreign cryptomethods and continued research in cryptologic areas. As the armed forces of Japan and Nazi Germany grew more formidable, greater effort was directed toward their encryption systems. Due to U.S. interests in the Pacific, SIS paid special attention to Japan as its empire expanded in the late 1930s.

Solomon Kullback was one of a group of young cryptanalysts who met the difficult challenges of the Japanese language and cryptosystems. Others who took on the challenge with William Friedman were Frank Rowlett and Abraham Sinkov. These men built upon Agnes Driscoll's pioneering foundation for cipher machine solving. It was largely Kullback and

Rowlett who unraveled the Japanese machine called Red by U.S. analysts. This breakthrough, in turn, was important in the solving of a machine called Purple, which was used for Japan's diplomatic traffic. The Friedman team launched an attack on Purple with the help of the U.S. Navy's OP-20-G staff directed by Laurence Safford.

Solomon was thus involved in one of the greatest decryption achievements of any era when this top-level diplomatic encryption system was broken in 1940, primarily with the help of a U.S. cipher machine, the Purple Analog. This breakthrough was to affect a number of cryptologic and military actions throughout the course of World War II.

During the war, Kullback directed the SIS cryptanalytic branch. With the U.S. victory in 1945, his employment shifted to the Armed Forces Security Agency. In 1952, when the AFSA became the National Security Agency, Kullback became chief of its Office of Research and Development. He retired in 1962 and later taught at George Washington University in Washington, D.C., Florida State University and Stanford University.

Lamphere, Robert (b. 1918)

Lamphere had worked with the FBI in Alabama before being transferred to New York City. After investigating violations of the Selective Service Act for three and a half years, he joined the FBI's Soviet Espionage Squad in early 1945. It was here that he played a primary role in the investigation of Gerhart Eisler, an Austrian-born member of Soviet intelligence who had taken advantage of Russia's wartime alliance with the United States to practice espionage.

By 1947 Robert Lamphere had become a supervisor in the FBI's Espionage Section in Washington, D.C. Here he studied files of partially deciphered telegrams between Moscow and sites in the United States, such as the Soviet consulate in New York City. The intercepts had been clandestinely gathered by U.S. military intelligence since 1939.

Lamphere used solutions of some of the telegrams to expose Judith Coplin, an employee in the Justice Department. She had had access to documents from the Foreign Agents Registration offices and was providing information to Valentin Gubitchev, a Russian employee of the United Nations. Both were arrested by the FBI and convicted of espionage conspiracy. However, Coplin's 15-year sentence was overturned because of technicalities, and Gubitchev was deported.

Lamphere persevered with his investigations and made even more important revelations, including the infiltration of the Manhattan Project by Soviet agent and physicist Klaus Fuchs in 1944. Lamphere and the Espionage Section were able to trace Fuchs to the Rosenberg ring of atomic bomb spies.

Robert Lamphere concluded his FBI employment in 1955. His later work included executive positions with the Veterans Administration and a major insurance company.

Mauborgne, Joseph (1874–1971)

Joseph Mauborgne's early claim to cryptanalytic achievement was his solving a Playfair cipher in 1914. At that time this encryption method (actually created by Englishman Charles Wheatstone) was considered a difficult one, and Mauborgne is credited with the first documented third-party solution of this standard British army field cipher. He further enhanced his reputation in cryptologic circles by writing a 19-page booklet in 1914 about the procedures he used, *An Advanced Problem in Cryptography and its Solution*, the first such material published by the U.S. government.

One of his primary contributions to cryptography was his important improvement in the security method devised by AT&T engineer Gilbert Vernam. Mauborgne saw a weakness in the Vernam security system, which used the Baudot code and paper tapes of different lengths to encipher messages on the printing telegraph or teletypewriter.

Mauborgne saw that heavy use of the system might bring about repetition of the chosen keys even though two loops of tape were used. This repetition in turn might make the cipher susceptible to analysis by superimposition, the cryptanalytic system developed by Auguste Kerckhoffs. Mauborgne eliminated the possibility of repetition by combining the random key of Vernam's tape system with the nonrepeating key developed by the Army Signal Corps. The combination was called the "one-time system," and barring an accidental repetition, the method was considered unsolvable.

His second major achievement was in his role as chief signal officer for the Army Signal Corps. As a major general, in 1937, he promoted an increased interest in and funding for U.S. cryptographic efforts. He arranged for the Signal Intelligence Service to have more training staff and eavesdropping stations, a difficult task during the Depression. From a few dozen often undecipherable intercepts, the amount of usable decrypted data increased under Mauborgne's careful leadership. He also deserves special credit for his encouragement and support of William Friedman, a master analyst. It was Friedman, SIS staffers and Navy OP-20-G members who made the crucial attack on the top-level Japanese diplomatic encryptions named Purple by U.S. experts. Though Mauborgne retired in autumn 1941, his legacy served the United States through World War II and beyond.

Myer, Albert (1827–1880)

Born in 1827 near West Point, New York, Myer's childhood hopes of attending the United States Military Academy were dashed when his mother passed away. An aunt residing in Buffalo took him in and Myer obtained gainful employment which allowed him to finance his education. Myer worked at the New York, Albany, and Buffalo Telegraph Company during the summers while he attended Hobart College in Geneva, New York. His experiences with a needle indicator telegraph and an early version of a Morse telegraph influenced his fascination with communications methods.

After graduating from Hobart in 1847, Myer attended the Buffalo Medical College. His thesis, "A New Sign Language for Deaf Mutes," proposed Morse-like touches on the face and hands to represent words. After earning his medical degree in

1851, Myer had a private practice for three years before marrying in 1854. In that same year he passed his U.S. Army examination and was commissioned an assistant surgeon.

Army life often called for grueling and dangerous tours of duty in remote areas of the expanding nation. In October 1854 Myer began serving at various frontier forts in Texas. He had a healthy respect for the Native Americans of the region, and learned from their sign language.

During his time in Texas, Myer developed his signaling system that he called "flag telegraphy." His message exchangers used single banners with which they conveyed letters and numbers with varied movements. In October 1856 while at Fort Duncan, Texas, Myer wrote to the secretary of war and offered his method for consideration. After a long period with no reply, he turned to the U.S. Navy in 1858, who also turned him down. In February 1859 the Army finally agreed to evaluate his signals.

Myer was asked to conduct field texts, which he did with the help of Second Lieutenant Edward P. Alexander. From October through December 1859, they signaled each other from sites in New Jersey, Staten Island and Brooklyn.

Myer's process was accepted, and in June 1860 he was appointed signal officer with the rank of major. He patented his method in January 1861, and in the spring he was ordered to conduct further field texts in New Mexico. By this time an unidentified Army general is said to have nicknamed the motions of flag telegraphy "wigwag," and the name eventually superseded Myer's choice.

The Civil War brought Myer's system into prominence. First used successfully by Alexander, who had joined the Confederate Army, at the first Battle of Bull Run in July 1861, wigwag soon served both sides for the duration of the conflict.

Aware that visible exchanges could be understood by experienced men like Alexander, Myer developed a cipher disk to accompany the wigwag system. The disk enabled exchangers to equate numerals with letters. When the numerals were sent through wigwag, only the intended recipient with an identical disk could decrypt the meanings.

As the war proceeded, Myer perfected other inventions including signals with disks for daytime use and torches or lanterns for night. His career advanced as the value of his methods gave him increased credibility and prestige. In March 1863 he was promoted to colonel and was named chief signal officer of the Signal Corps. Under his guidance, the once fledgling corps grew substantially and by October 1863 their official roll included Myer, 198 other officers and 814 enlisted personnel.

As the Signal Corps' leader, Myer had been in charge of the Field Telegraph Section of the Army, which accompanied the North's troops and a few Navy vessels. They maintained land telegraph connections and some shore-to-ship links. Myer's overall responsibilities covered the training of new men, their battle area operations, and the purchase and management of materiél.

A bureaucratic rivalry with Anson Stager, head of the War Department's Military Telegraph Service, hindered Myer's career. In October 1861 the Telegraph Service was established as a civilian section of the Quartermaster Corps and took over existing corporate and railroad telegraph lines as part of the Union's defensive strategy. These already established wires handled administrative and strategic communications until the end of the Civil War.

Stager and Myer disagreed about the expanding Signal Corps' field telegraph links and the broader territories they covered. During the dispute, Stager won the favor of Edwin Stanton, the secretary of war, who ordered Myer to give up his responsibilities and turn them over to the next ranking Signal Corps officer.

Myer was a loyal Unionist and grudgingly accepted his demotion. For a time he was posted to Memphis, Tennessee, as a signal officer. Accompanying General Edward Canby's land forces, Myer saw action in a combined land-sea operation with Admiral David Farragut's ship in the Battle of Mobile Bay in Alabama on August 4, 1864. During this time Myer approached many officials in an attempt to return to his former position.

Some timely postwar accounting helped rejuvenate Myer's career. Expenditures by the Military Telegraph Service had amounted to $2,655,500. In comparison, the Signal Corps had spent $1,595,257—a difference of $1,060,243. This discrepancy drew high-level attention to Myer's wigwag.

Backed by this favorable report, Myer petitioned the War Department and central military figures such as Ulysses Grant, William Sherman and Philip Sheridan for their support. While awaiting an official response, he began writing his influential *Manual of Signals*, a compilation of practical communication methods.

Ironically, it was Stanton who informed Myer that he had been reinstated as both a colonel and chief signal officer on October 30, 1866. He lived to see his *Manual of Signals* (1868) and an 1872 reprint published to nearly unanimous acclaim.

He rose to the rank of brigadier general in June 1880 and in early August he retired and closed his office in Washington, D.C. He died 21 days later at the age of 52.

Porta, Giovanni (1535–1615)

Born in Naples, Giovanni Porta was a true Renaissance man. Brought up by a caring uncle, young Giovanni could pen essays in Latin and Italian by his tenth birthday. As a young adult he traveled to various European capitals and sites of scientific and natural curiosities. He wrote his first book, *Magia naturalis* at age 22.

Porta opened his Naples residence to men of similar pursuits and they began to conduct experiments on various topics. They became the premier organization of scientific thinkers, calling themselves the Accademia Secretorum Naturae. Influenced by these meetings and his own boundless curiosity, Porta wrote texts about astronomy and its occult counterpart astrology, architecture, memory enhancement, and light refraction, as well as comedies and tragedies. Porta later expanded *Magia naturalis* to 20 volumes, including recipes for secret inks and methods of steganography.

In 1563 another book established his place in the pantheon of cryptology. *De Furtivis Literarum Notis* contained four sections tracing the history of ciphers up to his time, solution processes and specific language characteristics. It included one of the earliest classifications of transposition and substitution methods. He also suggested the use of cipher keys with lengthy terms or irrelevant words to confuse interceptors.

For encryptions, Porta developed a disk with two dials, a fixed one that contained an ordered alphabet and a mobile one with a series of symbols. Even more important was his tableau, which replaced a letter at the top of the table and the other at the side with a single symbol at their coordinating point. This was the first digraphic cipher in cryptologic history.

Although he did not create any new polyalphabetic systems, he did list the primary methods of Alberti, Belaso and Trithemius, synthesizing their findings for future cryptographers. He also published the first European solution of a monoalphabetic substitution that had neither real nor false word divisions (their presence in a missive had been an aid to earlier analysts).

Porta pioneered the solution method known as the probable word. If the communiqué was suspected to be about treaty dealings, words such as bargain, offer, negotiate, accept or propose might be expected to be present. Awareness of the spellings and likely sentence structure would aid analysts in finding them and in identifying other terms associated with them.

Porta also proposed methods to break polyalphabetic ciphers in the first issue of *De Furtivis* and in a chapter added to an edition published in 1602. He was apparently the first to notice that alphabetically ordered letters in a word, such as *s t u* in the Latin word *studium* resulted in repeating cipher letters or symbol substitutions when polyalphabets were used. This was also true when letters were reversed but still in alphabetical order, like the letters *p o n* in *pondus*. He had also discovered that repeating plaintext characters generated repeating ciphertext characters in a polyalphabetic cipher. He came close to discovering the link between repeating key and plaintext sections that create repetitions in the ciphertext, which was not fully defined until the Prussian army officer Friedrich Kasiski did so in 1863.

Cryptology historians consider Porta to be the foremost Renaissance scholar of the science. His various studies made him the equal of any of the scholars of the time.

Rawlinson, Sir Henry Creswicke (1810–1895)

Sir Henry Rawlinson was born in Chadlington, Oxfordshire, in 1810. Educated at Warrington and Ealing, he grew up in Great Britain's golden age. Although the 1812 war with the United States had been controversial and had cost Britain considerable prestige, the empire still held numerous colonies across the world.

At the age of 17 Rawlinson followed in the footsteps of many young men when he chose military service with the East Indian Company. He began his career as a cadet and served in India (1827–33), Persia (1833–39) and Afghanistan (1838–42), where he distinguished himself in the Afghan War of 1842. During this time he studied some of the primary languages of the region including Hindustani, Arabic and modern Persian.

In 1833 he became a cavalry major and rose to the position of military adviser to the Shah's brother. While helping reorganize the Shah's army, he became fascinated by Persian antiquities. While traveling through western Persia's Zagros Mountains, he was first shown the series of inscriptions regarding Darius the Great near Behistun (now Bisitun, Iran). Carved into the side of a mountain, the symbols were for three languages, Old Persian, Elamite and Babylonian.

In 1835 Rawlinson began a life-threatening effort to transcribe the inscriptions while working on a rock ledge. With the equally courageous help of a Kurdish youth they made papier-mâché molds of a number of the symbols, the project took a decade to complete. In 1846 Rawlinson succeeded in translating the Old Persian section of the Behistun inscriptions. In conjunction with the efforts of other epigraphers and archeologists, decryptions of the other scripts soon followed.

During this long period of transcription and translation, Rawlinson's military career advanced. In 1843 he had become a political agent in Turkish Arabia, and in 1843–44 he was appointed British consul at Baghdad and became consul general in 1851. He also won a grant from the British Museum to continue the Assyrian and Babylonian excavations begun by the noted archaeologist Henry Austen Layard.

In 1855 Rawlinson returned to England, where he was subsequently knighted and became a crown director of the East India Company. He served in Parliament (1858 and 1865–68), was created a baronet in 1891 and served as president of the Royal Asiatic Society (1878–81) and Royal Geographic Society (1871, 1874). His primary texts included *Persian Cuneiform Inscription at Behistun* (1846–51) and *Outline of the History of Assyria* (1852). He died in London in March 1895.

Rossignol, Antoine (1600–1682)

Antoine Rossignol was born in 1600 in Albi, a small town in southern France, but very little is known about the first years of the man who became France's first full-time cryptologist.

His professional career began in 1628 during religious violence between the Huguenots and Catholics. The Huguenot town of Réalmont was being besieged by forces led by Henry II of Bourbon. A Protestant courier had been captured and was found to possess an enciphered missive apparently meant for supporters elsewhere. Someone informed Henry II about a resident of Albi, a dozen miles from Réalmont, who was skilled at deciphering such secrets.

The encryption was sent to Rossignol and he discovered its contents with surprising ease. He reported that the

Huguenots' ammunition supply was very low. Henry II sent the message back to the Protestant defenders and, their weakness revealed, they quickly surrendered.

Rossignol then became the nemesis of the Huguenots, as he solved their dispatches captured at the sieges of La Rochelle and Hesdin. By this time Rossignol was working for the very influential Cardinal Richelieu, one of France's most prominent citizens. Soon favored by Louis XIII himself, Rossignol was able to afford to build a château near Paris.

He continued his full-time service to the subsequent king, Louis XIV, and Richelieu's successor, Cardinal Mazzarin. In addition to analysis, he contributed to better national security procedures. He recognized that the nomenclator, the era's primary concealment method, had a significant flaw. Both its plaintext words and their codeword replacements were arranged in alphabetical or numerical order. He knew such predictable patterns of certain terms preceding or following others had helped him break cryptic shields. He therefore developed the two-part nomenclator, wherein the encoding section's elements were in random arrangement alongside alphabetically ordered words. In the decoding section, codewords were alphabetized and plaintext was mixed.

He married at the age of 45 and his life was filled with important friends and elaborate banquets. Poems were written in his honor and he was praised in the memoirs of contemporaries. Rossignol's son and one of his grandsons carried on the family tradition into the 1700s. Their skills were imitated by other Frenchmen who later established France's organized Cabinet Noir.

Safford, Laurence (1893–1973)

A Massachusetts native, Laurence Safford graduated from the United States Naval Academy and was commissioned as an ensign in 1916. After serving in China, by 1924 he was officer-in-charge of the (Cryptologic) Research Desk of the Code and Signal Section of the Division of Naval Communications. With his tiny staff, Safford made his office in Room

1621 of the old Navy Department Building in Washington, D.C.

Under the guidance of Agnes Driscoll, Safford put to use his own aptitude for mathematics and mechanical devices. In the 1920s, the Office of Naval Intelligence (ONI) surreptitiously photographed codes kept in the Japanese consulate in New York. After decryption, this data was called the Red code (also Red book) because of the color of its binding. This material became a pillar upon which Safford and Lieutenant Joseph Rochefort built a full-fledged department staffed by well-trained personnel (nicknamed the On the Roof Gang), as well as a network of listening posts in major naval operations areas.

With these improvements in the late 1920s came a new name for Safford's office and staff, OP-20-G. *OP* was from OPNAV, the Office of the Chief of Naval Operations; *20* referred to the 20th division of OPNAV and *G* designated the Communications Security Section. OP-20-G was also known as Station Negat. However, with such expansion came division of responsibilities as well as personal rivalries for control of intelligence data.

Disputes over analysis and distribution of intelligence material erupted between ONI and OP-20-G in the early 1930s and in the late 1930s between OP-20-G and the Navy War Plans Division led by Rear Admiral Richmond K. Turner. The latter dispute continued while Safford and OP-20-G worked with William Friedman and his Army SIS staff to solve a Japanese diplomatic encryption given the name Purple. Amazingly, the Army group then developed a Purple analog and accomplished decryptions that became part of the United States' cryptanalytic magic.

Tragically, the full range of Purple solutions and their implications about potential Japanese actions were not conveyed to U.S. commanders at Pearl Harbor, and Safford became a central figure in the ongoing attempts to assign blame for the December 7, 1941, disaster. From the Congressional hearings in 1946 he will always be remembered as the man who believed that a crucial warning sign of war called the Winds code had been intercepted and should have been made known to the Hawaiian commanders.

Safford managed to survive the bitter debate about intelligence failures and coverups, though his career was permanently damaged. After the war he served as assistant director of Naval Communications for Cryptographic Research until 1949. He later held the positions of special assistant to the director of the Armed Forces Security Agency and special assistant to the head of the Security Branch in the Division of Naval Communications. His active duties ended on March 2, 1953.

Despite his controversial link to the Pearl Harbor disaster, Laurence Safford is remembered for his pioneering work in naval intelligence when it was needed most.

Trithemius, Johannes (1462–1516)

Ridiculed by a stepfather for his interest in learning, the young Johannes Trithemius nevertheless persevered, studying at the University of Heidelberg and becoming an abbot of a Benedictine order in Germany at a very young age. His energetic mind generated collections of sermons, histories and biographies, earning him the title "Father of Bibliography." He also became deeply involved in studies of the occult and the practices of alchemy. From this realm of mysticism and pseudoscience, his interest in cryptology developed.

In 1499 Trithemius began a series of volumes entitled *Steganographia* (Greek for "covered writing"). In its early sections, he wrote about methods of vowel-consonant substitution, and the cryptographic uses of nulls and nonsensical words. The text remained in manuscript form for more than a century because of its heretical contents about magic, otherworldly beings and secret powers. After it was finally printed in 1606, it was placed on the Catholic Index of Prohibited Books.

Trithemius's direct contribution to cryptology was in his *Polygraphia*, published by a friend in 1518, a year and a half after his death. A collection of six books, the *Polygraphia* contained columns of Latin terms, words equated with plaintext letters and the first square table in cryptographic writings, the foundation of polyalphabetic substitution. Trithemius also developed the original progressive key, whereby all the cipher alphabets were used before repeating any one of them.

Trithemius's influence as a scholar and author did much to promote an understanding of secret-writing techniques. His work laid the foundations for many who came after him, including Leon Battista Alberti of Florence and Giovanni Porta of Naples.

Ventris, Michael G. F. (1922–1956)

Born in Hertfordshire, England, Ventris became fascinated with classical history as a boy, and he studied Greek and Latin at school. He later began adding French, German, Polish and Swedish to the list of languages in his repertoire.

After hearing a lecture by Sir Arthur Evans, Ventris became curious about the decipherment of Linear B, the script that Evans had found at Knossos, Crete. By the time he was 18 years old Ventris's paper proposing a relationship between Linear B and Etruscan had been accepted by the *American Journal of Archaeology*.

Ventris had also shown an aptitude for architecture, and he pursued this interest rather than the more obscure decipherments. His studies were interrupted by World War II, but after honorable service in the Royal Air Force, Ventris completed his architectural degree. In 1949, he began a full-fledged effort to solve Linear B.

Using the articles of U.S. professor Alice Kober, Ventris developed a pivotal grid of vowels and consonants that brought more order to his research. He also shared his findings ("Work Notes") with others and the exchanges contributed significantly to his progress. While being interviewed on a British radio program in June 1952, he publicly announced for the first time that he believed Linear B was an archaic form of Greek.

The next month he and Cambridge linguist John Chadwick, a Greek scholar, began a collaboration that built upon Ventris's findings. In 1953 they published their conclusive paper "Evidence for Greek Dialect in the Mycenaean Archives." Their text *Documents in Mycenaean Greek* was published in 1956, just a few weeks after Ventris's fatal car accident in September near Hatfield, England.

Vigenère, Blaise de (1523–1596)

Born in April 1523 in the village of Saint-Pourçain just northwest of Vichy, Vigenère's small town origins were no handicap to his future success.

Little is known about his upbringing until his teenage years. His education brought him a secretarial position at the Diet of Worms, a major convocation of the Catholic Church held in 1521 in Germany. This employment brought him into contact with church representatives, embassies and emissaries. Under the tutelage of the duke of Nevers, Vigenère began his life's work in diplomacy. He first became acquainted with cryptology on missions to Rome, and soon began to study and practice it.

At the age of 47, Vigènere retired from diplomatic life and devoted himself to writing. He produced a variety of texts including a *Traicté des comètes,* which disproved certain myths about the appearance of comets.

Although his 1585 book *Traicté des chiffres* was a landmark text, Vigenère apparently believed that cryptology was a wasteful use of time and intelligence. Written after his marriage to a much younger woman, the *Traicté*'s contents included analyses of subjects from ciphers to alchemy and magic. He also discussed a series of polyalphabetic systems of his day.

Like his predecessors Leon Battista Alberti and Johannes Trithemius, Vigenère primarily focused on the use of tables with shifting alphabets in his discussion of polyalphabetic ciphers. He developed a square table similar to Trithemius's *tabula recta*, but created varied alphabets that ran along the top and side of the tableau. Vigenère also improved Girolamo Cardano's autokey system in two ways. One of his inventions was the nonrepeating priming key. His second advance was an autokey with a priming key that used the ciphertext itself to encrypt the rest of the message.

The *Traicté* and the key developments were important contributions to cryptology, yet the priming and autokeys fell into disuse and were lost for many decades. It was not until the latter 1800s that others independently discover the practical ideas Vigenère had proposed almost 300 years before. In the meantime, Vigenère's name became associated with a much less secure system which had a brief repeated keyword and a standard A to Z alphabet table without the variations he had recommended.

Wheatstone, Sir Charles (1802–1875)

From his birth in February 1802 through his formative years, Charles Wheatstone remained in his native Gloucester, England, where he had a private school education.

Pursuing his early interest in sound, he founded a stringed instrument business in London when he was 21. His first scientific experiments to receive professional attention were studies of motions affected by harmonics and observation of vibrating surfaces. He also wrote articles about phonetics and hypothetical speaking mechanisms.

In 1834 he expanded into other areas when he became a professor of experimental philosophy at King's College in London. A retiring man, he rarely lectured, but the scholastic settings and facilities enabled him to study various esoteric subjects. During this time he invented the stereoscope, a mechanism which contained two photographs of an object taken from slightly varied angles. When each was viewed side by side through two eyepieces, a three-dimensional image appeared.

Wheatstone also experimented with measurements of the velocity of electric current in different types of wire. This research may have influenced his studies with telegraphy. A collaboration with inventor William F. Cooke resulted in a type of electrical telegraph with needles that moved to indicate letters, an invention which actually preceded that of Samuel Morse. Historians disagree on the extent of its practicality or use before the Morse version, with its practical dots and dashes, made electrically sent signals popular worldwide. Wheatstone and Cooke later developed telegraphs with a dial letter indicator and a type-printer feature.

One of the inventions named for Wheatstone was in fact not his. The Wheatstone bridge, an instrument applied to determine electrical resistance, was actually the creation of Samuel H. Christie in 1833. Around 1843 Wheatstone began promoting the device, and his better-known name became linked with Christie's mechanism. Ironically, a nearly identical situation occurred with Wheatstone's cryptography hobby.

Sir Charles amused himself by solving newspapers' "agony columns." These personal advertisements popular in the mid-to-late 1800s were encrypted communications used by secret lovers. Discovering how weak some of the simple letter substitution or number replacement methods actually were, Wheatstone experimented with better systems of his own.

In 1854 he created a manual, polygraphic encryption system that was the first literal digraphic cipher. Wheatstone's friend Baron Lyon Playfair extolled the method's virtues among British officials, and it soon became known as Playfair's cipher. While Wheatstone did not succeed in persuading the military hierarchy to apply his process in the Crimean War, it did see service in the Boer War in South Africa and in both world wars.

Ever the experimenter, Wheatstone devised another system with a mechanism that he called a Cryptograph, which he first presented at a Paris exposition in 1867. The Cryptograph had outer and inner alphabet rings and two clock-type hands linked to gears. The gearing aspect made the hands move in a ratio that was supposed to create a smooth operation and make very good concealments. It accomplished the former but fell short of the latter. The alphabet differential proved to be only one letter, and this was no real barrier to experienced cryptanalysts.

Like any good scientist, Wheatstone moved on and more acclaim followed. Already a member of the Royal Society, he became a corresponding participant in many of the primary scientific organizations of Europe. He was knighted in 1868 and received honorary degrees from Cambridge and Oxford. While he had written no major texts, his numerous articles in professional journals were a fitting record of his eclectic achievements.

On his passing in Paris in October 1875, he was eulogized as an example of the scientific mind at its best.

Yardley, Herbert (1889–1958)

Herbert Osborn Yardley may very well be the number-one "character" of cryptology. While not as brilliant as his contemporaries William Friedman of the United States and Georges Painvin of France, Yardley's life was quite colorful, as well as important to some of the modern developments of U.S. codes, ciphers, and cryptanalysis.

Born in Worthington, Indiana, Yardley became first a railroad telegrapher and then a code clerk with the U.S. State Department, where he saw the expansion of communication in business and in worldwide diplomacy.

Yet as the United States flexed its industrial muscles and looked for new markets and alliances, its supposedly private embassy exchanges were not well protected. Yardley amused himself by solving some State Department codes and thereby enhanced his reputation as an expert.

As the United States entered World War I, Yardley persuaded his superiors to give him a captain's commission and the direction of MI-8, the new cryptologic section of U.S. military intelligence. His energy and creativity helped to build MI-8 into a very credible organization that developed new Army codes, broke international encryptions, and produced invisible inks, as well as the chemicals to reveal them. He continued with MI-8 until the November 1918 armistice and remained in France to direct the cryptology section of the U.S. delegations to the Peace Conference.

In 1919, back at the State Department code room, Yardley became concerned about maintaining a full-time cryptologic department amid peacetime financial cutbacks. He directed his efforts toward this cause and achieved it in 1919 with a new organization jointly funded by the State and War Departments. Due to government spending technicalities, it was decided to shift the staff and the equipment to New York City, where the new group came to be known as the American Black Chamber.

The group had some of its greatest difficulties and most brilliant successes with decryptions of intercepted Japanese telegrams. Once language problems were surmounted and it was ascertained that the Japanese codes used *kata kana*, Yardley and his team made important discoveries. Some of the most crucial involved a disarmament conference that began in Washington, D.C., in November 1921. The decryption of Japan's diplomatic exchanges helped U.S. negotiators to gain the advantage during the meetings. The resulting Five-Power Treaty gave the United States superior ship-tonnage ratios, especially in relation to Japan, a growing rival in the Pacific.

Yet in spite of this success and others (reputedly around 45,000 international telegraphs were decrypted), funding cuts and isolationistic political sentiment combined in the decisions that led to the closing of the chamber in October 1929.

Embittered by this and financially drained by the Depression, Yardley published his memoir, *The American Black Chamber*. Published in 1931 and serialized in the *Saturday Evening Post*, the book was a commercial success and created a furor at home and abroad. Its revelations alarmed Tokyo's cryptographers, who quickly adopted improved secrecy procedures. William Friedman of the U.S. Army Signal Corps and other U.S. government officials were aghast.

Yardley struck back at his critics in magazine articles indicating that his book was intended as a warning to the United States to strengthen its own encryptions. While his reputation suffered in some circles, Yardley's literary efforts produced two novels, *The Red Sun of Nippon* and *The Blonde Countess*. While neither book had widespread success, the latter was made into a film with the title *Rendezvous* in 1935.

With money from his books and the movie, Yardley made some New York City-area real estate deals. When these faltered, he signed on as a cryptanalyst for Chiang Kai-shek, whose China was already under assault by Japan. From 1938 to 1940, he mixed cryptography with his long-time interest in poker, as well as with a series of female companions and a joie de vivre that historians say masked a growing cynicism.

After leaving China in 1940, he was variously involved in a failed restaurant deal, the Canadian Department of External Affairs (cryptanalysis work), and the Office of Price Administration in Washington, D.C. Nonfiction rewarded him a second time in 1957 with his successful book *The Education of a Poker Player*. He died in 1958.

Despite his highly questionable revelations, Herbert Yardley is credited with stimulating public interest in cryptology to a degree not seen since the days of author and amateur cryptologist Edgar Allan Poe.

QUIZZES AND ANSWERS

Quizzes

QUIZ 1

Apply the Polybius checkerboard to reveal the famous sites behind the numbers.

1. 33, 11, 43, 11, 45, 23, 35, 34
2. 14, 15, 32, 41, 23, 24
3. 11, 45, 23, 15, 34, 44
4. 13, 54, 13, 32, 11, 14, 15, 44
5. 41, 11, 43, 45, 23, 15, 34, 35, 34
6. 32, 11, 13, 35, 34, 24, 11
7. 11, 13, 43, 35, 41, 35, 32, 24, 44
8. 41, 24, 43, 11, 15, 51, 44
9. 33, 45, 35, 32, 54, 33, 41, 51, 44
10. 41, 15, 32, 35, 41, 35, 34, 34, 15, 44, 51, 44

QUIZ 2

Use the Polybius checkerboard again to discover these historic names.

1. 23, 15, 43, 13, 51, 32, 15, 44
2. 41, 43, 24, 11, 33
3. 11, 22, 11, 33, 15, 33, 34, 35, 34
4. 32, 11, 35, 13, 35, 35, 34
5. 23, 15, 32, 15, 34, 35, 21 45, 43, 35, 54

6. 41, 11, 43, 24, 44
7. 11, 13, 23, 24, 32, 32, 15, 44
8. 45, 43, 35, 25, 11, 34 23, 35, 43, 44, 15
9. 23, 15, 13, 45, 35, 43
10. 13, 32, 54, 45, 15, 33, 34, 15, 44, 45, 43,
    11

QUIZ 3

In this version of Caesar's cipher, *d* = *a*. Use this cipher to uncover the people, places and events of Caesar's time.

1. u x e l f r q
2. f o h r s d w u g
3. d x j x v w x v
4. j d x o
5. e u x w x v
6. s r p s h b
7. o h j l r q v
8. d q w r q b
9. f d o s x u q l d
10. l g h v r i p d u f k

QUIZ 4

In this Caesar substitution *a* = *w*. Set up your alphabet accordingly and identify these names linked with the Roman Empire.

1. p n e x q j a
2. d w j j e x w h
3. y k h k o o a q i
4. o a j w p a
5. d q j o
6. a i l a n k n
7. y w n p d w c a

8. w p p e h w
9. p e x a n
10. r e o e c k p d o

QUIZ 5

During the Middle Ages, alchemy flourished for many years. Refer to the zodiac alphabet and learn this terminology linked with those who sought the philosopher's stone.

1. ♄ ☉ ♈ ♊ ♆ ♐ ♍ ♌
2. ♒ ☿ ☉ ♊
3. ♃ ♅ ☉ ♉ ♅ ♐
4. ⊸ ☉ ♓ ♅ ♐
5. ♎ ♅ ♑ ♓ ♊ ♅
6. ♑ ♄ ☉ ♊ ♅ ♑
7. ♁ ☿ ♐ ♅
8. ♋ ♍ ♐ ♓ ☉ ♐
9. ♊ ♅ ☉ ♆
10. ♄ ♂ ☉ ♐ ♄ ♍ ☉ ♊

QUIZ 6

Refer to the Alberti cipher disk and its letter alignment to determine the names of these Renaissance cities.

1. f o y p z
2. n p y v q z y q c y p i h o
3. et o y c n o
4. f k o y z b z
5. y z i h o v
6. e z k n o h p y z
7. k o c t v
8. s h p k o y n o
9. t z b k c b
10. z et c f y p y

QUIZ 7

Use the Alberti cipher disk once more to uncover the names of these prominent Renaissance leaders.

1. c v z e o h h z
2. i p h p
3. k e z n p y
4. n p k q o r
5. b z f z t z
6. s o k b c y z y b
7. b o h o p y
8. n o k et z y q o v
9. t z f o h h z y
10. n b o t o b c n c

QUIZ 8

The Porta disk will help you unravel these facts related to Renaissance explorers.

1. [symbols]

2. [symbols]

3. [symbols]

4. [symbols]

5. [symbols]

6. [symbols]

7. [symbols]

8. [symbols]

9. [symbols]

10. [symbols]

QUIZ 9

Using the Roman numerals on the Porta disk, discover these encrypted words relating to capitalism.

1. XVII I XIX IX XII VII
   XVII
2. X XII I XII
3. IIII V XII XI I IIII
4. VI XVI V V XVIII XVI
   I IIII V

5.	XI	V	XVI	III	VIII	I
	XII	XVIII				
6.	II	XIII	XII	IIII	XVII	
7.	XIIII	XVI	XIII	VI	IX	XVIII
8.	IX	XII	XIX	V	XVII	XVIII
	XIII	XVI				
9.	VI	IX	XII	IXII	III	IX
	I	X				
10.	XI	IX	IIII	IIII	X	V
	III	X	I	XVII	XVII	

QUIZ 10

During the time of Francis Bacon there was much interest in magic and the occult. Use his bi-literal cipher to discover these words.

1. babaa, abaaa, baaba, aaaba, aabbb
2. aaabb, aabaa, ababb, abbab, abbaa, baaab
3. abbba, aabaa, abbaa, baaba, aaaaa, aaaba, ababa, aabaa
4. baaab, abbab, baaaa, aaaba, aabaa, baaaa, babba
5. aabaa, baabb, abaaa, ababa aabaa, babba, aabaa
6. babaa, aaaaa, baaaa, ababa, abbab, aaaba, abbab
7. aaaaa, ababb, baabb, ababa, aabaa, baaba
8. aabbb, aabaa, babab
9. abbba, abbab, baaba, abaaa, abbab, abbaa
10. aaaab, ababa, aaaaa, aaaba, abaab, ababb, aaaaa, aabba, abaaa, aaaba

QUIZ 11

Mary, Queen of Scots was quite familiar with these people and place. Use her secret alphabet to learn about them.

1. *w a b R t k H o   w b b t k H*

2. *Ⅱ b k b ʊ ɯ f a b ʊ*

3. *g b g ʊ ɯ f t x ɯ   ∞ k x t*

4. *∞ a H k ʊ ∞ ∞ H b*

5. *g b k k b R m*

6. *b w x t k b ɯ m*

7. *∞ a ʊ k ʊ ∞   x ʊ   b ∞ b ʊ ɯ*

8. *H k ʊ ᴦ b g H t a*

9. *f ʊ ʊ ʊ x R m*

10. *ʊ x t a H R ʊ ɯ f a b o   w b b t k H*

QUIZ 12

Using the knock cipher, translate these sounds from the next cell to formulate your own escape plan. Note: Each section of knocks represents one word spelled vertically.

Knock, Knock, Knock Knock =
Knock Knock, Knock, Knock, Knock, Knock =
Knock, Knock, Knock, Knock, Knock Knock, Knock, Knock, Knock, Knock =

Knock, Knock Knock, Knock, Knock, Knock =
Knock, Knock, Knock Knock, Knock, Knock, Knock =

Knock, Knock, Knock, Knock, Knock Knock, Knock,
Knock =
Knock Knock =
Knock, Knock, Knock Knock, Knock =
Knock, Knock, Knock Knock, Knock =

Knock Knock, Knock, Knock =
Knock, Knock, Knock, Knock Knock, Knock, Knock =
Knock Knock, Knock, Knock, Knock, Knock =
Knock, Knock, Knock, Knock, Knock Knock, Knock =
Knock, Knock Knock, Knock, Knock, Knock =
Knock Knock, Knock, Knock =
Knock Knock, Knock, Knock, Knock, Knock =

QUIZ 13

Your benefactor in the next cell has heard you find the key.
Decipher his second aural message to complete your plan.

Knock Knock, Knock, Knock =
Knock Knock, Knock, Knock, Knock, Knock =
Knock, Knock, Knock Knock, Knock =
Knock, Knock, Knock Knock, Knock =
Knock Knock =
Knock, Knock, Knock, Knock Knock, Knock, Knock =

Knock, Knock, Knock, Knock, Knock Knock, Knock,
Knock =
Knock Knock, Knock, Knock, Knock, Knock =
Knock, Knock, Knock Knock, Knock =
Knock, Knock, Knock Knock, Knock =

Knock, Knock, Knock, Knock Knock, Knock, Knock, Knock =

Knock, Knock, Knock, Knock, Knock Knock, Knock, Knock =

Knock, Knock Knock, Knock, Knock, Knock, Knock =
Knock, Knock, Knock Knock, Knock, Knock =

Knock, Knock, Knock, Knock Knock, Knock, Knock, Knock, Knock =
Knock, Knock, Knock Knock, Knock, Knock, Knock, Knock =

Knock, Knock Knock =
Knock, Knock, Knock, Knock Knock, Knock, Knock =
Knock Knock, Knock, Knock, Knock, Knock =
Knock Knock, Knock, Knock, Knock, Knock =
Knock Knock, Knock, Knock, Knock =
Knock, Knock, Knock Knock, Knock, Knock, Knock, Knock =
Knock, Knock, Knock Knock, Knock, Knock =

QUIZ 14

Break the long-held cipher of the Rosicrucians to determine the names of these wise people.

1. ⌐ ▢ ⌐ ⌐∙ ⌐ ⌐⌐ ⌐

2. ⌐ ▢ ▢∙ ⌐ ⌐ ⌐ ⌐∙ ⌐⌐

3. ⌐ ⌐ ⌐ ⌐∙ ⌐ ⌐∙ ▢

4. ⌐ ▢ ⌐ ⌐∙ ⌐∙

5. ∙⌐ ⌐ ⌐ ⌐ ⌐∙ ⌐

6. ▢ ▢ ⌐ ⌐ ⌐∙ ⌐ ⌐ ⌐ ⌐∙ ⌐⌐

7. ⌐ ▢ ⊐ ⊏ ⊡ ▢ ⊡

8. ⊏ ▢ ⌐ ⌐ ⌐ ⊔ ⊔ ⌐

9. ▢ ⊔ ⊓ ⌐ ▢ ▢

10. ⊐ ⊔ ⊡

QUIZ 15

The Rosicrucians sought a balance of religion, science and philosophy for mankind's betterment. Find the names of the philosophers hidden below.

1. ⊏ ⊐ ⊔ ⌐ ▢

2. ⊔ ▢ ⊏ ⊔ ⊏ ⊡ ⊾ ⊔ ⌐ ⌐

3. ▢ ▢ ⊾ ⊔ ▢ ▢ ⊔ ⊔

4. ⊐ ⊡ ⊏ ⊓

5. ⊔ ⊏ ⊾ ⌐ ⌐ ▢ ⌐ ⊐ ⊔

6. ⌐ ⌐   ⌐ ⊾ ▢ ▢ ⊔ ⌐

7. ▢ ⊔ ⊡ ⊔ ⊔ ⊐

8. ⊔ ⊔ ⊐ ⊓ ⊾ ⊡

9. ⊾ ⊔ ⊏ ⊓ ⊔ ⊏

10. ⌐ ⌐   ⊔ ⊏ ⊔ ▢ ⊔ ⊾ ⌐

QUIZ 16

A number of U.S. presidents have been Masons. First discern the meaning of these ciphers, then name the president with whom the word is best associated.

1. ⊔ ∟ ∟ ∨ < ˃ ⌐

2. ⊓ ⊔ ⌐ < ∟ ∟ ☐ ⊡

3. ⊓ ⊔ ∧ ☐ < ˃ ⊡ ∟ ☐ ⊔ ☐

4. ∟ ☐ ˃ ⊡ ☐ < ⊔ ∟

5. ∨ ∟ ⊔ ⊡   ⊓ ⊔ ⌐ ⌐ ∧

6. ☐ ⊡ ∨   ⊔ ∟ ⌐ ⊡ ⌐ ☐ ∨

7. ⌐ ∟ ⌐ ☐ ⊓ ∨ ⊔ ☐

8. ∟ ⊔ ∧ ⌐ ⊏ ⌐ ☐ ⊓   ∟ ⊡ ⌐ ⊓ ∧

9. ⌐ ⌐ ⌐ ⊏ ⊡ ⌐ ⊔ ∟

10. < ⊏ ∟ ∟ < ☐ ⊡ ☐ < ⊏

QUIZ 17

Continue to break the clues to these presidents' identities.

1. ⊏ ⊔ ☐ ⊡ ∨ <

2. ∟ ⊔ ∧ ⌐ ⊏   ∟ ∟ ⊓ ⊡ ∟ ∨

3. ∨ ∟ ⌐ ☐ ☐ <

4. ∟⊓⌐▢⌐⅃⊏⊓▢▢<

5. ⅃∟>∟⅃ ∨⌐∟ ⊏▢∟∪

6. ⊓∟∨⌐∧<▢⊓ >∪<▢

7. ∨∪▢ ∪∪ ⌐∟▢∨∟⊓▢▢<

8. <▢∟⊓∨ ∨⌐⅃∟<

9. ∨∧⌐∟▢⊓▢ ⅃∪∧∟<

10. ⊓⌐∟⅃ ⊏∪∟∨▢

QUIZ 18

Use the one-handed alphabet of Abbe Charles de l'Epée to decipher these words.

1.

2.

3.

4.

5.

6.

7.

8.

9.

10.

QUIZ 19

Decipher these numbers signed in Epée's alphabet.

1.

2.

3.

4.

5.

6.

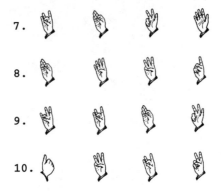

7.

8.

9.

10.

QUIZ 20

The Culper-Tallmadge alphabet is needed to solve these ciphers. Prepare your own alphabet chart using the Culper-Tallmadge example and place equated numbers above and below the line (e.g., $a = 1$; $e = 5$) to discover the names of these famous patriots.

1.	27	30	40	48	6	40	10	12	30	8
	42	32	30							
2.	10	26	30	32	50					
3.	30	14	46	14	30	14				
4.	40	6	12	6	27	40				
5.	46	32	30	40	42	14	44	8	14	30
6.	25	12	6	30	30	6	8	10		
7.	16	30	6	30	26	25	12	30		
8.	30	32	10	10	6	27	8	14	6	44
9.	30	10	6	25	14					
10.	27	34	12	42	10	10	14	30		

QUIZ 21

The Culper-Tallmadge alphabet will help you discern these terms and tools of Colonial spies. Once you've discovered how the numbers were created, the cipher will be easy to solve.

1. 399 45 21 216 45 440 27 224 165
2. 32 5 7 7 45 216
3. 440 216 5 288 32 255 255 216
4. 156 5 182 288 399 27 7 224
5. 7 255 12 45 440 575 45 45 224
6. 288 255 27 399 255 224
7. 16 27 32 32 45 224 288 255 21
   165 45 440
8. 156 255 224 7 7 156 5 399 399
9. 575 5 624 399 45 5 156
10. 156 156 255 575 12 483 156 156 45
    440

QUIZ 22

Focus your long glass on these banners and identify ten vessels famous for exploration or military glory.

1.

2.

3.

4.

5.

6. [flag symbols]

7. [flag symbols]

8. [flag symbols]

9. [flag symbols]

10. [flag symbols]

QUIZ 23

Refer to Chappe's signaling method to signify these people, places and events that affected the life of Napoleon Bonaparte.

1. ⌐ ↘ ⌐ ˢ ↑ ⌐ ⌣

2. ↗ ↘ ⌐ ↙ ✓ ↑ ↘ ↘

3. ↖ ↘ ˢ ⌐ ↘ ↘

4. ∟ ⌐ ↑ ↘ ⌐ ↑ ↘

5. ⌐ ∟ ↗ ⌣

6. ↘ ∟ ✓ ↘ ↘ ⌣ ⌐ ✓

7. ↗ ⌐ ∟ ↘ ↑ ↘ ↖

8. ↘ ⌐ ∟ ∟ ↑ ↘ ↘ ↗ ↘ ↘

9.) ⌣ ∕ [⌐ L ⟍⟍

10. S∕ ſ [L [⟍ ⌣

QUIZ 24

Morse Code has helped save many lives during national emergencies. The American Morse version will help you uncover these newsmaking events.

1.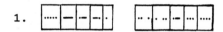

2.

3.

4.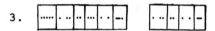

5.

6.

7.

8.

9.

10.

QUIZ 25

Use Louis Braille's dots and cells to determine which letters are between those represented here by the Braille dots. Add up their totals according to their position in the alphabet, follow the directions of the mathematical signs and discover the numerical answers.

1. ⠂   --   ⠒⠄ + ⠘⠢ -- ⠰⠒

2. ⠿   --   ⠡ − ⠈ -- ⠠⠌

3. ⠣   --   ⠢⠠ × ⠆ -- ⠰⠃

4. ⠇   --   ⠑⠈ + ⠑⠈ -- ⠦⠒

5. ⠈⠂   --   ⠠⠑ ÷ ⠂ -- ⠰⠃

6. ⠘⠘   --   ⠣⠄ − ⠣⠠ -- ⠡

7. ⠡⠠   --   ⠭⠭ × ⠇ -- ⠠⠌

8. ⠠⠈   --   ⠑⠈ ÷ ⠂ -- ⠒⠒

9. ⠒⠒   --   ⠑⠈ − ⠄⠆ -- ⠒⠒

10. ⠈⠠   --   ⠠⠆ × ⠿ -- ⠄⠈

QUIZ 26

Use William Moon's method to discover these constellations.

1. Λ И – Λ \ Γ ∕

2. C Λ \ I И Λ

3. Ɔ \ Λ C O

4. ☉ J Ɔ \ ◡ ∕

5. L ل \ Λ

6. O \ I O И

7. ⊂ Γ \ ∕ Γ ◡ ∕

8. ∕ Γ \ ⊂ Γ И ∕

9. — ◡ C Λ И Λ

10. V Γ ᠑ Λ

QUIZ 27

The Moon system continues to be a guide to these words regarding the universe.

1. ᠑ I L < Γ ل ⌐ Λ ل

2. O \ ᒪ I —

3. L I ᠑ ☉ — ل Γ Λ \

4. ᠑ Λ L Λ > ل

5. ᠑ Γ — Γ O \ ∕

6. Λ > I ∕

7. ⊂ ⊃ ⅂ ⌐ –

8. ⅂ \ ∧ ∨ ⌐ – ⌐

9. ∧ / – ⌐ \ ○ ⌐ ⊃ /

10. N ○ \ – ○   / – ∧ \

QUIZ 28

Using the keyword, *heart,* make your own square like that in the Wheatstone/Playfair example to decipher these messages from the "agony columns."

```
1.  kr  ah  rh  he  aw  de  cr
2.  or  aw  mo  da  he  ta  rh
3.  vs  st  zu  dk  aw  li  pv  qy
4.  hm  ca  bp  zu  ht  zy  qd  kt
5.  po  ay  ur  qi  tk  bv  st  qy
6.  hm  pm  he  aw  ht  ry  az
7.  lk  qy  qy  pr  dq  ac
8.  ha  qa  na  dq  aw  om  bt  qy
9.  wr  qy  sr  iq  wh
10. ik  bt  az  ca  mk  bt  az
```

QUIZ 29

In the heat of battle, two signalers have used their field glasses to view these number groups sent by your allies. Use Myer's numbers to interpret their meaning.

```
1.  221, 322, 221, 132, 111 3
2.  133, 331, 223, 223, 313, 332 3
3.  112, 331, 221 3
4.  132, 223, 222, 213, 322, 123 3
5.  322, 223, 311 33
6.  133. 312, 221, 213, 331 3
7.  331, 213, 123, 312, 133 3
```

8. 122, 231, 112, 322, 323 3
9. 112, 313, 313, 221, 112, 331, 332 3
10. 311, 221, 112, 323 333

QUIZ 30

Use your own field glasses to view the signals. Your skill in deciphering these messages may determine the result of the battle.

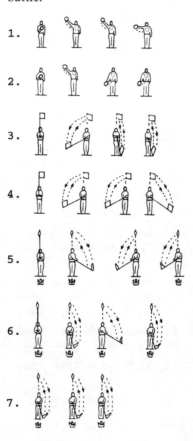

QUIZ 31

This secret list of weapons and supplies must be conveyed rapidly by wigwag to the next observation post. Send this message with Myer's alphabet.

1. cannon
2. rifles
3. mortars
4. bullets
5. powder
6. bayonets
7. hardtack
8. flour
9. salt
10. medicine

QUIZ 32

These messages have been sent to you using the rail fence cipher. Apply what you have learned to discern these letter groups.

1. frsm esat otut rtrx
2. pnnu acma gxei slra pinx
3. ioca sate rnld btlx
4. ioc uchl hhrh
5. atea cekx nitm rexx
6. uinl caex nobo kdxx
7. cacl osil flsx hnel rvle alxx
8. vcsu gigx ikbr seex
9. cvly hrex aarc agsx
10. alna undx tatb rexx

QUIZ 33

Reverse the cipher-solving process. Use the rail-fence method to conceal the names of these figures from the Civil War.

1. Clara Barton
2. Graxton Bragg
3. John Brown
4. Jubal Early
5. Horace Greeley
6. John Hood
7. Julia Howe
8. Mary Lincoln
9. William Seward
10. Harriet Stowe

QUIZ 34

These heliograph signals are defined as long flash (*lf*) and short flash (*sf*). Using the International Morse Code, equate *lf* with a dash and *sf* with a dot. A comma separates letters and a dash separates words.

1. sf lf lf, sf sf, sf lf sf sf, lf sf sf-lf sf sf sf, sf sf, sf lf sf sf, sf lf sf sf-sf sf sf sf, sf sf, lf sf lf sf, lf sf lf, lf lf lf, lf sf lf

2. sf sf sf, sf sf, lf, lf, sf sf, lf sf, lf lf sf-lf sf sf sf, sf sf lf, sf lf sf sf, sf lf sf sf

3. sf lf, lf sf, lf sf, sf sf, sf-lf lf lf, sf lf, lf sf lf, sf lf sf sf, sf, lf sf lf lf

4. lf sf sf sf, sf sf lf, sf sf lf sf, sf sf lf sf, sf lf, sf lf sf sf, lf lf lf-lf sf sf sf, sf sf, sf lf sf sf, sf lf sf sf-lf sf lf sf, lf lf lf, lf sf sf, lf sf lf lf

5. lf sf lf sf, sf lf, sf lf sf sf, sf lf, lf
    lf, sf sf, lf, lf sf lf lf-sf lf lf lf, sf lf,
    lf sf, sf

6. sf lf lf, lf sf lf lf, sf lf, lf, lf-sf, sf
    lf, sf lf sf, sf lf lf sf

7. lf sf lf sf, lf lf lf, lf sf lf sf, sf sf sf
    sf, sf sf, sf sf sf, sf

8. lf sf sf sf, sf, sf lf sf sf, sf lf sf sf,
    sf-sf sf sf, lf, sf lf, sf lf sf, sf lf sf

9. lf sf lf, sf sf, lf-lf sf lf sf, sf lf, sf
    lf sf, sf sf sf, lf lf lf, lf sf

10. lf lf sf, sf, sf, lf sf, lf lf lf, lf sf, sf
    sf, lf lf, lf lf lf

QUIZ 35

Terms regarding the American West are being signaled to you
by heliograph from a distant mesa.

1. lf sf lf sf, lf lf lf, sf lf lf, lf sf sf
    sf, lf lf lf, lf sf lf lf

2. sf lf sf sf, sf lf, lf sf, lf sf sf-lf lf
    sf, sf lf sf, sf lf, lf sf, lf

3. sf lf lf, sf lf, lf lf sf, lf lf lf, lf sf-
    lf, sf lf sf, sf lf sf sf, lf sf

4. sf lf sf sf, lf lf lf, lf sf, lf lf sf-sf sf
    sf sf, lf lf lf, sf lf sf, lf sf, sf sf sf

5. sf lf sf, sf lf, lf sf, lf lf sf, sf-sf lf
    lf, sf lf, sf lf sf

6. lf sf sf sf, sf sf lf, sf sf lf sf, sf sf lf
    sf, sf lf, sf lf sf sf, lf lf lf

7. sf lf lf, sf lf, lf, sf, sf lf sf-sf lf sf,
    sf sf, lf lf sf, sf sf sf sf lf, sf sf sf

8. sf sf sf, lf lf lf, lf sf sf-lf sf sf sf, sf
sf lf, sf sf sf, lf, sf, sf lf sf

9. lf sf sf sf, sf lf, sf lf sf, lf sf sf sf,
sf, lf sf sf-sf lf lf, sf sf, sf lf sf, sf

10. sf sf sf sf, lf lf lf, sf, sf sf sf, lf, sf,
sf lf, lf sf sf

QUIZ 36

Make use of the basic Esperanto vocabulary to translate these statements and questions.

1. La patrino visitas Sinjorino Aames.
2. Tiu esta Arla Aames.
3. La patro iras en la domo.
4. Brad esta la frato.
5. La fratino laboras en la gardeno.
6. Kiel estas la automobile?
7. Kiu mangas en la kuirejo?
8. Estas Brad Aames en la cambro?
9. Kie estas la sinjoro?
10. Kiu amas la knabino?

QUIZ 37

Use the Esperanto numbers to decode this combination of words and mathematic symbols.

1. tri + ses × ok ÷ kvar =
2. kvin − du × nau + tri =
3. sep × tri − dek + unu =
4. du + sep × tri − kvin =
5. ok × dek ÷ du + kvar =
6. nau × tri − unu × du =
7. tri × ses + ok − unu =
8. dek ÷ du × sep + du =
9. ses × nau ÷ du − ok =
10. kvar + sep × tri − kvin =

QUIZ 38

Use the Ardois chart to identify these ships and aircraft from fact and fiction, and then match their names with the means of their loss listed below (A to J).

1.  ▯▯▯▯▯▯

2.  ▯▯▯▯▯

3.  ▯▯▯▯▯▯▯

4.  ▯▯▯▯▯▯▯▯

5. 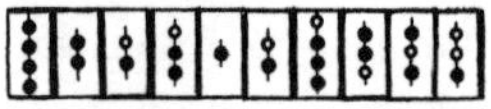 ▯▯▯▯▯▯▯▯

6.  ▯▯▯▯▯   ▯▯▯▯▯

7. 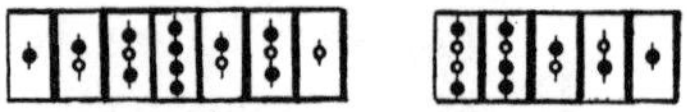 ▯▯▯▯▯▯▯   ▯▯▯▯▯

8.  ▯▯▯▯▯▯   ▯▯▯▯▯

9.  ▯▯▯▯▯▯▯▯

10. 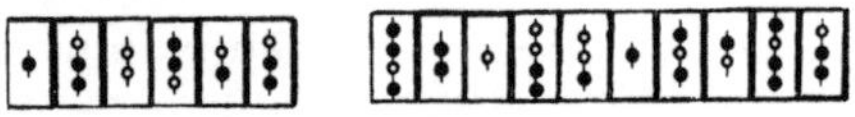 ▯▯▯▯▯▯   ▯▯▯▯▯▯▯▯▯

A. fire
B. iceberg
C. air disaster
D. torpedoes
E. lake mystery
F. giant wave
G. collision
H. harbor mystery
I. vanished
J. Moby Dick

■ 654

QUIZ 39

An agent has sent this terse phrase: $M = 15$, *alphabet sequential*. The chamber's files contain the key for her messages: 47351268903624015789. You have a series of ten number groups from another valid source. Use the Mauborgne system to discern the missing numbers that make the sum each time. Then match them with this particular alphabet to reveal the plaintext. You learn that $N = 16$ and $C = 5$ for your alphabet/number equations. But remember, Mauborgne's addition is not standard.

1.   47351268
     + _____
     41422762

2.   473512689036
     + _____
     574618649742

3.   4735
     + _____
     4136

4.   4735126890362401
     + _____
     4040198801312717

5.   47351268903624
     + _____
     44401661115725

6.   473512
     + _____
     523820

7.   47351268
     + _____
     67422973

8.   4735
+ _____
      5441

9.   473512689036
+ _____
      524228649337

10.   47351268903624
+ _____
      52423274014223

QUIZ 40

Recall the honored naval heroes of the world who are named in the following semaphore signals.

1.

2.

3.

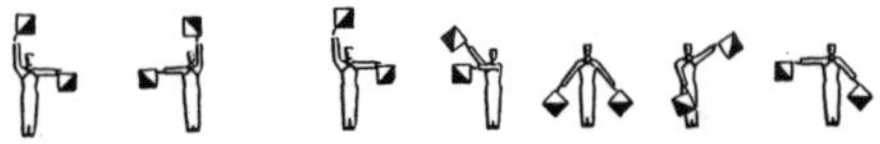

4.

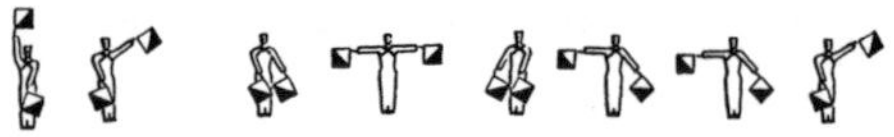

5.

6.

7.

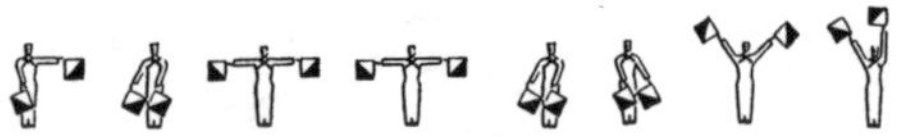

8.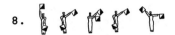

9.

10.

QUIZ 41

It is the Cold War. A Czech defector in Vienna has given the West vital facts from an apparently useless, crumpled piece of paper. On this KGB tear sheet are instructions for Russia's worldwide spokespersons. Add the correct column numbers (_ _ _ _ _) to the alphabet-equivalent numbers and solve their secret plans.

```
Bolshoi = 25 (standard reversed), columns =
vertical (from top left)

54321
65432
76543
87654
98765
12345
23456
34567
45678
56789
67891
78912
89123
91234
11234
```

11245
11256
11267
11278
11289
12234
12245
13356
12267

+	+	+	+
54328	65458	76558	87670

+	+	+	+	+
98776	12367	23482	34591	45700

+	+	+
56801	67897	78921

+	+	+
89131	91257	11252

+	+	+	+	+
11254	11278	11293	11301	11291

Answers

QUIZ 1

1. Marathon   2. Delphi   3. Athens   4. Cyclades
5. Parthenon   6. Laconia   7. Acropolis
8. Piraeus   9. Mt. Olympus   10. Peloponnesus

QUIZ 2

1. Hercules   2. Priam   3. Agamemnon   4. Laocoon
5. Helen of Troy   6. Paris   7. Achilles
8. Trojan Horse   9. Hector   10. Clytemnestra

QUIZ 3

1. Rubicon   2. Cleopatra   3. Augustus   4. Gaul
5. Brutus   6. Pompey   7. Legions   8. Antony
9. Calpurnia   10. Ides of March

QUIZ 4

1. Tribune   2. Hannibal   3. Collesseum
4. Senate   5. Huns   6. Emperor   7. Carthage
8. Attila   9. Tiber   10. Visigoths

QUIZ 5

1. cauldron   2. vial   3. beaker   4. water
5. pestle   6. scales   7. fire   8. mortar   9. lead
10. charcoal

QUIZ 6

1. Genoa   2. Constantinople   3. Venice
4. Grenada   5. Naples   6. Barcelona   7. Reims
8. Florence   9. Madrid   10. Avignon

QUIZ 7

1. Isabelle 2. Polo 3. R. Bacon 4. Cortez
5. da Gama 6. Ferdinand 7. de Leon
8. Cervantes 9. Magellan 10. C. (Catherine) de
Medici

QUIZ 8

1. Cape Horn 2. astrolabe 3. cartographer
4. shoreline 5. Asia passage 6. compass
7. St. Christopher 8. navigator 9. tides
10. Cape of Good Hope

QUIZ 9

1. savings 2. loan 3. demand 4. free trade
5. merchant 6. bonds 7. profit 8. investor
9. financial 10. middle class

QUIZ 10

1. witch 2. demons 3. pentacle 4. sorcery
5. evil eye 6. warlock 7. amulet 8. hex
9. potion 10. black magic

QUIZ 11

1. Chartley Castle 2. Walsingham 3. Babington
Ploat 4. Phelippes 5. Ballard 6. Scotland
7. Philip of Spain 8. Elizabeth 9. Gifford
10. Fotheringhay Castle

QUIZ 12

key in wall crevice

QUIZ 13

cellar well swim to freedom

QUIZ 14

1. Socrates 2. Confucius 3. Galileo 4. Locke
5. Buddha 6. Nostradamus 7. Solomon
8. Rousseau 9. Newton 10. Kant

QUIZ 15

1. Plato 2. Copernicus 3. Mohammed 4. Knox
5. Aristotle 6. St. Thomas 7. Mendel
8. Calvin 9. Harvey 10. St. Francis

QUIZ 16

1. First V.P.-J. Adams 2. Doctrine-Monroe
3. Mount Vernon-Washington 4. inventor-
Jefferson 5. wife-Dolly Madison 6. New
Orleans-Jackson 7. grandson-B. Harrison
8. Rough and Ready-Z. Taylor 9. bachelor-
Buchanan 10. thirteenth-Fillmore

QUIZ 17

1. honest-A. Lincoln 2. Rough Riders-T.
Roosevelt 3. silent-Coolidge 4. impeachment-
A. Johnson 5. Civil War hero-Grant 6. disputed
vote-Hayes 7. son of president-J.Q. Adams
8. terms split-Cleveland 9. Supreme Court-Taft
10. dark horse-J. Polk

QUIZ 18

1. faith 2. trust 3. charity 4. love
5. mercy 6. hope 7. peace 8. honor 9. truth
10. loyalty

QUIZ 19

1. 33 2. 21 3. 29 4. 53 5. 39 6. 48 7. 15
8. 31 9. 45 10. 24

QUIZ 20

CULPER-TALLMADGE ALPHABET CHART WITH NUMBER EQUIVALENTS

a	b	c	d	e	f	g	h	i
j	k	l	m	n	o	p	q	r
s	t	u	v	w	x	y	z	

1,	2,	3,	4,	5,	6,	7,	8,	9,
10,	11,	12,	13,	14,	15,	16,	17,	18,
19,	20,	21,	22,	23,	24,	25,	26	

5,	6,	7,	8,	9,	10,	1,	2,	3,
4,	15,	13,	14,	16,	17,	18,	11,	12,
21,	22,	23,	24,	25,	26,	19,	20	

e	f	g	h	i	j	a	b	c
d	o	m	n	p	q	r	k	l
u	v	w	x	y	z	s	t	

Note: Some letter/number equivalents have the same total. Look for the letters needed for a correct answer. As shown here, the letters' number equivalents have been added.

m	r	s
13	18	19
14	12	21

n	l	u
27	30	40

1. Mrs. Washington 2. H. Knox 3. Revere
4. S. Adams 5. Von Steuben 6. L. Darragh
7. Franklin 8. Rochambeau 9. N. Hale
10. M. (Molly) Pitcher

QUIZ 21

1. secret ink   2. dagger   3. trap door   4. lamp sign   5. go between   6. poison   7. hidden pocket 8. long glass   9. wax seal   10. hollow bullet

QUIZ 22

1. Pinta   2. Mayflower   3. Half Moon   4. Monitor 5. Merrimac   6. Arizona   7. Wasp   8. Nautilus 9. Calypso   10. Polaris

QUIZ 23

1. Corsica   2. Borodino   3. Moscow   4. Leipzig 5. Elba   6. Old Guard   7. Belgium   8. Wellington 9. Waterloo   10. St. Helena

QUIZ 24

1. plane crash   2. dam break   3. prison riot 4. flood watch   5. cyclone   6. toxic fumes 7. blizzard   8. forest fire   9. oil spill 10. ptomaine

QUIZ 25

Example solution for 1:

Braille dots read *a* to *f*; letters *between* are *b*, *c*, *d*, *e*. Their number equivalents total 14; + Braille dots read *v* to *z*; letters *between* are *w*, *x*, *y*, whose number equivalents total 72. Thus, 14 + 72 = 86

1. a to f (14) + v to z (72) = 86;   2. g to k (27) − a to e (9) = 18;   3. q to u (57) × b to d (3) = 171;   4. l to p (42) + s to y (110) = 152; 5. c to i (30) ÷ a to d (5) = 6;   6. m to r (62) − h to k (19) = 43;   7. u to x (45) × b to

ARGE

e (7) = 315;   8. i to o (60) ÷ a to c (2) = 30;
9. n to t (85) − w to y (24) = 61;   10. e to j
(30) × g to i (8) = 240

QUIZ 26

1. Antares   2. Carina   3. Draco   4. Hydrus
5. Lyra   6. Orion   7. Perseus   8. Serpens
9. Tucana   10. Vega

QUIZ 27

1. Milky Way   2. orbit   3. light year   4. galaxy
5. meteors   6. axis   7. comet   8. gravity
9. asteroids   10. North Star

QUIZ 28

Note: *x* is used as a null to make pairs in
diagraph stage.

1.   kr ah rh he aw de cr
me et at th ex ca fe
Meet at the cafe.

2.   or aw mo da he ta rh
sh ex is ax th re at
She is a threat

3.   vs st zu dk aw li pv qy
yo ur un ci ex kn ow sx
Your uncle knows.

4.   hm ca bp zu ht zy qd kt
ri de co un tr xy la ne
Ride country lane.

5.   po ay ur qi tk bv st qy
ou rx st ol en ho ur sx
Our stolen hours.

6. hm pm he aw ht ry az
ri sk th ex tr ys tx
Risk the tryst.

7. lk qy qy pr dq ac
ki sx sx se al ed
Kiss sealed.

8. ha qa na dq aw om bt qy
te lx lt al ex si gh sx
Telltale sighs.

9. wr qy sr iq wh
ye sx my lo ve
Yes my love.

10. ik bt az ca mk bt az
ni gh tx de li gh tx
Night delight.

QUIZ 29

Note: 3, 33, and 333 indicate end of word, end of sentence and end of message, respectively.

1. enemy 2. troops 3. are 4. moving 5. now
6. their 7. right 8. flank 9. appears
10. weak

Message: Enemy troops are moving now. Their right flank appears weak.

QUIZ 30

Position Motion Message

1.	1	1 1 2	A
2.	1	1 3 3	T
3.	1	1 3 3	T
4.	1	1 1 2	A

5. 1 2 1 1 C
6. 1 3 2 3 K
7. 3 3 3 End of message

QUIZ 31

1. 211, 112, 322, 322, 223, 322; 2. 331, 213,
122, 231, 221, 332; 3. 132, 223, 331, 133, 112,
331, 332; 4. 121, 233, 231, 231, 221, 133, 332;
5. 313, 223, 311, 212, 221, 331; 6. 121, 112,
111, 223, 322, 221, 133, 332; 7. 312, 112, 331,
212, 133, 112, 211, 323; 8. 122, 231, 223, 233,
331; 9. 332, 112, 231, 133; 10. 132, 221, 212,
213, 211, 213, 322, 221. 333

QUIZ 32

a. Divide cipher into equal halves:

frsm esat otut rtrx

b. Start with first letter on left (f), then first
letter on right side of line (o) and continue
alternating left to right across the dividing
line. X is a null.

1. fort sumter start 2. peninsula campaign
3. iron-clads battle 4. shiloh church
5. antietam creek 6. union blockade
7. chancellorsville falls 8. vicksburg siege
9. cavalry charges 10. atlanta burned

QUIZ 33

1. c a a a t n
l r b r o x
c a a a t n l r b r o x
c a a a t n l r b r o x

2. b a t n r g
r x o b a g
batnrgrxobag
batn rgrx obag

3. j h b o n x
o n r w x x
jhbonxonrwxx
jhbo nxon rwxx

4. j b l a l x
u a e r y x
jblalxuaeryx
jbla lxua eryx

5. h r c g e l y x
o a e r e e x x
hrcgelyxoaereexx
hrcg elyx oaer eexx

6. j h h o
o n o d
jhhoonod
jhho onod

7. j l a o e x
u i h w x x
jlaoexuihwxx
jlao exui hwxx

8. m r l n o n
a y i c l x
mrlnonayiclx
mrln onay iclx

9. w l i m e a d x
i l a s w r x x
wlimeadxilaswrxx
wlim eadx ilas wrxx

10. h r i t t w
a r e s o e
hrittwaresoe
hrit twar esoe

QUIZ 34

1. Wild Bill Hickok 2. Sitting Bull 3. Annie
Oakley 4. Buffalo Bill Cody 5. Calamity Jane
6. Wyatt Earp 7. Cochise 8. Belle Starr
9. Kit Carson 10. Geronimo

QUIZ 35

1. cowboy 2. land grant 3. wagon train
4. long horns 5. range war 6. buffalo
7. water rights 8. sod buster 9. barbed wire
10. homestead

QUIZ 36

1. The mother visits Mrs. Aames 2. That one is
Aria Aames. 3 The father goes in the house.
4. Brad is the brother 5. The sister works in
the garden. 6. How is the car? 7. Who eats in
the kitchen? 8. Is Brad Aames in the room?
9. Where is the gentleman? 10. Who loves the
daughter?

QUIZ 37

1. $3 + 6 \times 8 \div 4 = 18$ 2. $5 - 2 \times 9 + 3 = 30$
3. $7 \times 3 - 10 + 1 = 12$ 4. $2 + 7 \times 3 - 5 = 22$
5. $8 \times 10 \div 2 + 4 = 44$ 6. $9 \times 3 - 1 \times 2 = 52$
7. $3 \times 6 + 8 - 1 = 25$ 8. $10 \div 2 \times 7 + 2 = 37$
9. $6 \times 9 \div 2 - 8 = 19$ 10. $4 + 7 \times 3 - 5 = 28$

QUIZ 38

1. Pequod-J   2. Maine-H   3. Titanic-B
4. Lusitania-D   5. Hindenburg-C   6. Morro
Castle-A   7. Earhart plane-I   8. Andrea Doria-G
9. Poseidon-F   10. Edmund Fitzgerald-E

QUIZ 39

(Mauborgne's "difference" is noncarrying addition.)

1.   47351268
   +04171504 bomb
    41422762

2. +101106060716 hidden

3. +0401 by

4. +0315072011050316 American

5. +07150403212101 Embassy

6. +150318 map

7. +20171715 room

8. +1716 on

9. +151716060301 Monday

10. +15172016111609 morning

QUIZ 40

1. Drake   2. Magellan   3. J.P. Jones
4. de Grasse   5. Perry   6. Nelson   7. Farragut
8. Dewey   9. Halsey   10. Nimitz

QUIZ 41

(Note: standard reversed means $Z=1$, $Y=2$, $X=3$, etc.)

talk peace our sdi ready

APPENDIX

Note: The patterns of letter and word frequencies cannot be measured exactly in any language. The constant development of new terms make quantification an inexact science at best. However, a number of studies have been conducted on these topics, resulting in the general standards shown below.

ENGLISH LETTER AND WORD FREQUENCIES

SINGLE LETTERS (IN ORDER OF FREQUENCY):

						e	t	o
a	n	i	r	s	h	d	l	c
w	u	m	f	y	g	p	b	v
k	x	q	j	z				

DIGRAPHS:

	th	he	an	in	er	re	es
on	ea	ti	at	st	en	nd	or	to
nt	ed	is	ar	ou	te	of	it	ha
se	et	al	ri	ng			

TRIGRAPHS:

	the	and	tha	ent	ion	tio	for
nde	has	nce	edt	tis	oft	sth	men

COMMON REVERSALS:

	er-re	es-se	an-na	ti-it
on-no	in-ni	en-ne	at-ta	te-et	or-ro
to-ot	ar-ra	st-ts	is-si	ed-de	of-fo

TWO-LETTER WORDS:

	of	to	in	it	is	be
as	at	so	we	he	by	or	on	do
if	me	my	up	an	go	no	us	am

THREE-LETTER WORDS:

	the	and	for	are	but
not	you	all	any	can	had	her	was
one	our	out	day	get	has	him

FOUR-LETTER WORDS: that with have this your from
they know want been good much some time very
when come here just like long

THIRTY COMMON ENGLISH WORDS: the of and to
a in that is I it for as with
his he be not by but have you
are on or her had at from which

**SIXTY PATTERN WORDS (NUMBERS INDICATE ORDER OF LETTERS'
APPEARANCE AND REPETITION):**

121	did	eye				
122	add	all	bee	egg	off	see	too
1213	away	even	ever	nine	none	
1221	noon					
1223	been	book	cook	cool	deep	door	feed
	feel	feet	food			
1233	ball	bell	bill	call	fill	free	full
	hall	hill	knee	less		
12134	enemy		every		paper		usual
12234	allow		apple		offer	
12314	catch		clock		enter		taste
	truth					
12334	brook		carry		green		happy
	hurry		sleep			
122314	appear					
1234562	because		measure		service	
	through					
12345675	increase					

THIRTY NONPATTERN WORDS: an as at be by
do go he if in bow car day ear
far get has its joy led able boat cent
deal each fact gain have into join

GLOSSARY

Additive A digit or digits that are added to a codenumber to give the code additional concealment.

Algorithm A series of instructions whereby a mathematical formula is applied to the numeric representation of a message in order to encrypt or decrypt it.

Anagramming A method used by cryptanalysts who align a cryptogram's characters beneath a standard alphabet. Using this parallel reconstruction they hope to discern the correct letters' identities by comparing alphabet arrangements and similar positions.

Anthropomorphism The attribution of human shapes or characteristics to gods, animals and objects.

Argot (French: *argot*, "claw," "spur"). A form of speech described as the particular language of thieves or the specialized vocabulary of people in the same workplace.

ASCII The acronym for American Standard Code for Information Interchange, a widely used system for encoding letters, numerals, punctuation marks and signs as binary numbers.

Authentication The process of verifying the sender or receiver as well as the contents of a communication.

Autokey An early type of encryption control (key) that changes with each communication, thereby giving more protection than one that is repeated with several messages.

Bigram (Greek: *bi*, "having two," + *gram*, "what is written"). A letter pair. Bigrams are combined in various configurations when making or solving cryptosystems. Another word for bigram is *digraph*.

Biliteral Consisting of two letters. The term "biliteral" can refer to the use of letter pairs to encipher a message, or to the use of two different typefaces within a cipher system.

Binary Having two components or possible states, usually represented by ones and zeroes in varied patterns.

Binary code A system for representing things by combinations of two symbols, such as one and zero or true and false.

Bit The smallest unit of information in a computer. Equivalent to a single zero or one. The word *bit* is a contraction of *binary digit*.

Blocking The encrypting of a communication by segmenting it into blocks of letters or digits. These are arranged by preplanned formula and sent in the new arrangement.

Boustrophedon An ancient form of writing in which the lines run alternately left to right and right to left.

Byte A sequence of eight bits treated as a unit for computation or storage.

Cablegram A communication method associated with national and international cables, linked most closely with cryptology when used with commercial codes for business transactions.

Cell One of the rectangular units that compose the geometrical patterns used in some symbol ciphers and transposition ciphers.

Certificate The public key, name, expiration date and other pertinent information about an individual or entity that has been digitally signed by a Certificate Authority.

Chip An integrated circuit on a square of silicon or other semiconductor, made up of multiple thousands of transistors and other electronic components.

Cipher (Arabic: *sifr*, "nothing") A method of concealment in which the primary unit is the letter. The letters of the plaintext are substituted with other letters, numbers or symbols, or they can be transposed for other letters, letter pairs (bigrams or digraphs) and sometimes for many letters (polygrams).

Cipher alphabet An alphabet composed of substitutes for the normal alphabet or the particular alphabet in which the cipher is written.

Cipher device A manual mechanism used to encrypt and decrypt messages.

Cipher disk A disk used to arrange alphabets, numerals or symbols for encrypting and decrypting purposes.

Cipher machine A mechanism used to encrypt and decrypt messages. Although the term has been used synonymously with *cipher device*, it is more properly reserved for more mechanical, and later electrical, encryption machines.

Ciphertext The result of applying a cipher method to a given text (the plaintext). The new enciphered communication is the ciphertext.

Ciphony (*cipher* + *telephony*) A primary form of oral secrecy involving speech modification with scrambler devices.

Cleartext A communication sent without encoding or encryption. Such messages are also called *in clear* or, sometimes, *in plain language*.

Code (French, Latin: *codex*, "tree trunk," "writing tablet") A method of concealment that may use words, numbers or syllables to replace the original words and/or phrases of a message. Codes substitute whole words whereas cipher transpose or substitute letters or digraphs.

Codebook Either a collection of code terms or a book used to encode and decode messages.

Codenames Name concealments for a person, an object such as a secret weapon or a tactical operation.

Codewords In general, a word that masks another. Also a word used as a signal, exemplified by the jargon codes of D-Day.

Codenumbers Numbers that function like codewords when they replace the words of a plaintext message.

Codetext The result of encoding a given communication (the plaintext.) Similar to ciphertext, codetext differs mainly in that a code, rather than a cipher, conceals the text.

COMINT An acronym for *communications intelligence*, one of the primary categories of signals intelligence (SIGINT). It includes message interception; the location of enemy transmissions' sites; analysis of types of communications; and solution of ciphers and codes.

Compaction The process of reducing the length of an enciphered message in order to enhance transmission speed or to reduce its chance of being solved. For example, every fifth character or punctuation mark is removed, thereby compressing or compacting the remainder of the communication.

Compartmentation A term primarily applied to codes in government security classification listings. Compartmentation serves to restrict access to information that pertains to national security. Along with top-level secrecy designations, these codes also conceal subcategories within the primary groups.

COMSEC An acronym for *communications security* and a primary branch of signals security. It includes steganography (hiding the existence of a message); traffic security (fake communications, intermittent silence); and cryptography (ciphers, codes, secure phones and other protected transmissions mechanisms).

Construction tables Tables of letters used to form encryptions.

Contact chart A tool, used by a codebreaker or cipher solver, to analyze the frequency of certain letters or letter combinations in encrypted text.

Cryptanalysis (Greek: *kryptē* < *kryptós*, "secret, hidden," *ana*, "up, throughout," *lysys*, "a loosing"). The process used by third parties' cryptanalysts who attempt to intercept transmissions and reveal their concealed contents.

Cryptanalyst A unintended third party who deciphers or decodes without possessing the concealment method being used.

Cryptoeidography (Greek: *eidos*, "form," + *graphia* "writing"). It includes two methods of rendering pictures secret. One version, cifax, modifies electrical patterns to distort an image. The second system involves optical alterations.

Cryptogram (Greek: *kryptós*, "hidden" + *gram*, "written" or "drawn"). An encoded or enciphered message.

Cryptography (Greek: *kryptē* < *kryptós*, "secret, hidden," + *graphia*, "writing"). The processes of covering a message by codes (encoding) or ciphers (enciphering).

Cryptology (Greek: *kryptē* < *kryptós*, "secret, hidden," + *logos*, "word," or *ology*, "science"). The science that includes message concealment (cryptography), the solving of such a communication by the intended recipient (decoding and deciphering), and the solution of a message by an unauthorized third party (cryptanalysis).

Decipher (Latin: *de*, "away from" + cipher). The removal of a cipher concealment. Decipherment is achieved by the intended party using the correct key; cryptanalysis, the breaking of a cipher, is performed by an enemy party. Synonymous words include *decrypt*, *recover* and *solve*.

Decode The removal of a code concealment. The intended receiver does so by finding the code term's equivalent plainletter or word meaning in a list or codebook. Third parties try to do so by cryptanalysis. Synonymous words are *decrypt*, *recover* and *solve*.

Digital signature The electronic equivalent of a manual signature that consists of a digital message and a public key that uniquely binds the message to the sender.

Digraph (Greek: *di*, "twice," + *graphos*, "to write"). An encipherment in which the plaintext is written using letter pairs.

Digraphic An encryption process whereby two letters are enciphered so that both together affect the result.

Electronics security A primary branch of signals security, that tries to counteract the ELINT (electronics intelligence) efforts of other nations. It includes emissions protection such as shifting radar frequencies and counter-countermeasures such as "looking through" jammed radar.

ELINT An acronym for *electronics intelligence*, a major segment of signals intelligence. Within ELINT are the divisions of RADINT (radar intelligence) and TELINT (telemetry intelligence). As these acronyms imply, it includes eavesdropping on radar emissions as well as on telemetry diagnostic signals between rockets or unmanned aircraft. ELINT also involves countermeasures such as jamming enemy radar and emitting false radar echoes.

Encicode (*enciphered* + *code*). A coded message that has been further hidden by additional cipher methods like transposition or substitution. The process of providing such layers of protection is known as "superencipherment." Encicodes also include covers in which codewords are replaced by numbers.

Encryption A general term for enciphering or encoding. The specific methods are identified by the terms *encoded* or *enciphered*.

Expansion A type of concealing process that expands the plaintext by adding spaces or nonsensical syllables in order to hide the exact number of letters, numbers or characters. It is often combined with processes such as blocking or permutation.

Fist The particular touch or sending style of radiotelegraph staffpersons; such styles might be, light, quick or heavy.

Frequency With reference to code and cipher solutions, the number of times a letter or digit appears in the encryption. This is a primary clue for a cryptanalyst, because every legitimate language has discernible patterns for vowels, consonants and syllable pairs. With reference to electromagnetic or sonic waves, the number of waves that pass a fixed point each second.

Frequency tables Lists used by cryptanalysts that show frequency of letters, letter pairs, syllables and words in various languages.

Geometric patterns Configurations used to align, transpose or substitute alphabet letters with other letters, numerals or special forms such as those of symbol cryptography.

Grille A piece of paper with apertures which can be laid over a secret message to reveal only the relevent letters.

Hash A smaller fixed-length numerical representation of a larger message.

Homophones Multiple substitutes that replace a single letter. These may be a group of letters or numbers, but are always substituted for the same single letter. For example, *a* may be replaced by *l, m, n* or *12, 13* and *14.*

HFDF High-frequency direction finding, nicknamed "huffduff" during World War II. A means of tracing broadcast signals to their origin.

HUMINT An acronym for *human intelligence.* Intelligence gathered through human means, as contrasted with solely technological methods of garnering data.

Ideogram (Greek: *idea,* "idea" + *gramma* "written" or "drawn."). A graphic (written) symbol of an idea or object that represents an idea rather than the sounds that make its name. Also known as an ideograph.

Jargon codes Open methods of linguistic concealment. A type of open code, the jargon code is not hidden by symbols or transposed letters. Rather, an innocent word or words replaces another term in a sentence constructed in an innocuous fashion.

Kata kana A Japanese system of syllabic writing.

Key The directions that control the encryption and decryption of concealed communications as well as the variable aspects of a given code or cipher system. A *symmetric key* is the name for the long-existing practice in which two communicants share a secret key ideally known only to them. Two key, or public key, practices have existed publicly since the 1970s. This arrangement involves a publicly known and a private key combination.

Keynumber A series of digits that serves as a key for encryption and decryption purposes. A keynumber is often used to start a shifting process with a cipher alphabet. For example, the keynumber *8* could indicate that a cipher alphabet is shifted eight places from the standard alphabet.

Keyphrase A phrase that serves as a key.

Keyword A word used as a key for an encryption system. Keywords are chosen for their length and lack of repeating letters. A keyword can have its letters enumerated according to those letters' location in a given alphabet. For example, *vigil* could be represented as $v = 22, i = 9, g = 7, i = 9, l = 12$. Such digits help to arrange or shift groups of message characters aligned in columns beneath them.

Logogram (Greek: *logos,* "word" + *gramma,* "written" or "drawn"). A term used by scholars who believe that signs, whether they indicate syllables or other phonetic aspects, stand for words. In some texts, it is used instead of the words *ideogram* and *ideograph.*

Monoalphabetic A cryptomethod in which a single cipher alphabet is used to conceal the letters of a plaintext communication.

Nomenclator (Latin: *nomen* "name" + *calator,* "caller"). A collection of syllables, words and names similar to a code, with a separate cipher alphabet. The latter often had multiple replacements, known as homophones, for a given letter. It was the predominant concealment system from the 15th to the mid-19th century.

Nonsecret codes Codes that are used primarily for communicative rather than cryptographic purposes. This category includes a wide range of systems, from business codes to visual and audial signaling methods.

Null A meaningless letter, symbol or number inserted into a code list or a cipher alphabet. Nulls are often used to complete a pair from an otherwise uneven pattern of letters and numbers. They can also complicate the decryption efforts of unintended third parties by disrupting anticipated sentence patterns, word lengths and the frequency of syllable groups.

One-part code A code in which both the plaintext words or numbers and their concealment equivalents are listed alphabetically or sequentially.

Open code A code concealed in an apparently innocent message. Open codes are a branch of linguistically masked communications, which includes null ciphers, geometric methods and jargon codes.

Permutation A form of transposition in which symbols or letters are rearranged but retain their original identities.

Pictograph (Latin: *pictus,* "painted" + Greek: *graphein,* "to write"). Man-made marks formed into a recognizable image. Also known as a *pictogram,* pictographs can repreent both sounds and words.

Plaincode The code remaining after superenciphered codenumbers have been removed. Also known as the placode.

Plaintext A message either before it has been encoded or enciphered or after it has been decoded or deciphered.

Polyalphabetic A method of creating a cipher through the use of more than one replacement alphabet. This type of concealment method gives cryptographers several levels of letters within which to hide their original words and sentences and is generally more secure than a single (monoal-phabetic) substitution.

Polygraphic A concealment technique that encrypts two or more characters at one time. The digraphic involves an enciphered letter pair that together affect the result. The polygraph (or multigraph) used to consist of large awkward tables of equivalents, but around 1929 a tetragraphic form was developed algabraically wherein the keys and plaintext characters had a series of numerical values. In this system, when a matrix of letters is enciphered, the matrix is considered as a unit. If any part of this unit is altered, the whole encryption is changed.

Polyphone A cipher letter that represents a choice from among a set group of plaintext elements that are all equivalent to the same cipher letter, usually two or three at most. For example, q = x, from among the choices x, y, or z for the first message. In a second message, q would be y.

Rebus (Latin: *res* "thing"). A series of images in which the sounds of the names of the figures and symbols combine to create a word or phrase.

Route The order or sequence of transposition through a matrix.

Semagram (Greek: *sema*, "sign," + *gramma*, "written" or "drawn"). A form of steganography, wherein encryptions are made of arrangements of objects, images or symbols rather than by letters or numbers.

Semaphore (Greek: *sema* "sign" + *pherein* "to bear" or *phero*, "to carry"). A communications system whereby either hand-held flags or mechanical arms are moved into different positions to convey messages to an observer.

SIGINT The acronym for *signals intelligence*. It is the encompassing title for intelligence-gathering processes, including interception; signal origin direction-finding; traffic analysis (message numbers, sources and directions); cryptanalysis; electronic reconnaissance (eavesdropping on radar signals); and countermeasures (jamming and creating false radar echoes.)

Signals security The major signals' protection activities, including communications security (COMSEC) with steganography; traffic security; cryptography; electronics security such as radar frequency changes; and counter-countermeasures ("looking through" jammed radar).

Steganography (Greek: *steganos* "covered" + *graphein* "writing"). A primary form of communications security that conceals the physical presence of a secret message, which may or may not be additionally protected by a code or cipher.

Substitution A cryptographic method in which the plaintext is replaced by letters, numbers or other symbols while its original word or letter order remains unchanged.

Superencipherment A cryptographic process in which a code or cipher is itself enciphered by transposition (especially for numbers) or substitution (for words and numbers) in order to provide an extra layer of cryptographic protection. The result is called an encicode, also known as an enciphered code.

Symbol cryptography A form of cryptography in which the plaintext is replaced by a symbol, such as a design or a written or printed mark.

TELENT An acronym for *telemetry intelligence* (from Greek: *tele*, "far off," + *metron*, "a measure"). The self-diagnosis sent by a missile to tracking stations on the ground. This information is a primary target for enemy electronics intelligence (ELINT).

Transmissions security An electronic form of communications security similar to steganography. Transmissions security tries to hide the existence of secret messages in electrical exchanges, whether or not they are encrypted.

Transposition A system of cryptography whereby the letters or words of a message are rearranged. Permutation is a form of transposition.

Trigraph a group of three letters.

Two-key cryptography Also known as "public key." A security process which involves two keys, one obtainable from a public directory and the other kept private. The former is primarily applied to encrypt a message and the latter is mainly used to decrypt it.

Two-part code A code in which the code equivalents are randomized, while the plaintext is listed in alphabetical order. More complicated than a one-part code, it requires a second list or a book in which the code equivalents are complied in numerical or alphabetical order for decoding. Friendly communicators have both versions.

Unbreakable ciphers include one-time methods and unconditionally secure cryptosystems.

Variants Two or more letters that represent plaintext letters.

BIBLIOGRAPHY

BOOKS

Abernethy, Thomas Perkins. *The Burr Conspiracy*. New York: Oxford University Press, 1954.

Allen, Col. Robert S. *Lucky Forward*. New York: Vanguard Press, 1947.

Allen, Thomas B. *War Games*. New York: McGraw-Hill, 1987.

Andrews, Carol. *The Rosetta Stone*. Chicago: Ares Publishing, 1986.

Bahn, Paul, and Jean Vertut. *Images of the Ice Age*. New York: Facts on File, 1988.

Bakeless, John. *Turncoats, Traitors and Heroes*. Philadelphia: J. B. Lippincott Company, 1959.

Bamford, James, *The Puzzle Palace*. Boston: Houghton Mifflin, 1982.

Barker, Wayne, ed. *The History of Codes and Ciphers in the United States Prior to WWI*. Laguna Hills, CA: Aegean Park Press, 1978.

____. *The History of Codes and Ciphers in the United States during the Period Between the World Wars, Part II, 1930–39*. Laguna Hills, CA: Aegean Park Press, 1989.

Barnes, Howard R. *Report of Code Compilation Section, General Headquarters, American Expeditionary Forces, December 1917–November 1918*. Washington, DC: GPO, 1935.

Basham, A. L. *The Wonder That Was India*. New York: Grove Press, 1959.

Bates, David Homer. *Lincoln in the Telegraph Office*. New York: Century, 1907.

Bazeries, Etienne. *Les Chiffres secrets dévoilés*. Paris: Librairie Charpentier et Fasquelle, 1901.

Bearden, Bill, ed. *The Bluejacket's Manual*, 20th ed. Annapolis: U.S. Naval Institute Press, 1978.

Beker, Henry, and Fred Piper. *Cipher Systems*. New York: John Wiley, 1982.

Bennett, Ralph. *Ultra in the West: The Normandy Campaign, 1944–45*. London: Hutchinson, 1979.

Bernikow, Louise. *Abel*. New York: Trident Press, 1970.

Beymer, William G. *On Hazardous Service: Scouts and Spies of the North and South.* New York: Harper & Bros., 1912.

Billeter, Jean-François. *The Chinese Art of Writing.* New York: International Publications, Inc., 1990.

Bittner, William. *Poe: A Biography.* Boston: Little, Brown & Co., 1962.

Boatner, Mark M., III. *Encyclopedia of the American Revolution.* New York: McKay, 1974.

Bond, Raymond T. *Famous Stories of Code and Cipher.* New York: Rinehart, 1947.

Bonfonte, Larissa. *Etruscan.* Berkeley and Los Angeles: University of California Press, 1990.

Bowers, William M. *The Bifid Cipher*, Practical Cryptanalysis 2. American Cryptogram Association, 1960.

_____. *The Trifid Cipher*, Practical Cryptanalysis, 3. American Cryptogram Association, 1961

Boyd, Carl. *Hitler's Japanese Confidant: General Oshima Hiroshi and MAGIC Intelligence, 1941–1945.* Lawrence, Kansas: University of Kansas Press, 1993.

Bright, Charles. *Submarine Telegraphs: Their History, Construction, and Working.* London: Crosby Lockwood & Son, 1898.

Broome, Capt. Jack E. *Make a Signal!* London: Putnam, 1955.

Burton, Sir Richard. *Kama-sutra.* New York: E. P. Dutton & Co., 1962.

Catton, Bruce. *Glory Road.* New York: Doubleday & Co., 1952.

Cave Brown, Anthony. *Bodyguard of Lies.* New York: Harper & Row, 1975.

Chadwick, John. *Linear B and Related Scripts.* Berkeley and Los Angeles: University of California Press, 1987.

Chant, Christopher. *The Encyclopedia of Codenames of World War II.* New York: Methuen, 1988.

Claiborne, Robert. *The Birth of Writing.* Alexandria, VA: Time-Life Books, 1974.

Coe, Michael. *Breaking the Maya Code.* New York: Thames and Hudson, 1992.

Coggins, Jack. *Flashes and Flags: The Story of Signalling.* New York: Dodd, Mead & Company, 1963.

Cohen, Frederick B. *Protection and Privacy on the Information Superhighway.* New York: John Wiley & Sons, 1995.

Cook, B. F. *Greek Inscriptions.* Berkeley and Los Angeles: University of California Press, 1987.

Cookridge, E. H. (pseud.). *Spy Trade.* New York: Walker, 1971.

Corbett, Julian S. *Fighting Instructions, 1530–1816.* London: Naval Records Society, 1905; reprinted, London: Conway Maritime Press, 1971.

____. *Signals and Instructions, 1776 to 1794*. London: Naval Records Society, 1909; reprinted, London: Conway Maritime Press, 1971.

Costello, John. *The Pacific War*. New York: Rawson, Wade, 1981.

____. *Mask of Treachery*. New York: William Morrow, 1988.

Coulmas, Florian. *The Writing Systems of the World*. Malden, MA: Blackwell Publishers, 1989.

Cresswell, John. *Teach Yourself Esperanto*. New York: David McKay Co., 1968.

Currey, Cecil B. *Code Number 72: Ben Franklin: Patriot or Spy?* Englewood Cliffs, NJ: Prentice-Hall, 1972.

Dallin, David. *Soviet Espionage*. New Haven, CT: Yale University Press, 1955.

Davies, D. W., and W. L. Price. *Security for Computer Networks*. New York: John Wiley, 1984.

Davies, W. V. *Egyptian Hieroglyphs*. Berkeley and Los Angeles: University of California Press, 1987.

Deacon, Richard. *The Chinese Secret Service*. New York: Taplinger Publishing, 1974.

Deavours, Cipher A., and Louis Kruh. *Machine Cryptography and Modern Cryptanalysis*. Dedham, MA: Artech House, 1985.

DeFrancis, John. *The Chinese Language: Fact and Fancy*. Honolulu: University of Hawaii Press, 1984.

de Toledano, Ralph. *The Greatest Plot in History*. New York: Duell, Sloan and Pearce, 1963.

Donovan, James B. *Strangers on a Bridge: The Case of Colonel Abel*. New York: Atheneum, 1964.

Donovan, Robert J. *Tumultuous Years*. New York: W. W. Norton & Co., 1982.

Dorwart, Jeffery M. *The Office of Naval Intelligence: The Birth of America's First Intelligence Agency, 1865–1918*. Annapolis, MD: U.S. Naval Institute Press, 1979.

____. *Conflict of Duty: The U.S. Navy's Intelligence Dilemma, 1919–1945*. Annapolis, MD: U.S. Naval Institute Press, 1983.

Doyle, Arthur Conan. *A Treasury of Sherlock Holmes*. New York: Hanover House, 1955.

Dupuy, Ernest R. and Trevor N. Dupuy. *The Encyclopedia of Military History*. New York: Harper & Row, 1970.

Erickson, John. *The Soviet High Command: A Military-Political History, 1918–1941*. London: Macmillan, 1962.

Fagan, M. D., ed. *A History of Engineering and Science in the Bell System: National Service in War and Peace (1925–1975)*. Murray Hill, NJ: Bell Laboratories, 1978.

Farago, Ladislas. *The Broken Seal: "Operation Magic" and the Secret Road to Pearl Harbor*. New York: Random House, 1967.

____. *The Game of the Foxes: The Untold Story of German Espionage in the United States and Great Britain During World War II*. New York: David McKay Co., 1971.

Faulk, Odie B. *The Geronimo Campaign*. New York: Oxford University Press, 1969.

Fea, Allan. *Secret Chambers and Hiding-Places*. London: S. H. Bousfield, 1901.

Flexner, James T. *The Traitor and the Spy: Benedict Arnold and John André*. New York: Harcourt, Brace, 1953.

Folliot, Denise. *Marie Antoinette*. New York: Harper & Row, 1957.

Foote, Alexander. *Handbook for Spies*. Garden City, NY: Doubleday, 1949.

Frank, Thomas, and Edward Weisband, eds. *Secrecy and Foreign Policy*. New York: Oxford University Press, 1974.

Friedman, William F. *American Army Field Codes in the American Expeditionary Forces during the First World War*. Washington, DC: GPO, 1942.

____. *Cryptography and Cryptanalysis Articles*, 2 vols. Laguna Hills, CA: Aegean Park Press, 1976. Reprinted from "The Use of Codes and Ciphers in the World War and Lessons to Be Learned Therefrom." *Articles on Cryptography and Cryptanalysis*. Signal Corps Bulletin (Washington, DC: GPO, 1942.)

____. *Elementary Military Cryptography*. Laguna Hills, CA: Aegean Park Press, 1976.

____. *History of the Use of Codes*. Laguna Hills, CA: Aegean Park Press, 1977. Reprinted from *Report on the History of the Use of Codes and Code Language, the International Telegraph Regulations Pertaining Thereto, and the Bearing of This History on the Cortina Report*. International Radiotelegraph Conference of Washington, 1927, Delegation of the United States of America (Washington, DC: GPO, 1928).

____. *Military Cryptanalysis II*. Laguna Hills, CA: Aegean Park Press, 1984.

____. *Solving German Codes in World War I*. Laguna Hills, CA: Aegean Park Press, 1997.

Friedman, William F., and Elizebeth Friedman. *The Shakespearean Ciphers Examined: An Analysis of Cryptographic Systems Used as Evidence That Some Author Other Than William Shakespeare Wrote the Plays Commonly Attributed to Him*. Cambridge: Cambridge University Press, 1957.

Friedman, William, F. and Charles J. Mendelsohn. *The Zimmermann Telegram of January 16, 1917 and Its Cryptographic Background*. Laguna Hills, CA: Aegean Park Press, 1976.

Gaines, Helen F. *Cryptanalysis: A Study of Ciphers and Their Solution*. New York: Dover Publications, 1956.

Gardner, Martin, ed. *Oddities and Curiosities*. New York: Dover Publications,, 1957.

Gauer, Albertine. *A History of Writing* 3d ed. New York: Cross River Press, 1992.

Gibson, Hugh, ed. *The Ciano Diaries, 1939–1943*. Garden City, NY: Doubleday and Hudson, 1992.

Givierge, Marcel. *Cours de cryptographie*. Paris: Berger-Levrault, 1925. Translated as *Course in Cryptography* by John B. Hurt, Washington, DC: GPO, 1934. Reprinted by Aegean Park Press, Laguna Hills, CA, 1978.

Glover, Beaird. *Secret Ciphers of the 1876 Presidential Election*. Laguna Hills, CA: Aegean Park Press, 1991.

Gould, Robert. *The History of Freemasonry throughout the World*. New York: Charles Scribner, 1936.

Greene, Franklin. *Chronosemic Signals*. Washington, DC: GPO, 1864.

Groves, Leslie R. *Now It Can Be Told, The Story of the Manhattan Project*. New York: Harper and Row, 1962.

Gyldén, Yves. *The Contribution of the Cryptographic Bureaus in the World War*. Laguna Hills, CA: Aegean Park Press, 1978. Translated from the original *Chifferbyraernas Insatser I Världskriget Till Lands* (Stockholm, 1931).

Hamming, Richard W. *Coding and Information Theory*. 2d ed. New York: Prentice-Hall, 1986.

Handlin, Oscar, ed. *Eli Whitney and the Birth of American Technology*. Boston: Little, Brown & Co., 1956.

Harris, Roy. *The Origin of Writing*. Reprinted, Herndon, VA: Thoemmes Press, 1996.

Hinz, Walter. *The Lost World of Elam: Recreation of a Vanished Civilization*. New York: New York University Press, 1972.

Hitt, Parker. *Manual for the Solution of Military Ciphers*. Laguna Hills, CA: Aegean Park Press, 1976.

Hoehling, A. A. *Women Who Spied*. New York: Dodd, Mead & Co., 1967.

Holmes, W. J. *Double-Edged Secrets: U.S. Naval Intelligence Operations in the Pacific during World War II*. Annapolis, MD: Naval Institute Press, 1979.

Holzmann, Gerard J. and Björn Pehrson. *The Early History of Data Networks*. Los Alamitos, CA: IEEE Computer Society Press, 1995.

Horan, James D. *Confederate Agent: A Discovery in History*. New York: Crown, 1954.

Howe, Russell W. *Mata Hari*. New York: Dodd, Mead, 1986.

Howeth, Capt. Linwood S. *History of Communications-Electronics in the United States Navy*. Washington: DC: GPO, 1963.

Hoy, Hugh C. *40 O.B., or How the War Was Won*. London: Hutchinson, 1935.

Hutchinson, William T. and William M. Rachal, eds. *Papers of James Madison*. Chicago: University of Chicago Press, 1969.

Ind, Allison. *A Short History of Espionage*. New York: David McKay Co., 1963.

James, William. *The Codebreakers of Room 40*. New York: St. Martin's Press, 1956.

Jenks, Capt. Robert W. *The Brachial Telegraph*. New York: Henry Sanders & Co., 1852.

Johnson, Brian. *The Secret War*. London: British Broadcasting Corporation, 1978.

Kahn, David. *The Codebreakers: The Story of Secret Writing*. New York: Macmillan, 1967.

——. *Hitler's Spies: German Military Intelligence in World War II*. New York: Macmillan, 1978.

——. *Kahn on Codes*. New York: Macmillan, 1984.

Keegan, John. *The Price of Admiralty, The Evolution of Naval Warfare*. London: Penguin Books, 1988.

Keiser, Bernhard E., and Eugene Strange. *Digital Telephony and Network Integration*. New York: Van Nostrand Reinhold, 1985.

Kennen, George F. *Memoirs, 1923–1950*. Boston: Little, Brown, & Co., 1967.

Kessler, Ronald. *Moscow Station: How the KGB Penetrated the American Embassy*. New York: Charles Scribner's Sons, 1989.

Koch, Edward. *Cryptography or Cipher Writing*. 2d ed. Belleville, IL: Buechler Publishing, 1942.

Konheim, Alan G. *Cryptography: A Primer*. New York: John Wiley, 1981.

Koop, Theodore F. *Weapon of Silence*. Chicago: University of Chicago Press, 1946.

Krivitsky, Walter J. *In Stalin's Secret Service*. Frederick, MD: University Publications of America, 1985.

Ladd, James, Keith Melton, and Captain Peter Mason. *Clandestine Warfare*. London: Blandford Press, 1988.

Laffin, John. *Codes and Ciphers*. New York: Abelard-Schuman, 1964.

Lampalla, Princess de, *Secret Memoirs of the Royal Family of France during the Revolution*. Vol. 2 London: H.S. Nichols, 1895.

Lamphere, Robert J., and Tom Schachtman. *The FBI-KGB War: A Special Agent's Story*. New York: Random House, 1986.

Lange, André, and E. A. Soudart. *Treatise on Cryptography*. Laguna Hills, CA: Aegean Park Press, 1981.

Layton, Edwin T., with Roger Pineau and John Costello. *"And I Was There": Pearl Harbor and Midway—Breaking the Secrets*. New York: William Morrow, 1985.

Levitt, B. K., M. K. Simon, J. K. Omura and R.A. Scholtz. *Spread Spectrum Communications Handbook*. 2d rev. ed. New York: McGraw-Hill, 1994.

Lewin, Ronald. *Ultra Goes to War: The First Account of World War II's Greatest Secret Based on Official Documents*. New York: McGraw-Hill, 1978.

Lewis, Spencer. *Rosicrucian*. San José, CA: The Rosicrucian Press Ltd., 1929.

Lindsey, Robert. *The Falcon and the Snowman*. New York: Simon & Schuster, 1979.

Locard, Edmond. *Traité de criminalistique, 6.* "Les Correspondances secrètes," 831–931 at "Cryptoraphie à l'aide des objets," 901–3. Lyon, France: Joannès Desvignes, 1937.

Macalister, R. A. Stewart. *The Secret Languages of Ireland*. Cambridge: Cambridge University Press, 1937.

Macintyre, Donald. *The Battle of the Atlantic*. New York: Macmillan, 1961.

Marland, E. A. *Early Electrical Communication*. New York: Abelard-Schuman, 1964.

Mead, Peter. *The Eye in the Air*. London: Her Majesty's Stationery Office, 1983.

Meister, Dr. Aloys. *Die Anfänge der Modernen Diplomatischen Geheimschrift*. Paderborn: Ferdinand Schoningh, 1902.

____. *Die Geheimschrift im Dienste der Päpstlichen Kurie*. Paderborn: Ferdinand Schoningh, 1906.

Menezes, Alfred J., Paul C. van Oorschot and Scott A. Vanstone. *Handbook of Applied Cryptography*. New York: CRC Press, 1997.

Meyer, Carl H., and Stephen M. Matyas. *Cryptography: A New Dimension in Computer Data Security*. New York: John Wiley, 1982.

Moore, Dan Tyler, and Martha Waller. *Cloak and Cipher*. Indianapolis: Bobbs-Merrill, 1962.

Mosley, Leonard. *The Cat and the Mice*. New York: Harper & Brothers, 1958.

Murphy, John. *Book of Pidgin English*. New York: AMS Press, 1949.

Myer, Albert J. A. *Manual of Signals*. New York: D. Van Nostrand, 1868.

Nalder, Maj. Gen. R. F. H. *The Royal Corps of Signals*. London: Royal Signals Institution, 1958

Nanovic, John (Henry Lysing, pseud.) *Secret Writing: An Introduction to Cryptograms, Ciphers and Codes*. David Kemp & Company, 1936, reprinted New York: Dover Publications, 1974.

Newbold, William R. *The Cipher of Roger Bacon*. Philadelphia: University of Pennsylvania Press, 1928.

Noggle, Burl. *Teapot Dome: Oil and Politics in the 1920s*. Baton Rouge: Louisiana State University Press, 1962.

Norman, Bruce, *Secret Warfare*. Washington, DC: Acropolis Books, 1973.

Ore, Oystein. *Cardano, The Gambling Scholar*. With a translation from the Latin of Cardano's *Book on Games of Chance* by Sidney Henry Gould. Princeton, NJ: Princeton University Press, 1953.

O'Toole, G. J. A. *The Spanish War: An American Epic, 1898*. New York: W. W. Norton and Co., 1984.

____. *The Encyclopedia of American Intelligence and Espionage: From the Revolutionary War to the Present*. New York: Facts on File, 1988.

Parker, Donn B. *Fighting Computer Crime.* New York: Charles Scribner's Sons, 1983.

Parpola, Asko. *Deciphering the Indus Script.* Cambridge and New York: Cambridge University Press, 1994.

Peckham, Howard H. "British Secret Writing in the Revolution." *Quarterly Review of the Michigan Alumnus* 44 (winter 1938): 126–31.

Pei, Mario. *One Language for the World.* New York: Devin-Adair Co., 1958.

Pennypacker, Morton. *General Washington's Spies on Long Island and in New York.* Brooklyn: Long Island Historical Society, 1939.

Perrault, Gilles. *The Red Orchestra.* Translated by Peter Wiles. New York: Simon & Schuster, 1969.

Persico, Joseph E. *Piercing the Reich: The Penetration of Nazi Germany by American Secret Agents during World War II.* New York: Viking Press, 1979.

Pitt, Barrie, and the editors of Time-Life Books. *World War II: The Battle of the Atlantic.* Alexandria, VA: Time-Life Books, 1980.

Plaidy, Jean. *Mary Queen of Scots.* New York: G. P. Putnam, 1975.

Plum, William R. *The Military Telegraph During the Civil War in the United States.* 2 vols. Chicago: Jansen, McClurg, 1882.

Pope, Maurice. *The Story of Decipherment: From Egyptian Hieroglyphic to Linear B.* London: Thames and Hudson, 1975.

Postgate, J. N. *Early Mesopotamia: Society and Economy at the Dawn of History.* New York: Routledge, 1992.

Prados, John. *Combined Fleet Decoded: The Secret History of American Intelligence and the Japanese Navy in World War II.* New York: Random House, 1995.

Prange, Gordon W., with Donald M. Goldstein and Katherine V. Dillon. *At Dawn We Slept: The Untold Story of Pearl Harbor.* New York: McGraw-Hill, 1981.

Preble, Rear Adm. George Henry, USN. *History of the Flag of the United States of America*, 3d ed. Boston: James R. Osgood & Co., 1882.

Preston, Keith, trans. *De Furtivis Literarum Notis* by Giovanni Porta (1563). Fabyan Collection, Library of Congress.

Price, Derek J. *John Baptista Porta's Natural Magick.* New York: Basic Books, 1957.

Rachlis, Eugene. *They Came to Kill: The Story of Eight Nazi Saboteurs in America.* New York: Random House, 1961.

Raines, Rebecca Robbins. *Getting the Message Through: A Branch History of the U.S. Army Signal Corps.* Washington, DC: U.S. Army Center of Military History, 1996.

Richelson, Jeffrey T. *The U.S. Intelligence Community.* Cambridge, MA: Ballinger, 1985.

Robinson, Andrew. *The Story of Writing: Alphabets, Hieroglyphs, and Pictograms*. London: Thames and Hudson, 1995.

Robison, S. S., and Mary L. Robison. *A History of Naval Tactics from 1530 to 1930*. Annapolis, MD: U.S. Naval Institute, 1942.

Roskill, S. W. *The War at Sea: 1939–1945*. London: Her Majesty's Stationery Office, 1954.

Ross, Ishbel. *Rebel Rose: Life of Rose O'Neal Greenhow, Confederate Spy*. New York: Harper & Brothers, 1954.

Rowan, Richard W., and Robert Derndorfer. *The Story of Secret Service*. New York: Elsevier-Dutton, 1956.

Russell, Francis, and the editors of Time-Life Books. *World War II: The Secret War*. Alexandria, VA: Time-Life Books, 1981.

Ryan, Cornelius. *The Longest Day: June 6, 1944*. New York: Simon & Schuster, 1959.

Sacco, Luigi. *Manual of Cryptography*. Reprinted from a translation of *Manuale di Crittographia* (Rome, 1936). Laguna Hills CA: Aegean Park Press, 1977.

Sarton, George. *Six Wings: Men of Science in the Renaissance*. Bloomington: Indiana University Press, 1957.

Schachner, Nathan. *Aaron Burr: A Biography*. New York: Frederick Stokes Co., 1937.

Scott, Adm. Sir Percy. *Fifty Years in the Royal Navy*. New York: George H. Doran Co., 1919.

Senner, Wayne M., ed. *The Origins of Writing*. Lincoln: University of Nebraska Press, 1989.

Simkins, Peter. *Air Fighting, 1914–1918*. London: Imperial War Museum, 1978.

Smith, Laurence Dwight. *Cryptography: The Science of Secret Writing*. New York: Dover Publications, 1955.

Smith, Whitney. *Flags: Through the Ages and Across the World*. New York: McGraw-Hill Book Co., 1975.

____. *The Flag Book of the United States*. New York: William Morrow and Co., 1970.

Spector, Ronald H. *Eagle Against the Sun: The American War with Japan*. New York: Free Press, 1985.

____. *Listening to the Enemy*. Wilmington, DE: Scholarly Resources, 1988.

Stallings, William. *Data and Computer Communications*. New York: Macmillan, 1985.

Sterling, Comdr. Yates, USN. *Fundamentals of Naval Service*. Philadelphia: J. B. Lippincott Co., 1917.

Stern, Philip Van Doren. *Secret Missions of the Civil War*. New York: Crown, 1959.

Stevenson, William. *A Man Called Intrepid*. New York: Harcourt Brace Jovanovich, 1976.

Still, Alfred. *Communication through the Ages*. New York: Murray Hill Books, Inc., 1946.

Thayer, Charles W. *Diplomat*. New York: Harper & Brothers, 1959.

Thompson, George Raynor, and Dixie R. Harris. *The Signal Corps: The Outcome (Mid-1943 through 1945)* United States Army in World War II: The Technical Series. Washington, DC: GPO, 1966.

Thompson, George Raynor, Dixie R. Harris, Pauline Oakes, and Dulany Terrett. *The Signal Corps: The Test (December 1941 to July 1943)*; United States Army in World War II: The Technical Series. Washington, DC: GPO, 1957.

Thompson, James W., and Saul K. Padover. *Secret Diplomacy: Espionage and Cryptography*. New York: F. Ungar, 1963.

Thompson, Robert L. *Wiring a Continent*. Princeton, NJ: Princeton University Press, 1947.

Thorndike, Lynn. *A History of Magic and Experimental Science*. New York: Columbia University Press, 1926–1958.

Thorne, J. O., ed. *Chambers's Biographical Dictionary*. New York: St Martin's Press, 1969.

Time-Life Books. *Understanding Computers: Computer Basics*. Alexandria, VA: Time-Life Books, 1985.

____. *Understanding Computers: Military Frontier*, Alexandria, VA: Time-Life Books, 1988.

____. *Understanding Computers: Computer Security*. Alexandria, VA: Time-Life Books, 1988.

Tomkins, William. *Indian Sign Language*. New York: Dover Publications, 1968.

Tuchman, Barbara W. *The Zimmermann Telegram*. New York: Viking Press, 1958.

____. *The Guns of August*. New York: Macmillan, 1962.

U.S. Army Military History Institute. *Traditions of the Signal Corps*. Washington, DC: GPO, 1959.

Van Doren, Carl. *Secret History of the American Revolution,* New York: Viking Press, 1941.

Very, Edward W. "Signals, Marine." In *A Naval Encyclopedia*. Philadelphla: G. R. Hammersly & Co., 1881.

Voorhis, Harold V. B. "Masonic Alphabets." In *A History of Royal Arch Masonry*. Vol. 3 Lexington, KY: Royal Arch Masons, 1956.

Walker, C. B. F. *Cuneiform*. Berkeley and Los Angeles: University of California Press, 1989.

Way, Peter. *Undercover Codes and Ciphers*. London: Aldus Books, 1977.

Weber, Ralph E. *United States Diplomatic Codes and Ciphers, 1975–1938.* Chicago: Precedent Publishing, 1979.

——. *Masked Dispatches: Cryptograms and Cryptology in American History, 1775–1900.* Ft. Meade, MD: NSA's Center for Cryptologic History, 1993.

Welchman, Gordon. *The Hut Six Story: Breaking the Enigma Codes.* New York: McGraw Hill, 1982.

Wellman, Paul I. *The Indian Wars of the West.* 2 vols. New York: Doubleday and Co., 1954.

Whitehead, Don. *The FBI Story: A Report to the People.* New York: Random House, 1956.

Willoughby, Maj. Gen. Charles A. *Shanghai Conspiracy: The Sorge Spy Ring.* New York: E. P. Dutton & Co., 1952.

Willoughby, Malcolm F. *Rum War at Sea.* Washington, DC: GPO, 1964.

Wilson, George. *The Old Telegraphs.* Chichester, England: Phillimore, 1976.

Wise, David. *Molehunt: The Secret Search for Traitors That Shattered the CIA.* New York: Random House, 1992.

Wolfe, Jack M. *A First Course in Cryptanalysis.* Rev. 3 vols. Brooklyn: Brooklyn College Press, 1943.

Woods, David L., ed. *A History of Tactical Communication Techniques.* Orlando, FL: Martin-Marietta Corporation, 1965.

——. *Signalling and Communicating at Sea.* 2 vols. New York: Arno Press, 1985.

Wright, Louis B., and Marion Tinling, eds. *The Secret Diary of William Byrd of Westover 1709–1712.* Richmond, VA: Dietz Press, 1941.

Wright, Peter, with Paul Greengrass. *Spycatcher: The Candid Autobiography of a Senior Intelligence Officer.* Harrisonburg, VA: Donnelley & Sons, 1987.

Wrixon, Fred B. *Codes, Ciphers, and Secret Languages.* New York: Crown Publishers, 1989.

——. *Codes and Ciphers.* New York: Simon & Schuster, 1992.

Yardley, Herbert. *The American Black Chamber.* Indianapolis: Bobbs-Merrill, 1931.

Yates, Frances A. *Giordano Bruno and the Hermetic Tradition.* Chicago: University of Chicago Press, 1964.

Zauzich, Karl-Theodor. *Discovering Egyptian Hieroglyphs: A Practical Guide.* New York: Thames and Hudson, 1992.

PERIODICALS

Attansio, C. R., and R. J. Phillips. "Penetrating an Operating System: A Study of VM/370 Integrity." *IBM Systems Journal* 15, no. 1 (1976): 46.

August, David. "Cryptography and Exploitation of Chinese Manual Cryptosystems, Part I: The Encoding Problem." *Cryptologia* 13, no. 4 (October 1989): 289–302.

——. "Cryptography and Exploitation of Chinese Manual Cryptosystems, Part II: The Encrypting Problem." *Cryptologia* 14, no. 1 (January 1990): 61–78.

Beard, William W. "YIYKAEJRGZQSYWX." *U.S. Naval Institute Proceedings* 44, no. 8 (1918).

Bowers, William M. "F. Delastelle—Cryptologist," *The Cryptogram* 30 (March-April 1963): 79–82, 85.

——. *Cryptogram* 30 (May–June, 1963): 101, 106–9.

——. "Major F. W. Kasiski—Cryptologist." *Cryptogram* 31 (January–February 1964): 53–54, 58–60.

Broadhurst, George. "Some Others and Myself." *Saturday Evening Post* 199 (October 23, 1926): 42.

Chase, Pliny Earle. "Mathematical Holocryptic Cyphers." *Mathematical Monthly* I (March 1859): 194–96.

Church, Hayden. "A Sherlock Holmes of Secret War Codes." *New York Times Magazine* November 8, 1931, 17.

Cohan, Capt. Leon. "Uniform Signal System." *Marine Corps Gazette* (June 1966): 27.

Culenaere, N. A, "Cryptophotography." *International Criminal Police Review* 102 (November 1956): 284–90.

Davies, Donald W. "The Early Models of the Siemens and Halske T52 Cipher Machine." *Cryptologia*, no. 3 (July 1983): 235–53.

——. "Charles Wheatstone's Cryptograph and Pletts's Cipher Machine." *Cryptologia* 9, no. 2 (April 1985): 155–61.

Field, Alexander J. "French Optical Telegraphy, 1793–1855: Hardware, Software, Administration. *Technology and Culture* 35, no. 2 (April 1994): 315–347.

Fishel, Edwin C. "The Mythology of Civil War Intelligence." *Civil War History* 10, no. 4 (December 1964): 344–67.

Friedman, William F. "Edgar Allan Poe, Cryptographer." *American Literature* 8 (November 1936): 266–80.

Friedman, William F., and Charles Mendelsohn. "Notes on Code Words." *American Mathematical Monthly* (August–September 1932): 394–409.

Givierge, Marcel. "Questions de Chiffre." Parts 1 and 2. *Revue Militaire Français*, June 1, 1924, 398–417; July 1, 1924, 59–78. Translated as "Problems of Code" in *Articles on Cryptography and Cryptanalysis*. Reprinted from the *Signal Corps Bulletin*. Washington, DC: GPO, 1942.

Hardie, Bradford. "The Potus-Prime Connection." *Cryptologia* 11, no. 1 (January 1987): 40–46.

Helmick, Leah Stock. "Key Woman of the T-Men." *Reader's Digest* (September 1937): 51–55.

Hendricks, Dewayne and David R. Hughes. "Spread-Spectrum Radio." *Scientific American 278* (April 1998): 94–96.

Hill, Lester. "Cryptography in an Algebraic Alphabet." *The American Mathematical Monthly* 36 (June–July 1929): 306–12.

___. "Concerning Certain Linear Transformation Apparatus of Cryptography." *American Mathematical Monthly* 38 (March 1931): 135–54.

Hoover, J. Edgar. "The Enemy's Masterpiece of Espionage." *Reader's Digest* 48 (April 1946): 1–6.

Kahn, David. "Cryptology and the Origins of Spread Spectrum." *IEEE Spectrum* (September 1984): 70–80.

___. "The Wreck of the Magdeburg." *MHQ: The Quarterly Journal of Military History* 2, no. 2 (Winter 1990): 97–103.

___. "Pearl Harbor and the Inadequacy of Cryptanalysis." *Cryptologia* 15, no. 4 (1991): 273–94.

Kapany, Narinder S. "Picture Tube." *Time* (December 3, 1956): 69–70.

___. "Fiber Optics." *Scientific American* (November 1960): 72–81.

Keegan, John. "Jutland." *MHQ: The Quarterly Journal of Military History* 1, no. 2 (Winter 1989): 110–12.

Knight, Mary. "The Secret War of Censors vs. Spies." *Reader's Digest* 48 (March 1946): 79–83.

Kober, Alice. "The 'Adze' Tablets from Knossos,' *American Journal of Archeology* 48 (January–March 1944): 64:75.

___. "Evidence of Inflection in the 'Chariot' Tablets from Knossos." *American Journal of Archeology* 49 (April–June 1945): l43–151.

___. "Inflection in Linear Class B: 1—Declension." *American Journal of Archeology* 50 (April–June 1946): 268–276.

___. "The Minoan Scripts: Fact and Fancy." *American Journal of Archeology* 42 (January–March 1948): 82–103.

Koenig, Duane. "Telegraphs and Telegrams in Revolutionary France." *Scientific Monthly* 59 (December 1944) :431–37.

Kolodin, Irving. "What is the Enigma?" *Saturday Review* 36 (February 28, 1953): 53–55, 71.

Kruh, Louis. "The Genesis of the Jefferson/Bazeries Cipher Device." *Cryptologia* 5, no. 4 (1981): 193–208.

___. "The Mystery of Decius Wadsworth's Cipher Device." *Cryptologia* 6, no 3 (July 1982): 238–247.

Kruh, Louis, and Cipher Deavours. "The Typex Cryptograph." *Cryptologia* 7, no. 2 (April 1983): 145–165.

Leighton, Albert C., and Stephen M. Matyas. "The Search for the Key Book to Nicholas Trist's Book Ciphers." *Cryptologia* 7 (October 1983): 297–314.

Levine, Jack. "Some Elementary Cryptoanalysis of Algebraic Cryptography." *American Mathematical Monthly* 68 (May 1961): 411–18.

Levy, Steven. "Scared Bitless." *Newsweek*, June 10, 1996: 49, 50, 55.

Littlefield, Jack. "Melancholy Notes on a Cablegram Code Book." *New Yorker*, July 28, 1934.

Matthews, W. "Samuel Pepys, Tachygraphist." *Modern Language Review* 29 (October 1934): 397–404.

McKay, Herbert C. "Notes from a Laboratory." *American Photography* 60 (November 1946): 38–40, 50.

——. "Stereo Photography." *U.S. Camera* 13 (October 1950): 16.

McCracken, Lt. A. R. USN. "Signalling in the British Navy, 1800." *Proceedings of the United States Naval Institute* 58 (January 1932) :47–48.

Mead, Hilary. "Captain Frederick Marryat, Royal Navy." In *Proceedings of the United States Naval Institute*. Annapolis: U.S. Naval Institute, 1933.

Mead, Comdr. Hillary P., RN. "The Story of the Semaphore." Parts 1 and 2. *Mariner's Mirror* (January 1933): 331–33; (January 1934): 93–104.

——. "The History of the International Code." In *Proceedings* Annapolis: U.S. Naval Institute, 1934.

——. "The Admiralty Telegraphs and Semaphores." *Mariner's Mirror*. 24 (April 1938): 184–203.

Mendelsohn, Charles J. "Cardano on Cryptography." *Scripta Mathematica* 6 (October 1939): 157–68.

——. "Blaise de Vigenère and the 'Chiffre Carré.'" *Proceedings of the American Philosophical Society* 82 (March 22, 1940): 103–29.

Morris, C. Brent. "Fraternal Cryptography: Cryptographic Practices of American Fraternal Organizations." *Cryptologia* 7, no. 1 (January 1983): 27–36.

Morris, Robert. "The Hagelin Cipher Machine (M-209)." *Cryptologia* 2, no. 3 (July 1978): 267–89.

Niblack, Albert. "Proposed Day, Night and Fog Signals for the Navy with Brief Description of the Ardois Night System." *Proceedings of the United States Naval Institute*. Annapolis: U.S. Naval Institute, 1891.

Paltsits, Victor Hugo. "The Use of Invisible Ink for Secret Writing during the American Revolution." *Bulletin* (New York Public Library) 39 (May 1935): 361–64.

Post, Melville D. "German War Ciphers." *Everybody's Magazine* 38 (June 1918): 28–34.

Seideman, Tony. "Bar Codes Sweep the World." *American Heritage of Invention and Technology* 8, no. 4 (spring 1993): 56–64.

Simmons, Gustavus J. "Cryptology: The Mathematics of Secure Communication." *Mathematical Intelligencer* 1, no. 4 (1979): 233–46.

——. "Scanning the Issues on Cryptology." *Proceedings of the IEEE* 76, no. 5 (May 1988).

Smith, David E. "John Wallis as a Cryptographer." *Bulletin of the American Mathematical Society* 24 (1917): 83–96.

Smith, Richard A. "Business Espionage." *Fortune* 53 (May 1956): 118–26.

Tanaka, Jennifer. "Drowning in Data." *Newsweek,* April 28, 1997: 85.

Thomas, Evan. "Inside the Mind of a Spy." *Newsweek*, July 7, 1997: 34, 35.

Thomas, Evan, and Gregory L. Vistica. "The Spy Who Sold Out." *Newsweek* December 2, 1996: 35.

Vogel, Donald S. "Inside a KGB Cipher." *Cryptologia* 14, no. 1 (1990): 37–51.

ARCHIVAL AND UNPUBLISHED SOURCES

Data Encryption Standard. Federal Information Processing Standards. Publication 46–1. U.S. Department of Commerce. National Bureau of Standards. January 1988.

Eberhart and Taschner. "Criminal Codes and Ciphers," *FBI Bulletin*, January 1985.

Friedman, Elizebeth. "History of Work in Cryptanalysis, April 1927–June 1930." Washington, DC, National Archives, Record Group 26.

The General Signal Book of the United States Navy. Washington, DC: GPO, 1898.

The History of Army Strip Cipher Devices. U.S. War Department. Army Security Agency, 1948.

Influence of U.S. Cryptologic Organizations on the Digital Computer Industry. May 1977. National Security Agency.

Instructions for Using the Cipher Device M-94. U.S. War Department. Washington, DC: GPO, 1922.

"Interservice Instructions for Ground/Air Recognition and Identification 1943." The War Office, England. November 1943.

National Intelligence Reorganization and Reform Act of 1978. Senate Bill (s. 2525). U.S. Congress, Washington, DC.

O'Neill, E. F., ed. *A History of Engineering and Science in the Bell System-Transmission Technology, 1925–1975.* AT&T Bell Laboratories, 1985.

"Pioneers in U.S. Cryptology." Part 1 (1987): 9–11, and Part 2 (1987): 12–14. History and Publications Division. National Security Agency, Ft. George Meade, MD.

Record Group 59. Department of State Decimal File 411/421/424/ Record Group 457. SRH Histories: SRH 20, SRH 044, SRH 355. National Archives, Washington, DC.

VENONA Historical Monographs. No. 1–5. National Security Agency, Ft. George Meade, MD, 1995, 1996.

Woods, David L. "The Evolution of Visual Signals on Land and Sea." Ph.D. Dissertation. Ohio State University, 1976.

INDEX